T0325659

Luminescence and Luminescent Materials

MATERIALS RESEARCH SOCIETY
SYMPOSIUM PROCEEDINGS VOLUME 667

Luminescence and Luminescent Materials

Symposium held April 17–19, 2001, San Francisco, California, U.S.A.

EDITORS:

Peter C. Schmidt
Technische Universität Darmstadt
Darmstadt, Germany

Kailash C. Mishra
OSRAM Sylvania, Inc.
Beverly, Massachusetts, U.S.A.

Baldassare Di Bartolo
Boston College
Chestnut Hill, Massachusetts, U.S.A.

Joanna McKittrick
University of California-San Diego
La Jolla, California, U.S.A.

A.M. Srivastava
General Electric R&D Corporation
Niskayuna, New York, U.S.A.

Materials Research Society
Warrendale, Pennsylvania

CAMBRIDGE
UNIVERSITY PRESS

University Printing House, Cambridge CB2 8BS, United Kingdom

One Liberty Plaza, 20th Floor, New York, NY 10006, USA

477 Williamstown Road, Port Melbourne, VIC 3207, Australia

314-321, 3rd Floor, Plot 3, Splendor Forum, Jasola District Centre, New Delhi - 110025, India

79 Anson Road, #06-04/06, Singapore 079906

Cambridge University Press is part of the University of Cambridge.

It furthers the University's mission by disseminating knowledge in the pursuit of education, learning and research at the highest international levels of excellence.

www.cambridge.org
Information on this title: www.cambridge.org/9781558996038

Materials Research Society
506 Keystone Drive, Warrendale, PA 15086
http://www.mrs.org

First published 2001
First paperback edition 2013

Single article reprints from this publication are available through University Microfilms Inc., 300 North Zeeb Road, Ann Arbor, MI 48106

CODEN: MRSPDH

A catalogue record for this publication is available from the British Library

ISBN 978-1-558-99603-8 Hardback
ISBN 978-1-107-41222-4 Paperback

CONTENTS

*Invited Paper

POSTER SESSION

NANOCRYSTALLINE MATERIALS

SYNTHESIS AND PROCESSING

QUANTUM WELLS AND QUANTUM DOTS

DEVICES AND DEVICE APPLICATIONS

*Invited Paper

PREFACE

This volume contains papers from Symposium G, "Luminescence and Luminescent Materials," held April 17–19 at the 2001 MRS Spring Meeting in San Francisco, California. The symposium comprised seven sessions, six of them dedicated to the presentation of papers and one to posters. The participation and the attendance were very good, pointing to the continuous interest in luminescence phenomena.

The papers that appear in this volume include both invited and solicited papers from various countries on different aspects of luminescence. The main themes ranged from theory and modeling, characterization of luminescent materials, systems with confined structures such as nano-crystallites and quantum wells and dots, to synthesis and devices.

The great interest of the participants in the subject of our symposium reinforces our belief that luminescence is presently and will continue to be a challenging field of research in materials science, solid state physics and chemistry. Recent progress in opto-electronic and display technology will drive this field in the search for new luminescent materials. Demands on new procedures for synthesis, and understanding underlying luminescence processes in these materials will create new opportunities for both fundamental and applied research in luminescence. We sincerely hope that this volume will contribute to furthering our knowledge and interest in this area.

The editors would like to thank the individual contributors, and the referees whose contributions to the quality of this volume cannot be overstated. We also would like to thank Dr. M. Stephan for his assistance in editing this volume.

Peter C. Schmidt
Kailash C. Mishra
Baldassare Di Bartolo
Joanna McKittrick
A.M. Srivastava

May 2001

MATERIALS RESEARCH SOCIETY SYMPOSIUM PROCEEDINGS

Volume 633— Nanotubes and Related Materials, A.M. Rao, 2001, ISBN: 1-55899-543-9
Volume 634— Structure and Mechanical Properties of Nanophase Materials—Theory and Computer Simulations vs. Experiment, D. Farkas, H. Kung, M. Mayo, H. Van Swygenhoven, J. Weertman, 2001, ISBN: 1-55899-544-7
Volume 635— Anisotropic Nanoparticles—Synthesis, Characterization and Applications, S.J. Stranick, P. Searson, L.A. Lyon, C.D. Keating, 2001, ISBN: 1-55899-545-5
Volume 636— Nonlithographic and Lithographic Methods of Nanofabrication—From Ultralarge-Scale Integration to Photonics to Molecular Electronics, L. Merhari, J.A. Rogers, A. Karim, D.J. Norris, Y. Xia, 2001, ISBN: 1-55899-546-3
Volume 637— Microphotonics—Materials, Physics and Applications, K. Wada, P. Wiltzius, T.F. Krauss, K. Asakawa, E.L. Thomas, 2001, ISBN: 1-55899-547-1
Volume 638— Microcrystalline and Nanocrystalline Semiconductors—2000, P.M. Fauchet, J.M. Buriak, L.T. Canham, N. Koshida, B.E. White, Jr., 2001, ISBN: 1-55899-548-X
Volume 639— GaN and Related Alloys—2000, U. Mishra, M.S. Shur, C.M. Wetzel, B. Gil, K. Kishino, 2001, ISBN: 1-55899-549-8
Volume 640— Silicon Carbide—Materials, Processing and Devices, A.K. Agarwal, J.A. Cooper, Jr., E. Janzen, M. Skowronski, 2001, ISBN: 1-55899-550-1
Volume 642— Semiconductor Quantum Dots II, R. Leon, S. Fafard, D. Huffaker, R. Nötzel, 2001, ISBN: 1-55899-552-8
Volume 643— Quasicyrstals—Preparation, Properties and Applications, E. Belin-Ferré, P.A. Thiel, A-P. Tsai, K. Urban, 2001, ISBN: 1-55899-553-6
Volume 644— Supercooled Liquid, Bulk Glassy and Nanocrystalline States of Alloys, A. Inoue, A.R. Yavari, W.L. Johnson, R.H. Dauskardt, 2001, ISBN: 1-55899-554-4
Volume 645E—Thermal Barrier Coatings—2000, M.A. Antelo, N.P. Padture, B. Pint, S. Sampath, D.J. Wortman, 2001, ISBN: 1-55899-555-2
Volume 646— High-Temperature Ordered Intermetallic Alloys IX, J.H. Schneibel, S. Hanada, K.J. Hemker, R.D. Noebe, G. Sauthoff, 2001, ISBN: 1-55899-556-0
Volume 647— Ion Beam Synthesis and Processing of Advanced Materials, D.B. Poker, S.C. Moss, K-H. Heinig, 2001, ISBN: 1-55899-557-9
Volume 648— Growth, Evolution and Properties of Surfaces, Thin Films and Self-Organized Structures, S.C. Moss, 2001, ISBN: 1-55899-558-7
Volume 649— Fundamentals of Nanoindentation and Nanotribology II, S.P. Baker, R.F. Cook, S.G. Corcoran, N.R. Moody, 2001, ISBN: 1-55899-559-5
Volume 650— Microstructural Processes in Irradiated Materials—2000, G.E. Lucas, L. Snead, M.A. Kirk, Jr., R.G. Elliman, 2001, ISBN: 1-55899-560-9
Volume 651— Dynamics in Small Confining Systems V, J.M. Drake, J. Klafter, P. Levitz, R.M. Overney, M. Urbakh, 2001, ISBN: 1-55899-561-7
Volume 652— Influences of Interface and Dislocation Behavior on Microstructure Evolution, M. Aindow, M. Asta, M.V. Glazov, D.L. Medlin, A.D. Rollet, M. Zaiser, 2001, ISBN: 1-55899-562-5
Volume 653— Multiscale Modeling of Materials—2000, L.P. Kubin, J.L. Bassani, K. Cho, H. Gao, R.L.B. Selinger, 2001, ISBN: 1-55899-563-3
Volume 654— Structure-Property Relationships of Oxide Surfaces and Interfaces, C.B. Carter, X. Pan, K. Sickafus, H.L. Tuller, T. Wood, 2001, ISBN: 1-55899-564-1
Volume 655— Ferroelectric Thin Films IX, P.C. McIntyre, S.R. Gilbert, M. Miyasaka, R.W. Schwartz, D. Wouters, 2001, ISBN: 1-55899-565-X
Volume 656E—Materials Issues for Tunable RF and Microwave Devices II, S.C. Tidrow, W.D. Wilber, S. Streiffer, J. Levy, J. Talvacchio
Volume 657— Materials Science of Microelectromechanical Systems (MEMS) Devices III, M. deBoer, M. Judy, H. Kahn, S.M. Spearing, 2001, ISBN: 1-55899-567-6
Volume 658— Solid-State Chemistry of Inorganic Materials III, M.J. Geselbracht, J.E. Greedan, D.C. Johnson, M.A. Subramanian, 2001, ISBN: 1-55899-568-4

MATERIALS RESEARCH SOCIETY SYMPOSIUM PROCEEDINGS

Volume 659— High-Temperature Superconductors—Crystal Chemistry, Processing and Properties, U. Balachandran, H.C. Freyhardt, T. Izumi, D.C. Larbalestier, 2001, ISBN: 1-55899-569-2
Volume 660— Organic Electronic and Photonic Materials and Devices, S.C. Moss, 2001, ISBN: 1-55899-570-6
Volume 661— Filled and Nanocomposite Polymer Materials, A.I. Nakatani, R.P. Hjelm, M. Gerspacher, R. Krishnamoorti, 2001, ISBN: 1-55899-571-4
Volume 662— Biomaterials for Drug Delivery and Tissue Engineering, S. Mallapragada, R. Korsmeyer, E. Mathiowitz, B. Narasimhan, M. Tracy, 2001, ISBN: 1-55899-572-2
Volume 664— Amorphous and Heterogeneous Silicon-Based Films—2001, M. Stutzmann, J.B. Boyce, J.D. Cohen, R.W. Collins, J. Hanna, 2001, ISBN: 1-55899-600-1
Volume 665— Electronic, Optical and Optoelectronic Polymers and Oligomers, G.E. Jabbour, B. Meijer, N.S. Sariciftci, T.M. Swager, 2001, ISBN: 1-55899-601-X
Volume 666— Transport and Microstructural Phenomena in Oxide Electronics, D.S. Ginley, M.E. Hawley, D.C. Paine, D.H. Blank, S.K. Streiffer, 2001, ISBN: 1-55899-602-8
Volume 667— Luminescence and Luminescent Materials, K.C. Mishra, J. McKittrick, B. DiBartolo, A. Srivastava, P.C. Schmidt, 2001, ISBN: 1-55899-603-6
Volume 668— II-VI Compound Semiconductor Photovoltaic Materials, R. Noufi, R.W. Birkmire, D. Lincot, H.W. Schock, 2001, ISBN: 1-55899-604-4
Volume 669— Si Front-End Processing—Physics and Technology of Dopant-Defect Interactions III, M.A. Foad, J. Matsuo, P. Stolk, M.D. Giles, K.S. Jones, 2001, ISBN: 1-55899-605-2
Volume 670— Gate Stack and Silicide Issues in Silicon Processing II, S.A. Campbell, C.C. Hobbs, L. Clevenger, P. Griffin, 2001, ISBN: 1-55899-606-0
Volume 671— Chemical-Mechanical Polishing 2001—Advances and Future Challenges, S.V. Babu, K.C. Cadien, J.G. Ryan, H. Yano, 2001, ISBN: 1-55899-607-9
Volume 672— Mechanisms of Surface and Microstrucure Evolution in Deposited Films and Film Structures, J. Sanchez, Jr., J.G. Amar, R. Murty, G. Gilmer, 2001, ISBN: 1-55899-608-7
Volume 673— Dislocations and Deformation Mechanisms in Thin Films and Small Structures, O. Kraft, K. Schwarz, S.P. Baker, B. Freund, R. Hull, 2001, ISBN: 1-55899-609-5
Volume 674— Applications of Ferromagnetic and Optical Materials, Storage and Magnetoelectronics, W.C. Black, H.J. Borg, K. Bussmann, L. Hesselink, S.A. Majetich, E.S. Murdock, B.J.H. Stadler, M. Vazquez, M. Wuttig, J.Q. Xiao, 2001, ISBN: 1-55899-610-9
Volume 675— Nanotubes, Fullerenes, Nanostructured and Disordered Carbon, J. Robertson, T.A. Friedmann, D.B. Geohegan, D.E. Luzzi, R.S. Ruoff, 2001, ISBN: 1-55899-611-7
Volume 676— Synthesis, Functional Properties and Applications of Nanostructures, H.W. Hahn, D.L. Feldheim, C.P. Kubiak, R. Tannenbaum, R.W. Siegel, 2001, ISBN: 1-55899-612-5
Volume 677— Advances in Materials Theory and Modeling—Bridging Over Multiple-Length and Time Scales, L. Colombo, V. Bulatov, F. Cleri, L. Lewis, N. Mousseau, 2001, ISBN: 1-55899-613-3
Volume 678— Applications of Synchrotron Radiation Techniques to Materials Science VI, P.G. Allen, S.M. Mini, D.L. Perry, S.R. Stock, 2001, ISBN: 1-55899-614-1
Volume 679E—Molecular and Biomolecular Electronics, A. Christou, E.A. Chandross, W.M. Tolles, S. Tolbert, 2001, ISBN: 1-55899-615-X
Volume 680E—Wide-Bandgap Electronics, T.E. Kazior, P. Parikh, C. Nguyen, E.T. Yu, 2001, ISBN: 1-55899-616-8
Volume 681E—Wafer Bonding and Thinning Techniques for Materials Integration, T.E. Haynes, U.M. Gösele, M. Nastasi, T. Yonehara, 2001, ISBN: 1-55899-617-6
Volume 682E—Microelectronics and Microsystems Packaging, J.C. Boudreaux, R.H. Dauskardt, H.R. Last, F.P. McCluskey, 2001, ISBN: 1-55899-618-4
Volume 683E—Material Instabilities and Patterning in Metals, H.M. Zbib, G.H. Campbell, M. Victoria, D.A. Hughes, L.E. Levine, 2001, ISBN: 1-55899-619-2
Volume 684E—Impacting Society Through Materials Science and Engineering Education, L. Broadbelt, K. Constant, S. Gleixner, 2001, ISBN: 1-55899-620-6
Volume 685E—Advanced Materials and Devices for Large-Area Electronics, J.S. Im, J.H. Werner, S. Uchikoga, T.E. Felter, T.T. Voutsas, H.J. Kim, 2001, ISBN: 1-55899-621-4

Prior Materials Research Society Symposium Proceedings available by contacting Materials Research Society

Theory, Modeling and Luminescence Phenomena

Mat. Res. Soc. Symp. Proc. Vol. 667 © 2001 Materials Research Society

Electronic Structures and Nature of Host Excitation in Gallates

P. C. Schmidt[+], J. Sticht[+], M. Stephan[+], V. Eyert[++] and K. C. Mishra*
[+] Institut für Physikalische Chemie,Technische Universität Darmstadt, Darmstadt, Germany
[++] Institut für Physik, Universität Augsburg, Augsburg, Germany
[*] Research and Development, OSRAM SYLVANIA, Beverly, MA

Dedicated to Dr. Franz Kummer on the Occasion of his 60th Birthday

ABSTRACT

It is an interesting exercise in materials science to explore simple rules relating the electronic properties of ternary systems to those of their binary constituents. In the present work, we have investigated the electronic structures of the large band gap gallates MGa_2O_4 (M=Mg, Ca, Ba and Zn) and the corresponding binary oxides MO and Ga_2O_3. Using first-principles band structure methods, we find that the metal atoms in MO control the width of the O 2p-like valence band and the size of the optical band gap. Covalent metal-oxygen bonding is much more pronounced in Ga_2O_3 and leads to characteristic structure in the valence band density of states. These basic features are retained in the ternary compounds where the covalent admixture to the chemical bond is largest between Ga and O, and the transitions across the band gap involve the Ga_2O_3 sublattice.

INTRODUCTION

There has been resurgence of interest in gallates, particularly zinc gallate, for application in thin film electro luminescent devices (TFEL), vacuum fluorescent displays (VFD), field emission display (FED) utilizing low-voltage cathode luminescence [1]. Gallates in M_2O-Ga_2O_3 and MO-Ga_2O_3 systems activated by divalent manganese was first investigated by Hoffman and Brown [2]. Magnesium gallate activated by manganese was used for a while as a green phosphor in specialty lamps. Recently zinc gallate has gained much attention as a blue emitter and manganese activated zinc gallate as a green emitter for VFDs and TFEL devices [1].

Understanding the electronic properties of the ternary oxides like $MgGa_2O_4$, $ZnGa_2O_3$ and $CaGa_2O_4$ from those of their binary building blocks CaO, MgO, ZnO and Ga_2O_3 is of importance from the perspective of materials design for specific application. This issue has already been the subject of our previous studies on a class of ternary oxides containing anionic groups with strong intra-atomic covalent bonding [3-5]. There we showed that the band gap transitions in the ternary compounds do not necessarily involve the highest occupied molecular orbitals (HOMO) and the lowest unoccupied molecular orbitals (LUMO) of the anionic groups.

In the present work, we report results from electronic structure calculations of the ternary gallates MGa_2O_4 (M=Mg, Ca, Ba and Zn) and their binary building blocks, MO and Ga_2O_3. In particular, we investigate how the HOMO-LUMO transitions change on going from the binary compounds MO and Ga_2O_3 to the ternary compounds. A systematic understanding of the correlations among the binary oxides and higher order oxides is helpful for predicting physical and chemical properties of more complex systems, and for engineering materials for specific purpose.

THEORY

The band structures of the compounds under study are computed using the augmented spherical wave (ASW) method [6] in a new scalar-relativistic implementation [7]. The calculations are based on density functional theory and the local density approximation. Since the ASW method uses the atomic sphere approximation with spherically symmetric potentials within Wigner-Seitz spheres of radii R_{AS} centered at the nuclei, we used empty spheres to fill the interatomic space in the open crystal structures associated with these oxides. These empty spheres can be considered as pseudo atoms without a nucleus, and are useful in modeling the correct shape of the crystal potential within large voids.

The ASW calculations were complemented by calculations using the new full potential (NFP) method [8,9]. Results from both approaches were found very similar. In both the ASW and NFP programs we used a minimal basis set (Mg (2s,2p), Ca (4s,4p,3d), Ga(4s,4p,3d), Zn(4s,4p,3d), Ba(6s,6p,5d) and O(2s,2p)) for the valence electrons comprising only one principal quantum number for each angular momentum state.

CRYSTAL STRUCTURES

MgO, BaO and CaO crystallize in the NaCl structure [10], and ZnO in the hexagonal wurtzite structure [11]. The β-form of gallium sesquioxide crystallizes in a monoclinic lattice with space group C2/m (No. 12) [12]. There are two kinds of Ga sites located in octahedral (Ga_2) and tetrahedral (Ga_1) sites. In contrast, oxygen atoms are arranged in a distorted close-packed structure. Metal-oxygen distances range from 1.80 to 1.85 Å in the tetrahedra and from 1.95 to 2.08 Å in the octahedra.

Several gallates crystallize in the (partially) inverse or normal spinel structure with two different cation sites, which are either tetrahedrally or octahedrally coordinated. $MgGa_2O_4$ is believed to have a partially inverse spinel structure [13,14]. However, in order to simplify the calculations, we have assumed a complete inverse spinel structure (space group Fd-3m, Nr. 227) using the structure data of Casado and Rasines [14] with Ga_1 on the tetrahedral and Mg and Ga_2 on the octahedral site.

The structure of $CaO\text{-}Ga_2O_3$ is more complicated than that of the corresponding Mg compound [15,16]. Beside many other phases, two stable $CaGa_2O_4$ phases exist, the orthorhombic o-$CaGa_2O_4$ phase and the monoclinic m-$CaGa_2O_4$ phase [16], which crystallize in silicate type of structure. In both the structures, one finds distorted GaO_4 tetrahedra, which form six-member rings. The two structures are distinguished by the orientation of the tetrahedra within the rings.

$ZnGa_2O_4$ crystallizes in the normal spinel structure [17] with Zn-O and Ga-O distances of 1.973 Å and 1.990 Å. Finally $BaGa_2O_4$ crystallizes in a hexagonal structure [18] with Ga in a tetrahedral coordination with d(Ga-O) = 1.726 – 1.819 Å and Ba in a 6+2 surrounding with d(Ba-O) = 2.845 – 3.100 Å.

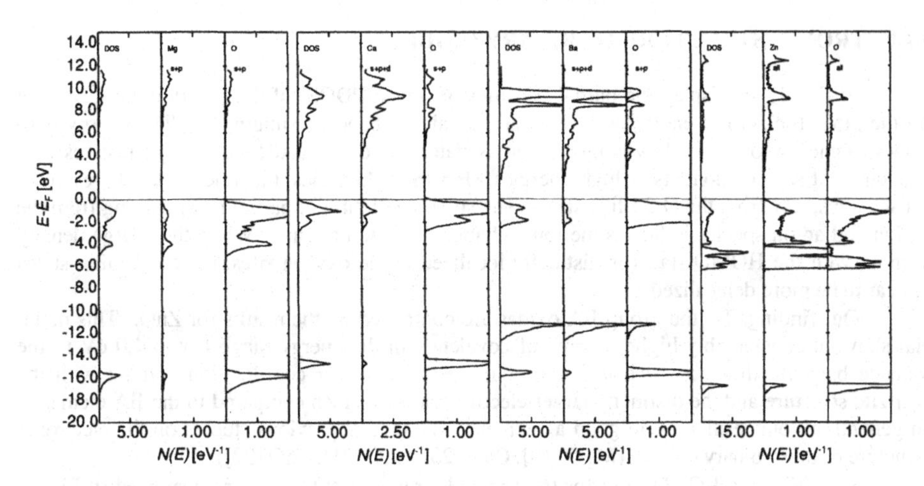

Figure 1: Density of States (DOS) and Partial Density of States (PDOS) of MgO, CaO, BaO and ZnO.

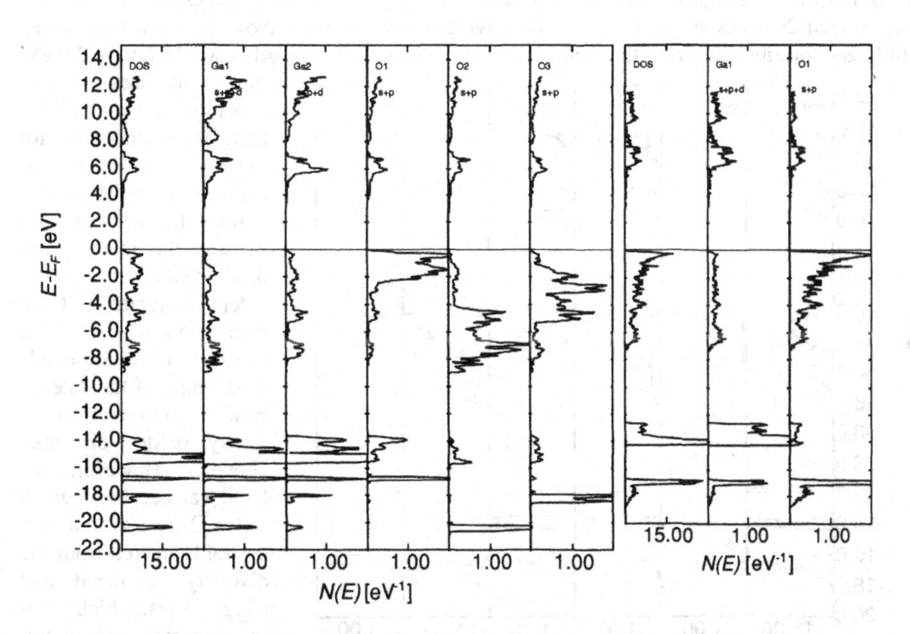

Figure 2: Density of States (DOS) and Partial Density of States (PDOS) of β-Ga_2O_3 (left) and α-Ga_2O_3 (right).

ELECTRONIC STRUCTURE OF BINARY OXIDES

In Fig. 1, we display the partial densities of states (PDOS) of the binary oxides, MO. In all the plots, the zero of energy was fixed at the valence band maximum (E_F). The energy window was chosen to include low lying oxygen 2s states as well as mixed metal-oxygen states with predominant s-p components at high energies. For the IIA oxides, the valence bands result almost exclusively from the O 2p-like states. The small admixture of partial DOS coming from the cation and anion species indicates the ionic character of the bonding. While the charge density coming with the HOMO states is distinctly localized at the oxygen sites, the conduction states appear to be more delocalized.

Our findings for the group IIA oxides are contrasted by the results for ZnO. This material shows a considerably higher degree of covalency in the energy range from –4.0 eV to the valence band maximum, which is a consequence of the lower coordination coming with the wurtzite structure and the distinctly larger electronegativity of Zn compared to the IIA elements. In general, all our results are in good agreement with several previous studies on the electronic structure of these binary oxides (MgO [19], CaO [20], BaO [21], ZnO [22]).

The DOS for β-Ga_2O_3 and for the trigonal structure (α-Ga_2O_3) are displayed in Fig. 2. As we will see below, taking both structures into consideration allows to discuss the differences arising from the non-equivalent atomic sites in the β-phase. For both phases, we observe low lying oxygen 2s states at –17.0 eV and high lying mixed sp states above the optical band gap with larger contributions from the metal states. In the energy range between –16.0 to –12.0 eV, the Ga 3d states give rise to a high partial DOS. Finally, the valence band for the lower valence states is made from similar contributions from the oxygen 2p and the metal states this indicating a large covalent character of these metal-oxygen states. It causes a considerable broadening of the valence band as compared to the binary oxides discussed above. However, the large valence band width in Ga_2O_3 results to a considerable degree from the diversity of metal and oxygen sites, which leads to a variety of metal-oxygen distances and coordinations. This situation leads, in particular, to the differences in oxygen

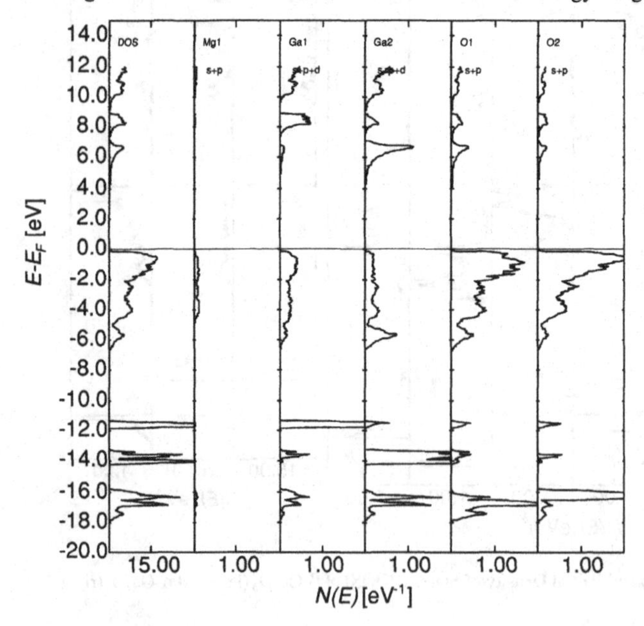

Figure 3: Density of States (DOS) and Partial Density of States (PDOS) of $MgGa_2O_4$.

partial DOS in β-Ga_2O_3, where O_1 and O_2 dominate in the upper and lower valence band region, respectively.

ELECTRONIC STRUCTURE OF GALLATES

The electronic structures of the gallates MGa_2O_4 may be essentially regarded as a superposition of the electronic states of the binary building blocks MO and Ga_2O_3, which we outlined in the previous section. While the main features apparently can be attributed to gallium sesquioxide, the change of the group II metal causes slight but distinct differences in the series MGa_2O_4. They trace back to the different spacings induced by the different sizes of the metal atoms and are rather independent from the details of the crystal structures. As it turns out, the optical properties and the character of the conduction band of the gallates can be predicted by the relative size of the band gaps of MO and Ga_2O_3. The partner with the smaller optical band gap eventually determines the optical band gap of the gallates.

In Fig. 3 we show the DOS of $MgGa_2O_4$. The O 2s and the Ga 3d like states are located below –10.0 eV. Because of the different sites of Ga, the position of the corresponding 3d like states differ by about 2.0 eV and are found at about –14.0 and –12.0 eV. Above these states we find the O 2p like states. The width of these bands of about 6.0 eV is slightly smaller than in Ga_2O_3. As expected from MgO the magnesium contribution in this energy range is very small indicating again a predominantly Mg-O ionic bonding. In contrast Ga-O bonding shows some covalent character. The lower conduction band states are very similar to those of Ga_2O_3. The amplitudes of these states are concentrated at O and Ga whereas we find almost no contribution of Mg. As mentioned above this finding is because the optical band gap of MgO is about 3.0 eV larger than that of Ga_2O_3.

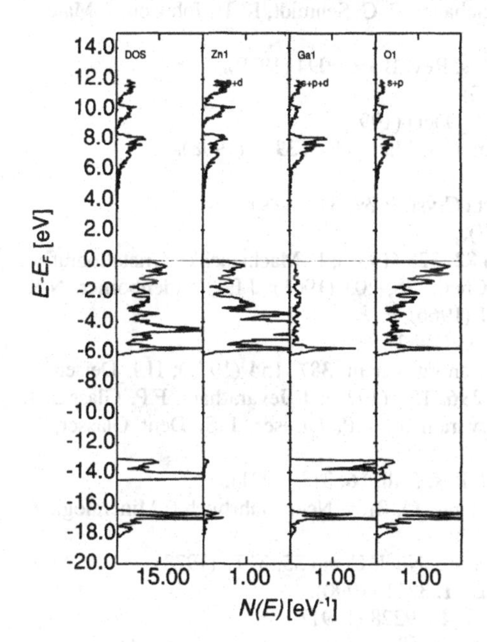

Figure 4: Density of States (DOS) and Partial Density of States (PDOS) of $ZnGa_2O_4$

Apart from the fact that the crystal structure of $CaGa_2O_4$ is more complicated, the gross features of its electronic structure are similar to those of $MgGa_2O_4$. However due to the higher number of non-equivalent sites, the DOS shows a large number of peaks. The situation is similar in $BaGa_2O_4$. The ordering of the partial DOS appears to be a superposition of the PDOS of the corresponding binary compounds. The band structure of both, $CaGa_2O_4$ and $BaGa_2O_4$, will be published elsewhere.

Finally, we display in Fig. 4 the partial DOS of $ZnGa_2O_4$. Since ZnO has a smaller optical band gap than Ga_2O_3 and the electronegativities of Ga and Zn are almost the same, the conduction band comprises similar contributions from all atoms. However, differences from the other three gallates stem essentially from the composition of the valence bands, which are dominated by the Zn 3d states. Only in the near gap region these states are superseded by the O 2p states. As a consequence, the charge density is not concentrated in the Ga_2O_3 sublattice but distributed over all three kinds of atoms.

CONCLUSION

In this report, we have discussed the electronic properties of the group II gallates MGa_2O_4 and related them to the binary building blocks MO and Ga_2O_3. The electronic structures of the ternaries arise from superposition of the electronic states of the respective binary partners. The nature of the chemical bond and the HOMO-LUMO transitions can be roughly predicted from the electronic properties of the binary counterparts.

REFERENCES

1. T. Minami, S. Takata, Y. Kuroi, T. Maeno, Soc. Inf. Dis. Sym. Digest **26**, 724 (1995), see references therein.
2. C.W.W. Hoffman, J.J. Brown, J. Inorg. Nucl. Chem. **30**, 63 (1968).
3. P.C. Schmidt, J. Sticht, V. Eyert, K.C. Mishra, Mat. Res. Soc. Symp. Proc. **560**, 323 (1999)
4. K.C. Mishra, B.G. DeBoer, P.C. Schmidt, I. Osterloh, M. Stephan, V. Eyert, K.H. Johnson, Ber. Bunsenges. Phys. Chem. **102**, 1772 (1998).
5. K. C. Mishra, I. Osterloh, H. Anton, B. Hannebauer, P. C. Schmidt, K. H. Johnson, J. Mater. Res. **12**, 2183 (1997).
6. A.R. Williams, J. Kübler, C.D. Gelatt Jr., Phys. Rev. B **19**, 6094 (1979).
7. V. Eyert, Int. J. Quant. Chem. **77,** 1007 (2000).
8. M. Methfessel, NFP Manual, IHP, Frankfurt (Oder) (1997).
9. E. Bott, M. Methfessel, W. Krabs, P.C. Schmidt, J. Math. Phys. **39**, 1 (1998).
10. W. Gerlach, Z. Phys. **9**, 184 (1922).
11. J. Albertsoon, S.C. Abrahams, A. Kvick, Acta Cryst. B **39**, 34 (1989).
12. S. Geller, J. Solid State Chem. **20**, 209 (1977).
13. T.F.W. Barth, E.Z. Pasnjak, Kristallografiya **82**, 325 (1932); F. Machatschki, Kristallografiya **82**, 348 (1932); H. Schmalzried, Z. Phys. Chem. **28**, 203 (1961); J.E. Weidenborner, N.R. Stemple Y. Okaya, Acta Crystallogr. **20**, 761 (1966).
14. P.G. Casado, I. Rasines, Z. Kristallogr. **160**, 33 (1982).
15. H.J. Deiseroth, H.K. Müller-Buschbaum, Z. anorg. Chem. **387**, 154 (1972); H.J. Deiseroth, H.K. Müller-Buschbaum, Z. anorg. Chem. **396**, 157 (1973); J. Jevaratham, F.P. Glasser, J. Amer. Ceram. Soc. **44**, 563 (1961); J. Jevaratham, F.P. Glasser, L.S. Dent Glasser, Z. Kristallogr. **118**, 257 (1963).
16. S. Ito, S. Banno, T. Kawano, K. Suzuki, Mat. Res. Bull. **16**, 313 (1981).
17. M. Wendschuh-Josties, H.St.C. O'Neill, K. Bente, G. Brey, Neues Jahrbuch f. Mineralogie **6**, 273 (1950).
18. H.J. Deiseroth, H.K. Müller-Buschbaum, J. Inorg. Nucl. Chem **35**, 3177 (1973).
19. Q.S. Wang, N.A.W. Holzwarth, Phys. Rev. B **41**, 3211 (1990).
20. R. Pandey, J.E. Jaffe, A.B. Kunz, Phys. Rev. B **43**, 9228 (1991).
21. M. Springborg, O.E. Taurian, J. Phys. C **19**, 6347 (1986).
22. W.F. Brinckmann, T.M. Rice, B. Bill, Phys. Rev. **8**, 1570 (1973).

Mat. Res. Soc. Symp. Proc. Vol. 667 © 2001 Materials Research Society

ADVANCES IN THE DEVELOPMENT OF QUANTUM SPLITTING PHOSPHORS

A.A. Setlur[1], H.A. Comanzo[1], A.M. Srivastava[1], W.W. Beers[2], W. Jia[3], S. Huang[4], L. Lu[4], X. Wang[4], and W.M. Yen[4]
[1]GE Corporate Research and Development, Niskayuna, NY 12309
[2]GE Lighting, Cleveland, OH 44110
[3]Department of Physics and Astronomy, University of Puerto Rico, Mayaguez, PR 00681
[4]Department of Physics and Astronomy, University of Georgia, Athens, GA 30602

ABSTRACT

Quantum splitting phosphors (QSPs) are phosphors that could convert VUV radiation into more than one visible photon. These phosphors have the potential to improve the efficacy of current Hg fluorescent lamps and/or Xe lamps by reducing the Stokes shift energy loss after VUV excitation (λ_{exc}=185 nm for Hg lamps or 147 nm and 172 nm for Xe lamps provided the emission color of the phosphor matches the eye sensitivity. The current technology in QSPs and their potential limitations will be discussed in this paper. At GE-CRD, we have discovered and developed QSPs that meet the requirements for use in current Hg based fluorescent lamps. The steady state and time resolved optical properties of one of these phosphors, $SrAl_{12}O_{19}:Pr^{3+},Mg^{2+}$, has been measured to estimate the maximum quantum efficiency and onset of concentration quenching in this phosphor. The maximum quantum efficiency for $SrAl_{12}O_{19}:Pr^{3+},Mg^{2+}$ has been calculated to be ~125-135% for 185 nm excitation with an upper bound on the Pr^{3+} doping level of ~1%.

INTRODUCTION

Phosphors which have quantum efficiencies for visible light greater than unity are well known for cathode ray or x-ray excitation. However, the efficiency of phosphors that convert ultraviolet (UV) or vacuum ultraviolet (VUV) light into visible light is typically less than unity. If phosphors are developed which emit more than one photon for each incident UV or VUV photon absorbed, a quantum splitting phosphor (QSP), this could significantly improve the efficacy of current Hg or Xe based fluorescent lamps. The current focus of QSP research is develop visible quantum splitting phosphors to minimize energy loss upon VUV excitation, either from the 185 nm line from Hg lamps or the 172 nm/147 nm lines from Xe lamps.

One potential route to developing QSPs is to use energy transfer and cross-relaxation between various rare earth ions for the emission of two visible photons after VUV excitation. This has been clearly demonstrated in $LiGdF_4:Eu^{3+}$ which potentially has a quantum efficiency of 190% upon excitation into the ^6G_J levels of Gd^{3+} at ~200 nm (Figure 1) [1]. After VUV excitation, the $^6G_J \rightarrow {}^6P_J$ transition is resonant with the $^7F_I \rightarrow {}^5D_0$ transition, leading to cross-relaxation and energy transfer to a Eu^{3+} ion. The Eu^{3+} ion then emits a orange-red photon, the first photon in this scheme. After the cross-relaxation between Gd^{3+} and Eu^{3+}, the ^6P_J level of Gd^{3+} is populated and this energy can then migrate along the Gd^{3+} sub-lattice until it reaches a Eu^{3+} ion, giving the second visible photon. Other quantum splitting schemes involving Er^{3+}-Gd^{3+}-Tb^{3+} in $LiGdF_4$ also have been demonstrated [2], but their efficiency has been estimated to be less than unity.

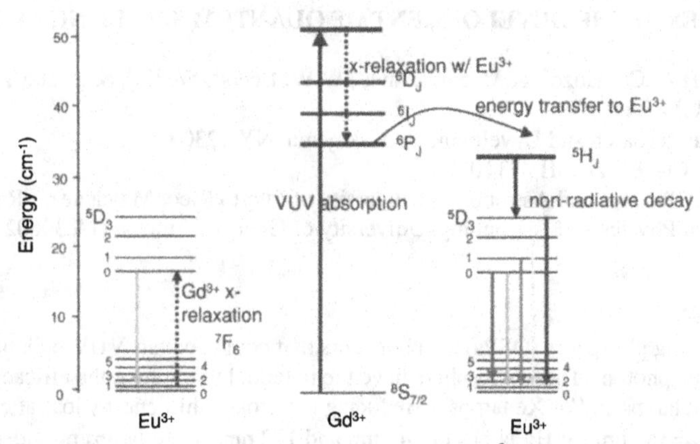

Figure 1. Schematic of quantum splitting process in $LiGdF_4:Eu^{3+}$ (drawn from Ref. 1)

Another potential set of quantum splitting phosphors are based upon Pr^{3+} doped hosts [3]. The energy levels of Pr^{3+} are such that two visible photons can be emitted after excitation with one VUV photon (Figure 2), a quantum splitting process. This process occurs by Pr^{3+} absorbing VUV radiation through the allowed 4f→5d transition, non-radiative decay into the $^1S_0$ state at ~46500 cm^{-1}, emission to the $^1I_6$ level (405 nm radiation), non-radiative decay into the $^3P_0$ state, and $^3P_0$ emission (480-700 nm). This process requires the onset of the allowed $4f^2$→4f5d absorption transition to be at high energy and above the $^1S_0$ state. If the $4f^2$→4f5d absorption bands are below the $^1S_0$ state, the emission after VUV excitation will be either UV 4f5d→$4f^2$ emission bands [4] or $^3P_0$ or $^1D_2$ emission which can be directly fed by the 4f5d band [5]. In either case, the quantum efficiency of these phosphors can never exceed 100%.

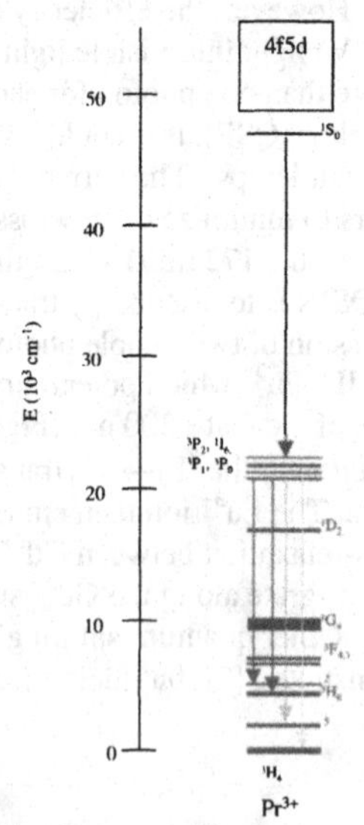

Figure 2. Schematic of the Pr^{3+} quantum splitting process. Note that the 4f5d band has to be above the $^1S_0$ level for quantum splitting to occur.

The initial discovery of Pr^{3+} QSPs was made at GE-CRD and Philips in 1974 in $YF_3:Pr^{3+}$ (Figure 3) [3]. The workers at GE-CRD found that the visible (400-700 nm) QE of $YF_3:Pr^{3+}$ was 140% and the total QE (with UV transitions) was 156% under 185 nm excitation. Other fluoride phosphors doped with Pr^{3+} were also found to be quantum splitting [3]. As detailed above, there have been many significant discoveries of quantum splitting phosphors using fluoride host lattices. However, fluoride phosphors cannot be implemented into Hg low pressure fluorescent lamps for the following reasons:

- Fluorides would have significant environmental and safety issues with their synthesis.
- There are many potential reactions for fluorides during the lamp making process such as reactions with H_2O in suspension and stability during the lehring process.
- Fluorides have a high affinity for Hg, leading to blackening of the phosphor, destroying the brightness.
- Under UV radiation and ion bombardment, fluorides would tend to reduce and damage, forming color centers, which would reduce the QE significantly.

It is unlikely that fluoride host lattices will be amenable to the many requirements of low pressure Hg fluorescent lamp manufacturing. Consequently, oxide QSPs need to be developed. This presents a problem for Pr^{3+} QSPs in that oxides generally have stronger crystal fields and covalency effects compared to fluorides, leading to a lower energy onset of the $4f \rightarrow 5d$ absorption band. Minimizing crystal field and covalency effects requires hosts with long Me-O bond lengths and high coordination number (crystal field strength α $1/R^5$). Using this as an initial guideline, GE-CRD was able to discover three new hosts for Pr^{3+} quantum splitting: $SrAl_{12}O_{19}$, LaB_3O_6, and $LaMgB_5O_{10}$ [6]. These discoveries give an opportunity to use Pr^{3+} quantum splitting in current Hg fluorescent lamp technologies. Note that the main emission peak in Pr^{3+} based QSPs is at ~405 nm, where the eye sensitivity is low. Optimal use of Pr^{3+} based QSPs will also require energy transfer schemes to convert these photons into useful visible light.

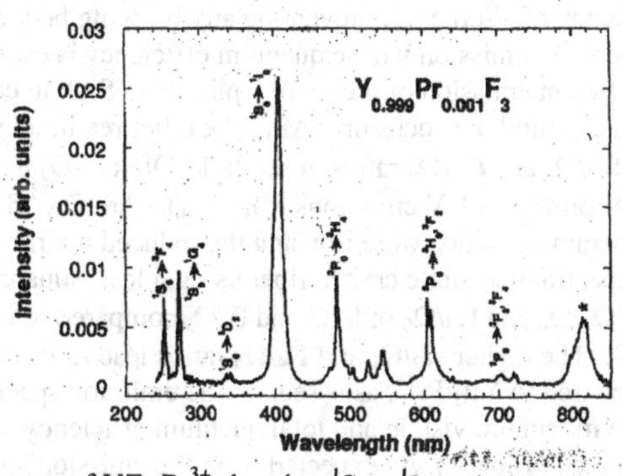

Figure 3. Emission spectra of $YF_3:Pr^{3+}$ showing $^1S_0 \rightarrow ^1I_6$ emission, the signature of Pr^{3+} quantum splitting.

EXPERIMENTAL METHODS

Solid state synthesis methods were used to make $SrAl_{12}O_{19}:Pr^{3+},Mg^{2+}$ (SAP), LaB_3O_6, and $LaMgB_5O_{10}$ (6). For SAP, $SrCO_3$, Al_2O_3, Pr_6O_{11}, and MgO were stoichiometrically mixed and fired at 1400°C in a reducing atmosphere. The borate phosphors were made using solid state synthesis using stoichiometric mixtures of La_2O_3, Pr_6O_{11}, MgO, and H_3BO_3 fired at 900-1000°C in reducing atmospheres. Single crystals fibers of SAP were grown using the laser heated pedestal growth method [7]. Single crystal samples were polished on both ends to optical quality.

Absorption spectra of single crystal fibers were measured using a Cary-14R UV/visible spectrometer. Emission spectra were measured using a Spex Fluoromax 2 spectrofluorometer with a resolution of 0.3 nm. Time resolved measurements from 80-300 K were also made in this work. Samples in a cryostat were excited with the 193 nm line of a PSX-100 excimer laser (MPB Technologies) using ArF as the laser medium. Fluorescence was measured through an optical fiber coupled to a Spex 500 M monochromator. The signal was detected by either a CCD (Princeton LN/CCD-1340/100-E-1) or by a PMT (Hamamatsu R-636-10) coupled with a boxcar (Stanford SR-250) for lifetime measurements.

RESULTS AND DISCUSSION

Comparison of the emission spectra of $SrAl_{12}O_{19}:Pr^{3+}$, Mg^{2+} (SAP) and $LaMgB_5O_{10}:Pr^{3+}$ (Figures 4a and 4b) shows the presence of the $^1S_0 \rightarrow ^1I_6$ emission line for both phosphors with $^3P_0 \rightarrow ^3H_4$ emission from SAP and weak $^1D_2 \rightarrow ^3H_4$ emission for $LaMgB_5O_{10}:Pr^{3+}$. This clearly shows that emission from the $^1S_0$ state does not necessarily mean that the phosphor has a total quantum efficiency (QE) significantly greater than unity. In borate QSPs, the maximum phonon frequency is large enough (~1400 cm^{-1}) so that the gap between the $^3P_0$ and $^1D_2$ levels is efficiently bridged by multiphonon relaxation. Due to energy migration and cross-relaxation between Pr^{3+} ions, the QE of red emission from the $^1D_2$ level is much lower than the QE of $^3P_0$ emission which reduces the overall QE for $LaMgB_5O_{10}:Pr^{3+}$ as seen in the emission spectra. This adds another requirement for efficient Pr^{3+} QSPs: the maximum phonon frequency must be low to prevent bridging of the gap between the $^3P_0$ and $^1D_2$ levels. Aluminates such as SAP have a maximum phonon frequency of ~700 cm^{-1}, preventing any bridging between the $^3P_0$ and $^1D_2$ states. This leads to efficient $^3P_0$ emission whose quantum efficiency is estimated to be close to unity (see below). The efficient emission of the second photon in SAP in comparison to the borate QSPs is the rationale behind our focus on SAP. The other requirements for a high visible QE in Pr^{3+} QSPs are low Ω_2/Ω_6 and Ω_4/Ω_6 ratios of the Judd-Ofelt (J-O) parameters which will give greater $^1S_0 \rightarrow ^1I_6$ emission versus UV emissions from $^1S_0$ to the $^1D_2$, $^1G_4$, and $^3F_4$ levels [9].

Judd-Ofelt (J-O) parameter ratios were fit using the reduced matrix elements for Pr^{3+} [10] to the visible absorption spectra of a single crystal fiber using a least squares method. We determined J-O ratios for Ω_2/Ω_6 and Ω_4/Ω_6 of 0.23 and 0.59, compared to Ω_2/Ω_6=0.013 and Ω_4/Ω_6=0.070 for $YF_3:Pr^{3+}$. The higher Ω_2/Ω_6 and Ω_4/Ω_6 ratios lead to much more intense UV $^1S_0$ emissions in SAP compared to $YF_3:Pr^{3+}$, as is seen in the emission spectra (Figures 3 and 4). Using these J-O ratios, the maximum visible and total quantum efficiency (QE) for SAP under VUV excitation was estimated (Table 1). As expected from the emission spectra, the maximum visible and total QE for SAP is lower than that of $YF_3:Pr^{3+}$. These calculations assume that the $^1S_0$ quantum efficiency is 100%, and that there is no concentration quenching of either the $^1S_0$ or $^3P_0$ luminescence. To determine the onset of concentration quenching, the decay times of the $^1S_0$ and $^3P_0$ emissions for various compositions and temperature in SAP were measured (Tables 2 and 3).

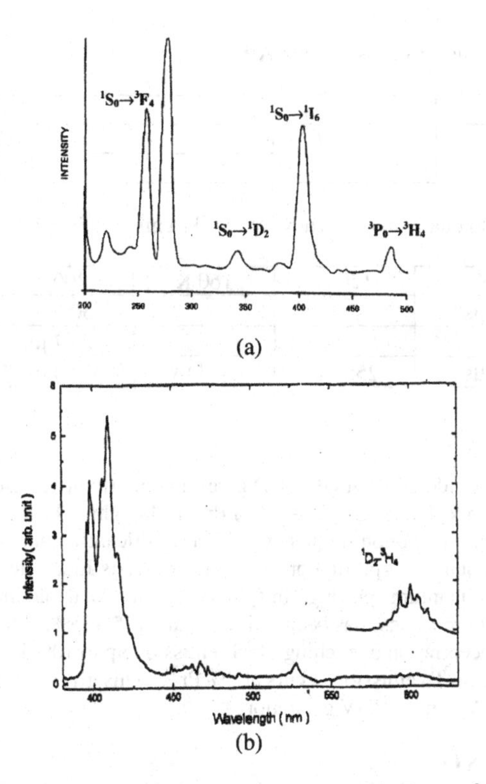

Figure 4. Emission spectra under 185 nm excitation for (a) $SrAl_{12}O_{19}:Pr^{3+}$, Mg^{2+} and (b) $LaMgB_5O_{10}:Pr^{3+}$. The red emission from $^3P_0$ in SAP is not shown due to instrumental limitations.

Table 1. Calculated quantum efficiencies for $SrAl_{12}O_{19}$: Pr^{3+}, Mg^{2+} and $YF_3:Pr^{3+}$

	Ω_2/Ω_6	Ω_4/Ω_6	$^1S_0 \rightarrow ^1I_6$ branching ratio	Visible QE	Total QE
Our J-O ratios	0.23	0.59	30%	53%	122%
Previously published J-O ratios [8]	0.12	0.23	45%	80%	135%
$YF_3:Pr^{3+}$ [3]	0.013	0.07	79%	135%	155%

Decay times were obtained by fitting decay curves to a single exponential function. Concentration quenching does not occur until 10% Pr^{3+} doping for the $^1S_0$ emission; however, it appears that concentration quenching occurs at <5% Pr^{3+} for $^3P_0$ emission. This is confirmed by measuring the quantum efficiency of the $^3P_0$ emission using the quantum splitting process. After population of the $^1S_0$ level, the only path to feed the $^3P_0$ state is the quantum splitting process. Consequently, the ratio of the total visible emission intensity from $^3P_0$ to the $^1S_0 \rightarrow ^1I_6$ emission intensity gives the QE for $^3P_0$ emission (Table 3). The onset of $^3P_0$ concentration quenching determined by decay time measurements corresponds well with the measurements of $^3P_0$ emission QE as well as theoretical estimates of the $^3P_0$ QE.

Table 2. Decay times for the $^1S_0$ emission in SAP

Pr^{3+}/Temperature	80 K	300 K
1%	673 ns	658 ns
10%	613 ns	591 ns

Table 3. Decay times and quantum efficiency for the $^3P_0$ emission in SAP

Pr^{3+}/T	80 K	120 K	160 K	300 K	$^3P_0$ QE
1%	31.7 μs			30.0 μs	71%
5%				23.7 μs	47%
10%	27.5 μs	25.2 μs	24.3 μs	18.3 μs	31%

CONCLUSIONS

The discovery of oxide QSPs at GE-CRD gives an opportunity to use quantum splitting phosphors in current lamp technologies. $SrAl_{12}O_{19}:Pr^{3+}$ is thought to be the most efficient oxide QSP due to its low maximum phonon frequency, which significantly improves the efficiency of the second photon in the quantum splitting process. However, its Judd-Ofelt parameters are less than optimal for efficient quantum splitting. In spite of this, the total maximum quantum efficiency (including UV emissions) has been estimated at ~125-135%. We have established that there is minimal concentration quenching of $^1S_0$ emission up to 10% Pr^{3+} and that concentration quenching of $^3P_0$ emission occurs at 5% Pr^{3+}. This gives the initial boundaries for maximizing the QE of SAP under VUV excitation.

ACKNOWLEDGEMENTS

This work was partially supported by the Department of Energy under contract DE-FC26-99FT40632.

REFERENCES

1. R. T. Wegh, H. Donker, K. D. Oskam, and A. Meijerink, Science **283**, 663 (1999).
2. R. T. Wegh, E. V. D. van Loef, and A. Meijerink, J. Lumin. **90**, 111 (2000).
3. W. W. Piper, J. A. DeLuca, and F. S. Ham, J. Lumin. **8**, 344 (1974); J.L. Sommerdijk, A. Bril, and A. W. de Jager, J. Lumin., **8**, 341 (1974).
4. K. H. Yang and J.A. DeLuca, Phys. Rev. B **17**, 4246 (1974).
5. H. E. Hoefdraad and G. Blasse, Phys. Stat. Sol. A **29**, K95 (1975).
6. A. M. Srivastava and W. W. Beers, J. Lumin. **71**, 285 (1997); A. M. Srivastava, D. A. Doughty, and W. W. Beers, J. Electrochem. Soc. **143**, 4113 (1996); A. M. Srivastava, D. A. Doughty, and W. W. Beers, J. Electrochem. Soc. **144**, 190 (1997).
7. W. M. Yen, J. Alloy and Comp. **193**, 175 (1993).
8. L. D. Merkle, B. Zandi, R. Moncorge, Y. Guyot, H. Verdun, and B. McIntosh, J. Appl. Phys. **79**, 1849 (1996).
9. R. Pappalardo, J. Lumin. **14**, 159 (1976).
10. M. J. Weber, J. Chem. Phys. **48**, 4774 (1968).

Mat. Res. Soc. Symp. Proc. Vol. 667 © 2001 Materials Research Society

Development of an organic dye solution for laser cooling by anti-Stokes fluorescence

Jarett L. Bartholomew, Peter A. DeBarber, Bauke Heeg, and Garry Rumbles[1]
MetroLaser, Inc.
Irvine, CA 92614, U.S.A.
[1]Center for Electronic Materials and Devices,
Imperial College, London, U.K.

ABSTRACT

Several independent groups have observed optical cooling by means of anti-Stokes luminescence in condensed media. The most promising materials are grouped into two categories: ion-doped glasses and organic dye solutions. It is this latter group that we focus our efforts on. Recent studies by our group show that irradiating a solution of rhodamine 101 in the long wavelength wing of the absorption spectrum results in the observation of optical cooling. To improve upon the initial observation of a few degree drop in temperature requires a better understanding of the conditions and phenomena leading to anti-Stokes luminescence in dye solutions. We develop a thermal lensing experiment to obtain fluorescence quantum yields of various dye solutions. The importance of concentration, choice of solvent, deuteration, and acidification are discussed.

INTRODUCTION

It is easy to see why the concept of cooling by anti-Stokes luminescence is so counter-intuitive. Not only does our experience with high power lasers teach us that laser light is very capable of heating, cutting, drilling, and vaporizing, early debate concerning the validity of anti-Stokes fluorescence generated much heat. Arguments for the existence of anti-Stokes fluorescence began when Pringsheim rebuffed Lenard's statement that anti-Stokes fluorescence violates the second law of thermodynamics [1]. This excited exchange heated up in the 1940s concerning the possibility of cooling by anti-Stokes fluorescence [2-4]. Landau settled the argument when he addressed the entropy change that occurs in the process of photoluminescence. Landau showed that the increase in the entropy for photoluminescence lies in the photon gas; the emitted photons are incoherent and occur over 4π steradians, thus accounting for the necessary increase in entropy.

Today, several groups have observed cooling by anti-Stokes luminescence [5-9]. In this paper, we examine specific factors affecting the anti-Stokes cooling performance including concentration, choice of solvent, deuteration, and acidification. We first present a brief synopsis of the theory underlying the phenomenon of cooling by anti-Stokes luminescence. This is followed by development of a thermal lensing experiment to measure fluorescence quantum yields for various dye solutions. The results of the thermal lensing experiments are presented and discussed. And finally, we conclude by making recommendations for developing optical refrigeration based on dye compounds.

BACKGROUND THEORY

The physics behind laser cooling by anti-Stokes luminescence in dyes relies on excitation into the low energy absorption wings of the dye spectrum. The excited states are deactivated by a combination of decay paths. It is the non-radiative pathways that lead to an increase in the sample's thermal energy. Dye molecules with near unit fluorescence quantum yield have negligible non-radiative decay paths and are therefore good candidates for laser cooling. If the resultant fluorescence is blue-shifted from the initial excitation wavelength, the emission is termed anti-Stokes fluorescence. The anti-Stokes fluorescence is attributable to excitation of high vibrational states of the ground electronic state. For each photon absorbed, a local cooling effect takes place by the removal of vibrational energy.

The power available, P_{th}, for altering a sample's thermal state by laser irradiation is given by

$$P_{th} = P_{abs}\left(1-\eta_{sol}\frac{\lambda_0}{\lambda^*}\right) \quad (1)$$

where P_{abs} is the absorbed power, η_{sol} is the fluorescence quantum yield of the solution, λ_0 is the wavelength of the pump light, and λ^* is the average wavelength of fluorescence emission. If $\lambda_0 > \lambda^*$ and η_{sol} is sufficiently large, P_{th} will be negative, and cooling will occur. If P_{th} is positive, then heating occurs. This relationship is evident in figure 1.

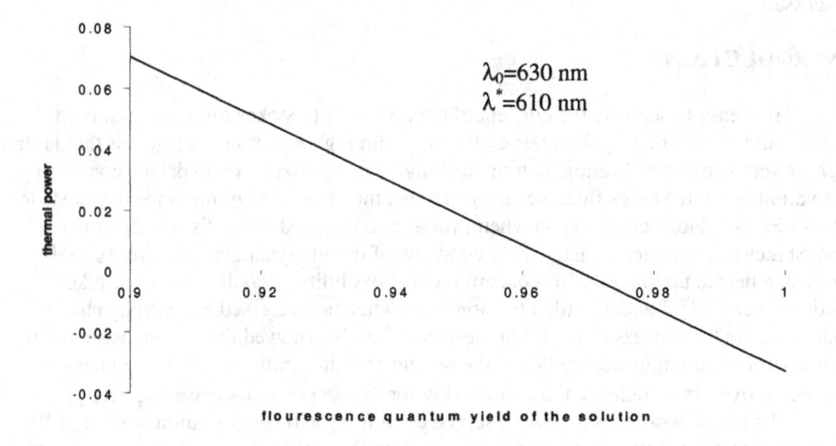

Figure 1. Plot showing the relationship between available thermal power and the fluorescence quantum yield.

EXPERIMENT

To optimize a solution for the highest possible η_{sol}, we developed a thermal lensing technique that allows us to measure the absolute fluorescent quantum yields of various dye solutions. As a laser beam is focused into a partially absorbing solution, the index of refraction in the solution decreases in the illuminated volume forming a divergent thermal lens. This is schematically shown in figure 2. As a result of the nascent thermal lens, the beam diverges (dashed line in figure) and the signal on the detector decreases until it reaches a steady state. Since the rate at which the thermal lens is formed is proportional to the power absorbed by the solution, a method of measuring the fluorescent quantum yield becomes possible.

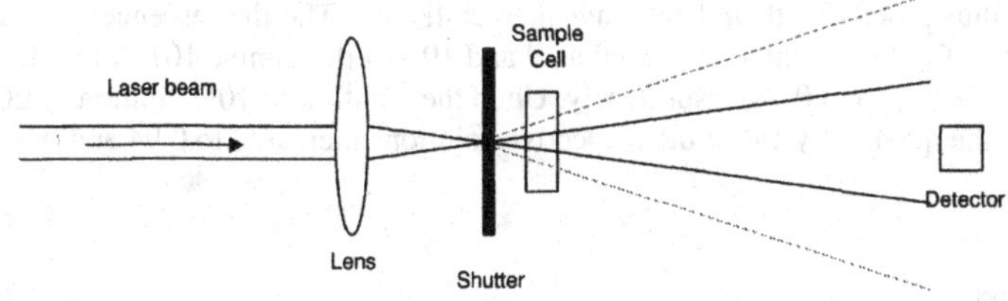

Figure 2. The thermal lens experiment is setup with four simple components, a lens, a shutter, a sample cell, and a detector.

The intensity on the detector obeys the expression

$$I(t) = I_0\left[1-\theta\left(1+\frac{t_c}{2t}\right)^{-1}+\frac{1}{2}\theta^2\left(1+\frac{t_c}{2t}\right)^{-2}\right]^{-1} \tag{3}$$

where the parameter t_c is the characteristic time of diffusion. The diffusion time is a function of the beam waist, the density, the heat capacity, and the heat conductivity of the solution. The quantity θ is directly proportional to P_{th}, the thermal power deposited in the sample. P_{th} may be calculated from the thermo-optic constants; however, this approach is flawed because the thermo-optic constants are not known to arbitrary precision. To avoid this problem, we use a reference compound with a nearly zero fluorescence quantum yield dissolved in the solvent [10,11]. By taking the ratio of P_{th} for the dye solution and P_{th} for the reference the thermo-optic constants cancel out. The fluorescence quantum yield of the dye, η_{dye}, is obtained by measuring the intensity absorbed by the dye, I_0, and the slope of the thermal lensing signal, m, caused by the intrinsic absorption of the solvent. Assuming constant laser power, we derived an expression for η_{dye}:

$$\eta_{dye} = \frac{\lambda^*}{\lambda_0}\left[1-\left(\frac{I_0^r}{I_0^s}\right)\left(\frac{m^s - m^{solvent}}{m^r - m^{solvent}}\right)\left(\frac{I_0^{solvent} - I_0^r}{I_0^{solvent} - I_0^s}\right)\right]. \tag{4}$$

We use the superscripts s and r to denote the sample with the dye and the sample with the reference, respectively.

RESULTS AND DISCUSSION

Several factors affect the optical cooling performance of the dye solution. To further our understanding of the cooling mechanism, we examined the effects of dye concentration and choice of solvent on the fluorescence quantum yield. Figure 4 displays the fluorescence quantum yield as a function of concentration for several solvent and dye combinations. The trend towards higher quantum yields with lower concentration is due to quenching of the luminescence. Quenching occurs when an excited state interacts with another molecule to open a non-radiative decay channel. Efficient quenching is caused by collisions with adjacent dye molecules. The effect is a function of dye concentration. Although high concentrations boost the effective absorption, at approximately 10^{-4} M, quenching collisions begin to take over. One strategy to minimize quenching is to use solution additives that form micelles. Micelle structures form cages around the dye molecules, thus protecting them from unwanted collisions. The fluorescence quantum yield of 7.5×10^{-5} M rhodamine 101 in ethanol and 10^{-4} M rhodamine 101 in DMSO was measured to be 0.89 and 0.78, respectively. Upon the addition of 10% ammonyx LO and 23% water, the quantum yield of the respective solutions increased to 0.94 and 0.93.

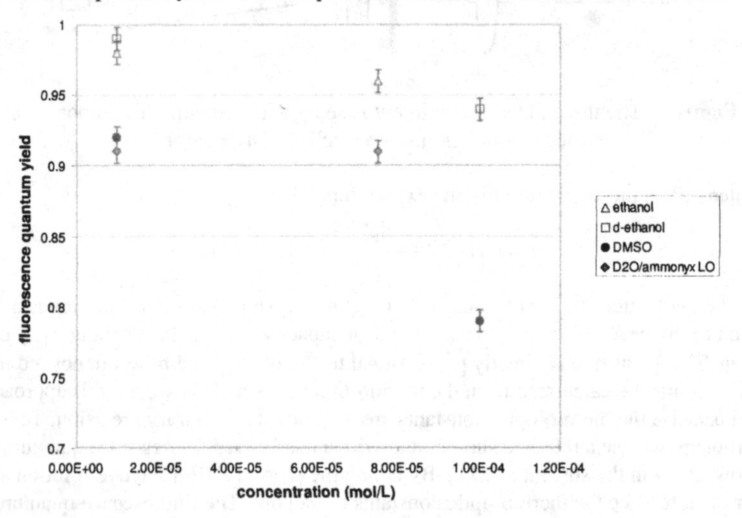

Figure 3. Fluorescence quantum yield versus dye concentration for several solvents.

Another strategy to minimize quenching is to inhibit the diffusion rate by using a viscous solvent [12,13] or to embed the dye in a solid host. An interesting example of this phenomenon is found in rhodamine B. This molecule has two mobile diethylamino groups that cause the fluorescence quantum yield to be 42% in ethanol at room temperature. As the temperature is decreased, the quantum yield rises to nearly 100%. A similar effect is seen at room temperature by changing solvent viscosity. The mobile groups are hindered in the viscous solvent glycerol, again resulting in a quantum yield of nearly 100%. In the case of rhodamine 101, the diethylamino groups are inherently rigid, thus its fluorescence quantum yield is virtually 100%, independent of the temperature and solvent viscosity.

Hydrogen vibrations are another common cause of fluorescence quenching in organic dyes. Dye molecules that have hydrogen atoms directly connected to the chromophore experience fluorescence quenching. Replacing the hydrogen with deuterium reduces the rate of nonradiative decay and increases the quantum yield [12]. Dissolving the dye in a deuterated solvent initiates hydrogen replacement. When rhodamine 6G is dissolved in monodeuterated ethanol, the fluorescence quantum yield rises to 98.0% from the value of 95.5% when dissolved in ethanol. Rhodamine 101 does not have hydrogen attached to the chromophore and is not subject to fluorescence quenching by hydrogen vibrations.

The formation of dye aggregates in certain solvents or at high concentrations can strongly quench fluorescence. One solvent notorious for inducing the formation of aggregates is water. It has been shown that the addition of surfactants to aqueous dye solutions results in complete deaggregation of the rhodamine 6G [14]. The fluorescence quantum yield of 10^{-4} M rhodamine 101 in aqueous solution was measured to be 24%. Upon the addition of 2% (w/v%) SDS, the quantum yield of the solution increased to 93%. A similar effect has also been seen upon the addition of 10% ammonyx LO.

One of the difficulties in developing a solution with unit quantum yield is the problem of non-fluorescent absorption of the solvent. Equation (5) shows the relationship between the quantum yield of a solution and the fractional absorption of the solvent.

$$\eta_{sol} = \frac{A_{dye}\eta_{dye}}{A_{dye} + A_{solvent}} \quad (5)$$

The absorption of laser dyes drops off sharply in the anti-Stokes region. As this happens, the cooling power of the dye is overcome by the heating power of the solvent. Thus, cooling is possible only over a limited range of wavelengths.

The choice of solvent is crucial for laser cooling since the photoluminescent properties including the fluorescence quantum yield of a dye solution is dependent upon the solvent. Rhodamine dyes with a free COOH group can exist in several forms depending on the polarity and the pH of the solvent [12]. In a polar solvent, rhodamine 101 takes part in a typical acid-base equilibrium. The cation is prevalent in acidic solution, and the zwitterion prevails in basic solution. Both forms will be found in neutral solution. The dominant ionic form can be determined spectroscopically since the absorption and emission spectra shift 10 nm to the blue as the equilibrium shifts to favor the zwitterion.

In non-polar solvents, the zwitterion is unstable. An inner lactone form of the dye is formed instead. Thus, in non-polar solvents, rhodamine 101 participates in an acid-base equilibrium between lactone and cation forms. The lactone form is colorless and therefore undesirable for this application. If a rhodamine dye is used in a non-polar solvent, acid should be added to assure that the cation prevails.

Another critical consideration involving the choice of solvent to be used in laser cooling is absorption. Equation (5) stresses the importance of extremely transparent solvents and shows that the intrinsic absorption of the solvent cannot be neglected. Taking this into account, carbon tetrachloride would be ideal for its transparency, but it is not polar enough to dissolve rhodamine dyes. A compromise is to use deuterated solvents such as monodeuterated ethanol or heavy water. In addition to the advantage that deuteration gives rhodamine 6G as discussed earlier, deuterated solvents are more transparent in the spectral region of interest.

CONCLUSIONS

A thermal lensing experiment was designed to examine the characteristics of various organic dye solutions for laser cooling by anti-Stokes fluorescence. Several important considerations for optimizing the fluorescence quantum yields and cooling performance in this application were investigated. Of primary importance are dye concentration and choice of solvent. Quenching is minimized for concentrations less than approximately 10^{-4} M. The addition of surfactants was also shown to minimize quenching. Deuterated and acidic solvents show some benefits.

ACKNOWLEDGMENTS

The research reported in this document/presentation was performed in connection with Contract DAAD17-00-C-0001 with the U.S. Army Research Laboratory. The views and conclusions contained in this document/presentation are those of the authors and should not be interpreted as presenting the official policies or position; either expressed or implied, of the U.S. Army Research laboratory or the U.S. Government unless so designated by other authorized documents. Citation of manufacturer's or trade names does not constitute an official endorsement or approval of the use thereof. The U.S. Government is authorized to reproduce and distribute reprints for government purposes notwithstanding any copyright notation hereon.

REFERENCES

1. P. Pringsheim, *ZS. f. Phys.*, **57**, 739 (1929).
2. P. Pringsheim, *Journal of Physics*, **10**, 495 (1946).
3. S. Vavilov, *Journal of Physics*, **9**, 68 (1945).
4. L. Landau, *Journal of Physics*, **10**, 503 (1946).
5. Zander, C.; Drexhage, K. H. *Advances in Photochemistry*; Neckers, D. C., Volman, D. H., Von Bünau, G. Eds.; Wiley: New York, 1995; p 59.
6. R.I. Epstein, M.I. Buchwald, B.C. Edwards, T.R. Gosnell, and C.E. Mungen, *Nature*, **377**, 500 (1995).
7. Clark, J. L.; Rumbles, G. *Phys. Rev. Lett.*, **76**, 2037 (1996).
8. T.R. Gosnell, *Opt. Lett.*, **24**, 1041 (1999).
9. C.W. Hoyt, M. Sheik-Bahae, R.I. Epstein, B.C. Edwards, and J.E. Anderson, *Phys.Rev. Lett.*, **85**, 3600 (2000).
10. Drexhage, K.H., Structure and properties of laser dyes., *Dye Lasers*, 2nd ed. (edited by F. P. Schäfer),144-179,269-287. Springer-Verlag, New York (1977).
11. J. H. Brannon, D. Magde, *J. Phys. Chem.*, **82**, 705 (1978).
12. J. H. Brannon, D. Magde, *J. Phys. Chem.*, **83**, 696 (1979).
13. M. Fischer, J. Georges, *Spectrochimica Acta Part A*, **53**, 1419-1430 (1997).
14. D. Magde, G. E. Rojas, and P. G. Seybold, *Photochemistry and Photobiology*, **70**(5), 737-744 (1999).

Mat. Res. Soc. Symp. Proc. Vol. 667 © 2001 Materials Research Society

Computer Modeling of Luminescence in ABO_3 Perovskites

R. I. Eglitis[1], E. A. Kotomin[1,2] and G. Borstel[1]
[1]Universität Osnabrück, Fachbereich Physik, D- 49069 Osnabrück, Germany
[2]Institute for Solid State Physics, University of Latvia, Kengaraga str. 8, Riga LV-1063, Latvia

ABSTRACT

We suggest theoretical interpretation to a long-debated discussion on a nature of the intrinsic "green" luminescence observed in many ABO_3 perovskites. For this purpose we performed quantum chemical calculations using the Intermediate Neglect of the Differential Overlap combined with the Large Unit Cell periodic model. Triplet exciton which is very likely responsible for the "green" luminescence is shown to be in a good approximation a pair of nearest Jahn-Teller electron and hole polarons (a bipolaron).

INTRODUCTION

Many ABO_3 perovskites reveal photoluminescence in the visible range ("green" luminescence) peaking around 2.2-2.3 eV in $KTaO_3$ and $KNbO_3$ (see Figure 1 and Figure 2) [1-3]. The origin of this luminescence has been discussed more than once. Suggested mechanisms include donor-acceptor recombination [4], recombination of electron and hole polarons [2], charge transfer vibronic exciton (CTVE) [5,6], transitions in MeO_6 complexes [7], etc. In this paper, we perform modelling of the triplet excitons in $KNbO_3$ and $KTaO_3$ and calculate their luminescence energies. Solution of this problem needs also a study of self-trapped electrons in perovskite crystals. An existence of small radius *polarons* in ionic solids was predicted theoretically by L. Landau in 1933 [8]. Strict experimental (ESR) proof of self-trapped *holes* has been given for alkali halides by Känzig in 1957, a quarter of century later [8]. Since then for a long time it was believed that the electron self-trapping is not energetically favourable in ionic solids due to a large energy loss necessary for an electron localization on a single cation, which is the first stage of the trapping process, and is not compensated by the energy gain due to crystal polarization, at the second stage of the self-trapping. However, in 1993 the first ESR evidence appeared [9] for the *electron* self-trapping in $PbCl_2$ crystals, and one year later- in $LiNbO_3$ perovskite crystals [10]. Lastly, very recently, existence of the self-trapped electrons was discussed in $KNbO_3$ [11]. We start this paper with theoretical modeling of the electron self-trapping in $KNbO_3$ and $KTaO_3$ perovskites as the first stage of the triplet exciton formation.

METHOD

We have used the semi-empirical, quantum chemical method of the Intermediate Neglect of the Differential Overlap (INDO) [12]. The modification of the standard INDO method for ionic solids is described in detail in Ref. [13-15]. This method is based on the Hartree-Fock formalism and allows self-consistent calculations of the atomic and electronic structure of pure and defective crystals. In the last decade the INDO method has been used to the study of bulk solids and defects in many oxides [13-18] and semiconductors [19,20]. This method has been earlier applied to the study of phase transitions and frozen phonons in pure $KNbO_3$ [21], pure and Li-doped $KTaO_3$ [22], point defects in $KNbO_3$ [23-25], and solid perovskite solutions KTN ($KNb_xTa_{1-x}O_3$) [26,27]. More details about the INDO method and relevant computer program CLUSTERD are given in references [12-15]. In present calculations we use periodic,

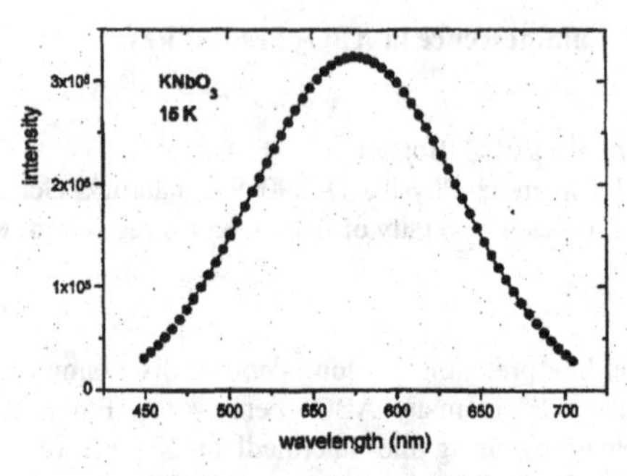

Figure 1. Luminescence spectrum of $KNbO_3$ under X-ray irradiation at 15 K [1]

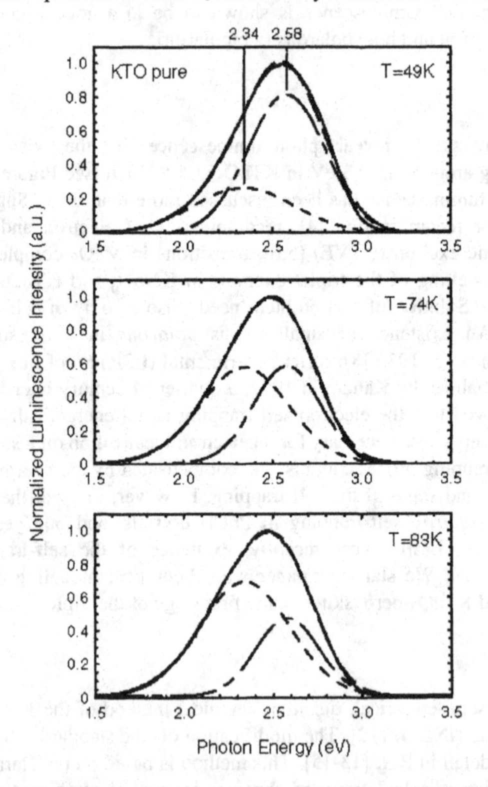

Figure 2. The "green" luminescence spectrum of nominally pure $KTaO_3$ and its temperature dependence. It was assumed in [2] that the band consists of two components.

the so-called *large unit cell* (LUC) model [28]. In this model the electronic structure calculations are performed for an extended unit cell at the wave vector **k**=0 in the narroved Brillouin zone (BZ) which is equivalent to the band-structure calculations at several special points of the normal BZ, transforming to the narrow BZ center after the corresponding extension of the primitive unit cell. In ABO_3 crystals under study with the unit cell containing five atoms the 2x2x2 extended LUC used in our calculations consists of 40 atoms.

The detailed analysis of the development of the INDO parametrization for pure $KNbO_3$ and $KTaO_3$ is given in Refs. [21,22]. The INDO method reproduced very well both available experimental data and results of *ab initio* LDA - type calculations. In particular, this method reproduces the effect of a ferroelectric instability of $KNbO_3$ due to off-center displacement of Nb atoms from the regular lattice sites, as well as the relative magnitudes of the relevant energy gains for the [100], [110] and [111] Nb displacements, which are consistent with the order of the stability of the tetragonal, orthorombic and rhombohedral ferroelectric phases, respectively, as the crystal's temperature decreases. This is a very non-trivial achievement since the typical energy gain due to the Nb off-center displacement is of the order of several mRy per unit cell.

The calculated frequencies of the transverse-optic (TO) phonons at the Γ point in the BZ of cubic and rhombohedral $KNbO_3$ and the atomic coordinates in the minimum energy configuration for the orthorhombic and rhombohedral phases of $KNbO_3$ are also in good agreement with experiment thus indicating that a highly successful INDO parametrization has been achieved. The frozen-phonon calculation for T_{1u} and T_{2u} modes of cubic $KTaO_3$ are also in good agreement with experiment. An appreciable covalency of the chemical bonding is seen from the calculated (static) effective charges on atoms (calculated using the Löwdin population analysis): 0.62 *e* for K, 2.23 *e* for Ta and –0.95 *e* for O in $KTaO_3$, which are far from those expected in the purely ionic model (+1 *e*, +5 *e* and –2 *e*, respectively) often used. These charges show slightly higher ionicity in $KTaO_3$ as compared with the relevant effective charges calculated for $KNbO_3$: 0.54 *e* for K, 2.02 *e* for Nb and –0.85 *e* for O.

MAIN RESULTS

In our recent calculations for the hole polarons in $KNbO_3$ [25] we found that both one-site (atomic) and two site (molecular) polarons bound to K vacancy could co-exist. A hole is either localized by a single O atom displaced toward the vacancy, or it is shared by a couple of nearest O atoms. The relevant absorption band was observed recently experimentally by means of the transient optical spectroscopy [29].

As an extension of the hole polaron study, we have modeled here first of all the *electron* polaron. As before, we used the 2x2x2 extended cubic cell with the LUC containing 40 atoms. We allowed six nearest oxygen atoms in the octahedron around a central Nb atom in $KNbO_3$ to relax, in order to find the energy minimum of the system. All other Nb and K atoms, as well as the rest of O atoms were kept fixed at their perfect lattice sites. According to our INDO calculations, initially the ground state is three-fold degenerate (t_{2g}). This degeneracy is lifted as a result of the combination of the breathing mode and the Jahn-Teller effect: 1.4 % a_0 (a_o is lattice constant) outward displacement of four nearest equatorial O atoms (with the energy gain of 0.12 eV) and *inwards* 1 % a_o relaxation of the two oxygens along the *z* direction (an additional gain of 0.09 eV). That is, the total lattice energy gain is 0.21 eV. A similar JT electron polaron was observed recently in $BaTiO_3$ [30].

As a result, a considerable (0.5 *e*) electron density is localized on the central Nb atom producing three closely spaced energy levels in the band gap. They consist mainly of the *xy*, *xz*

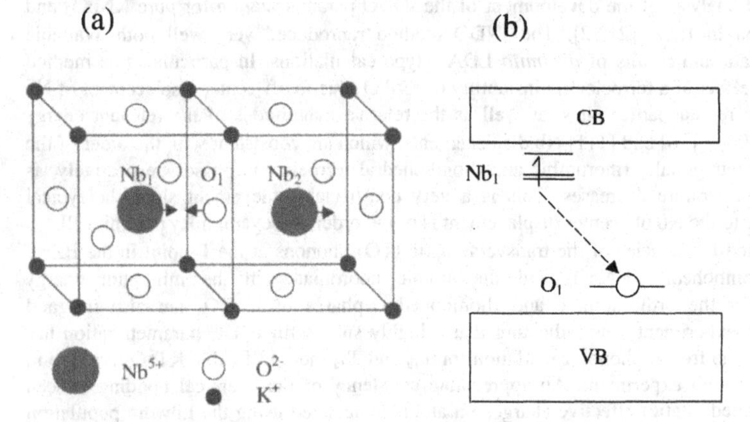

Figure 3. Schematic view of the charge transfer vibronic exciton in perovskites (a) and its luminescence (b)

Table 1. Calculated properties of excitons (see Figure 3.a for atom labeling). Numbers in brackets are experimental values

	Relaxation energy (eV)	Displacement (% a_o)	Effective charge (e)	Charge transfer (e)	Luminescenc e energy (eV)
$KNbO_3$	2.37				2.17 (2.2)
Nb1		2.9	1.55	-0.47	
Nb2		4.3	1.92	-0.10	
O1		4.9	-0.20	0.65	
$KTaO_3$	2.71				2.14 (2.3)
Ta1		3.1	1.74	-0.49	
Ta2		4.5	2.12	-0.11	
O1		5.2	-0.24	0.71	

and *yz* Nb *4d* atomic orbitals (splitted t_{2g} energy level in an isolated ion) whereas another two empty levels are located close to the conduction band bottom. The electron polaron absorption energy was estimated by means of ΔSCF to be 0.78 eV. The relevant absorption process corresponds to an electron transfer to the nearest Nb atom. This absorption band has been indeed recently observed under picosecond laser excitation [11].

As to the energetics of the self-trapped electron formation, previous *ab initio* LMTO calculations [21] have demonstrated that the bottom of $KNbO_3$ conduction band has a narrow subband which probably permits the electron localization energy to be small enough for a positive total self-trapping energy balance.

In similar calculations for $KTaO_3$ we obtained optical absorption of 0.75 eV, with the following ground state relaxation: 1.7 % a_o for four O equatorial atoms and 1.2 % a_o inward displacement of two other O atoms along the z axis. This results in the total lattice relaxation energy of 0.27 eV. This is larger than in the $KNbO_3$ case because of higher $KTaO_3$ ionicity.

Next, we have calculated the triplet state of the excitons in the two crystals. Preliminary calculations for $KTaO_3$ [6] have demonstrated that the triplet exciton is a triad center containing *one* active O atom (O1) and two Nb atoms located on the opposite sites from this O atom (see Figure 3. a. The total energy of the system is lowered by the combination of Coulomb attraction between electron and hole and the vibronic effect in this charge transfer vibronic exciton (CTVE) [5,6]. In order to find the CTVE energy minimum, we performed the self-consistent geometry optimization for 40 atom LUC (See Figure 3.a). According to our calculations, the oxygen ion O1 in $KNbO_3$ is displaced by 4.9 % a_o towards the Nb1 ion (Figure 3.a) which is active in the CTVE formation. Simultaneously, this Nb1 ion reveals a displacement of 2.9 % a_o towards the O1 ion, whereas another Nb2 ion (which is located on the other side from the oxygen ion O1 along the CTVE-axis) has a repulsion from the oxygen O1 and is displaced outwards by 4.3 % a_o. We obtained that the total energy reduction in CTVE in $KNbO_3$ due to the lattice relaxation of O1, Nb1 and Nb2 – three main atoms - is quite appreciable - 2.37 eV. The charge redistribution between ions is –0.47 e on the Nb1, 0.65 e on the O1, and –0.1 e on the Nb2.

A strong lattice distortion caused by the CTVE induces the local energy levels in $KNbO_3$ band gap. Namely, O1 energy level with a hole is located 0.9 eV above the VB top, its wave function consists mainly of its $2p_x$ atomic orbital directed towards the Nb1 atom. On the other hand, two closely located Nb1 energy levels (one of them is two-fold degenerate) appear at 0.7 eV below the CB. They have t_{2g} symmetry and consists mainly of 4*dxy* atomic orbitals of Nb1 and to a smaller extend Nb2 ions, with admixture of atomic orbitals of other nearest Nb atoms surrounding the CTVE (Figure 3.b). The luminescence energies (Figure 3.b) calculated using ΔSCF method are close to the experimentally observed values. Results of the exciton calculations in $KNbO_3$ and $KTaO_3$ are collected in Table 1. As one can see, the atomic displacements and relaxation energy in $KTaO_3$ is larger, due to more ionic nature of this crystal. However, this practically does not affect the luminescence energy.

CONCLUSIONS

Our calculations give a strong support to the "green" luminescence in these crystals as a result of the recombination of the electrons and holes forming the *charge transfer vibronic exciton* rather than due to the electron transitions in MeO_6 complex. Our results also demonstrate that well-parametrized quantum chemical methods are very efficient tool for the study of optical properties of advanced materials.

ACKNOWLEDGEMENTS

R. I. Eglitis and G. Borstel gratefully acknowledge DFG for the financial support, whereas E. A. Kotomin was supported by DAAD (Germany) and Latvian National Programme for New Materials in Opto - and Microelectronics. The authors are indebted to L. Grigorjeva, D. Millers, A. Popov, V. Trepakov, V. Vikhnin and R. Williams for many valuable discussions.

REFERENCES

1. A.I. Popov and E. Balanzat, *Nucl. Instr. Meth.* **B166-167,** 305-308 (2000).
2. V.S. Vikhnin, R.I. Eglitis, E.A. Kotomin, S.E. Kapphan, and G. Borstel, in: Proc. Williamsburg workshop on Fundamental Physics of Ferroelectrics, AIP, 2000, in press.
3. V. Pankratov, L. Grigorjeva, D.K. Millers, G. Corradi, and K. Polgar, *Ferroelectrics,* 2000, in press; L. Grigorjeva, D.K. Millers, A.I. Popov, E.A. Kotomin, and E.S. Polzik, *J. Lum.* **72-74,** 672-676 (1997).

4. G.Koshek and E. Kubalek, *Phys. Stat. Sol.* **A79**,131-145 (1983).
5. V.S.Vikhnin and S.E. Kapphan, *Phys. Sol. State* **40,** 834-840 (1998).
6. V.S. Vikhnin, H. Liu, W. Jia, S. Kapphan, R.I. Eglitis, and D.Usvyat, *J.Lum.* **83-84,** 109-120 (1999).
7. C. Blasse, *Mater. Res. Bull.* **18,** 525-530 (1983); L. G.de Haart, A.J. de Vires, and C. Blasse, *J. Sol. St. Chem.* **59,** 291-299 (1985).
8. K.S. Song and R.T. Williams, *Self-trapped Excitons* (Springer-Verlag, Berlin, 1993).
   A.L. Shluger and A.M. Stoneham, *J. Phys.: Condens. Matter* **5,** 3049-3082 (1993).
9. S.V. Nistor, E. Goovaerts, and D. Schoemaker, *Phys. Rev.* **B43,** 9575-9586 (1993).
10. B. Faust, H. Müller, and O.F. Schirmer, *Ferroelectrics* **153,** 297-304 (1994).
11. R.T. Williams, K.B. Ucer, G. Xiong, H.M. Yochum, L. Grigorjeva, D. Millers, and G. Corradi, Rad. Eff.& Def. in Sol., (2000), in press
12. J.A. Pople and D.L. Beveridge, *Approximate Molecular Orbital Theory* (McGraw-Hill, New York, 1970).
13. A.L. Shluger, *Theoret. Chim. Acta* **66,** 355-365 (1985).
14. E. Stefanovich, E. Shidlovskaya, A.L. Shluger, and M. Zakharov, *Phys. Stat. Sol.* **B160,** 529-539 (1990).
15. A.L. Shluger and E.V. Stefanovich, *Phys. Rev.* **B42,** 9664-9675 (1990).
16. A. Stashans, E.A. Kotomin, and J.-L. Calais, *Phys. Rev.* **B49,** 14854-14860 (1994).
17. E.A. Kotomin, A. Stashans, L.N. Kantorovich, A.I. Livshitz, A.I. Popov, I.A.Tale, and J.-L. Calais, *Phys. Rev.* **B51,** 8770-8785 (1995).
18. A. Stashans, S. Lunell, R. Bergstrom, A. Hagfeldt, and S.-E. Lundqvist, *Phys. Rev.* **B53,** 159-167 (1996).
19. E.V. Stefanovich and A.L. Shluger, *J. Phys.: Condens. Matter* **6,** 4255-4266 (1994).
20. A. Stashans and M. Kitamura, *Solid State Commun.* **99,** 583-587 (1996).
21. R.I. Eglitis, A.V. Postnikov, and G. Borstel, *Phys. Rev.* **B54,** 2421- 2430 (1996).
22. R.I. Eglitis, A.V. Postnikov, and G. Borstel, *Phys. Rev.* **B55,** 12976-12986 (1997).
23. R.I. Eglitis, N.E. Christensen, E.A. Kotomin, A.V. Postnikov, and G. Borstel. *Phys. Rev.* **B 56,** 8599-8607 (1997).
24. E.A. Kotomin, R.I. Eglitis, and A.I. Popov, *J. Phys.: Condens. Matter.* **9,** L315-320 (1997).
25. E.A. Kotomin, R.I. Eglitis, A.V. Postnikov, G. Borstel, and N.E. Christensen, *Phys. Rev.* **B 60,** 1-5 (1999).
26. R.I. Eglitis, E.A. Kotomin, and G. Borstel, *J. Phys.: Condens. Matter* **12,** L431-436 (2000).
27. R.I. Eglitis, E.A. Kotomin, G. Borstel, and S. Dorfman, *J. Phys.: Condens. Matter* **10,** 6271-6280 (1998).
28. R.A. Evarestov and V.A. Lovchikov, *Phys. Stat. Sol.* **B93,** 469-480 (1977).
29. E.A. Kotomin, R.I. Eglitis, G. Borstel, L. Grigorjeva, D. Millers, and V. Pankratov, *Nucl. Inst. Meth.* **B166-167,** 299-304 (2000).
30. S. Köhne, O.F. Schirmer, H. Hesse, T.W. Kool, and V.S. Vikhnin, *J. of Supercond.*, **12,** 193-200 (1999).

Characterization

Mat. Res. Soc. Symp. Proc. Vol. 667 © 2001 Materials Research Society

ANALYSIS OF TUNGSTATES AND SESQUIOXIDES, TWO OF THE BEST Yb^{3+}-DOPED LASER CRYSTALS ACCORDING TO DIFFERENT EVALUATIONS.

Georges Boulon, Alain Brenier, Laetitia Laversenne, Yannick Guyot,
Christelle Goutaudier, Marie-Thérèse Cohen-Adad, Gérard Métrat, Noelle Muhlstein,

Physical Chemistry of Luminescent Materials, Claude Bernard/Lyon1 University,
UMR 5620 CNRS, Bât. A.Kastler,10 rue Ampère, Domaine Scientifique de la Doua,
69622 Villeurbanne, France.

ABSTRACT

The development of reliable InGaAs laser diode pump sources emitting in the 900-980 nm spectral range is stongly influencing the field of lasers based on Yb^{3+}-doped solid state crystals. $Ca_5(PO_4)_3F$ (C-FAP) and S-FAP ($Sr_5(PO_4)_3F$) were soon recognized to be favourable hosts for Yb^{3+} lasing in the nanosecond pulse regime. This fact was supported by an evaluation of the spectroscopic properties of several Yb^{3+}-doped crystals useful for laser action. This evaluation is based on two parameters known from spectroscopy, the emission cross-section at the laser wavelength and the minimum pump intensity required to achieve transparency at the laser wavelength.We think there is a need of a new evaluation of Yb^{3+}-doped crystals in order to predict the laser efficiency in a more realistic manner in different kinds of regimes. We present here the main spectroscopic properties of two Yb^{3+}-doped laser crystals which are grown in our Group: (i) $KY(WO_4)_2$ double tungstates by the Floating Crystal method and (ii) Y_2O_3 sesquioxides by the Laser Heated Pedestal Growth method. The approach, based on a quasi-three level laser model, leads to compare all known Yb^{3+}-doped crystals in a two-dimensional diagram considering the laser extracted power and the slope efficiency. We shall show that tungstates and sesquioxides belong to the highest laser crystal potential in CW-end pumping configuration.

INTRODUCTION

The most promising ion that can be used in a non-Nd laser in the same range of emission wavelength is Yb^{3+}. The Yb^{3+} ion have some advantages over the Nd^{3+} ion as laser emitting center due to its very simple energy level scheme, constituted of only two levels : the $^2F_{7/2}$ ground state and the $^2F_{5/2}$ exited state. There is no excited state absorption reducing the effective laser cross-section, no up-conversion, no concentration quenching. The intense Yb^{3+}absorption lines are well suited for laser diode pumping near 980 nm and the small Stokes shift between absorption and emission reduces the thermal loading of the material during laser operation. The disadvantage of Yb^{3+} is that the final laser level of the quasi three-level system is thermally populated, increasing the threshold.

Among new directed searches for novel laser crystals,one important is the use of Yb^{3+} active ion in an inertial-fusion-energy diode pumped solid state laser. $Ca_5(PO_4)_3F$ (C-FAP) and $Y_3Al_5O_{12}$ (YAG) were soon recognized to be favorable hosts for Yb lasing in the nanosecond pulse regime. This fact was supported by an evaluation of the spectroscopic properties of several Yb-doped crystals useful for laser action [1]. This evaluation is based on two parameters known from spectroscopy : the emission cross-section σ_e at the laser wavelength and the minimum pump intensity I_{min} required to achieve transparency at the laser wavelength:

$$(1)\quad I_{\min} = \frac{\sigma_a}{\sigma_e + \sigma_a} \frac{h\upsilon}{\sigma_{ap}\,\tau}$$

where σ_a is the absorption cross-section at the laser wavelength, σ_{ap} is the the absorption cross-section at the pump wavelength and τ is the $^2F_{5/2}$ lifetime. σ_e and I_{min} were used in [1] as figure-of-merit to classify the hosts, in a two-dimensional diagram. In the diagram drawn in Fig.1 with the help of ,both, initial data already mentioned in [1] and recent data of sesquioxides, tungstates [2-5], YAB, GdCOB, YCOB, GGG, LNB [6-15] ,C-FAP ($Ca_5(PO_4)_3F$) and S-FAP ($Sr_5(PO_4)_3F$) appear to be exceptionally good, and YAB ($YAl_3(BO_3)_4$), GdCOB ($CaGd_4(BO_3)_3O$) and YCOB ($CaY_4(BO_3)_3O$) appear modest. This is somewhat in contradiction with experimental laser tests in which these latter materials are revealed very efficient : 73% slope efficiency in YAB, 77% in GdCOB, 78% in KYW ($KGd(WO_4)_2$), 72% in KGdW ($KGd(WO_4)_2$), to be compared to 71 and 79% for S-FAP and C-FAP respectively.

The purpose of this work is to present the spectroscopic properties of both Yb^{3+}-doped tungstates [4] and sesquioxides [5] and a new evaluation model based on a quasi-three level laser model, checked to be close to experimental laser data for CW longitudinally pumped lasers [2], [3].

SPECTROSCOPIC TECHNIQUES

Excitation of the Yb^{3+} fluorescence was performed with a Laser Analytical Systems dye laser (0.04 cm^{-1} resolution) pumped by a frequency-doubled Nd :YAG laser from BM Industries. The visible beam of the dye laser was sent through an hydrogen cell raman shifter to generate a new beam in the 950-980 nm range. The detection of the luminescence was carried out with a Jobin-Yvon monochromator equipped with a 1 µm blazed grating (2.4 nm resolution/mm slit) and a R1767 Hamamatsu photomultiplier. The signal was sent into a Stanford boxcar averager SRS 250. The decay kinetics were recorded with a Lecroy 9410 digital oscilloscope. The absorption spectra were obtained with a Carry 2300 spectrophotometer.

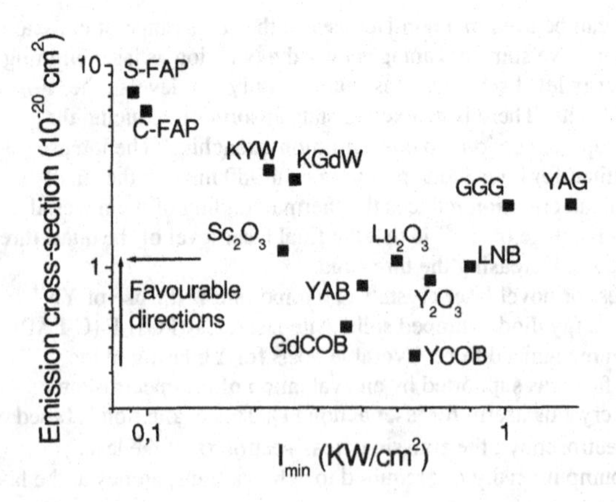

Figure 1. Figure-of-merite evaluated for several promising Yb^{3+}-doped laser crystals according to the approach of [1].

GROWTH BY THE TOP NUCLEATED FLOATING CRYSTAL METHOD AND SPECTROSCOPIC PROPERTIES OF Yb^{3+}-DOPED $KY(WO_4)_2$.

Crystal growth

The nucleation of a floating crystal was induced by adding a very small amount (10 µg) of WO_3 powder to the liquid surface.The small resultant crystal was quickly detected by image analysis when its size became larger than 50 µm. The solution temperature was then raised to reduce the growth rate (0.5 mm/h). During the growth, the crystal was floating freely. When it became too heavy, it must be kept up by a pumping device to avoid sinking. The crystals are unstrained because mechanical stresses and internal thermal gradients are lowered in the growing solid.The crystals are most frequently obtained as hexagonal looking platelets with two large faces perpendicular to b-axis. The space group is C2/c [4-18].

Spectroscopic properties

The absorption and emission spectra were recorded along the b-axis, not with a polarization along the c and a crystallographic axis as in [7] but in the two orthogonal directions corresponding to the two N_1 and N_2 axis of the optical indicatrix [18].Because there is a large overlap between emission and absorption, we used for emission spectra and decay time the 0.5 % (in the melt) Yb concentrated sample in order to reduce the deformation of the emission spectra in the short wavelengths range. The partial reabsorption of the emission in high Yb concentrated samples is clearly demonstrated by measuring the decay time of the fluorescence : we found 320 µs in 0.5% doped sample and the use of a concentrated sample is probably at the origin of the longer lifetime (460 µs) due to trapping effect or radiative transfer.For the absorption spectra (Fig.1) we used the 5 % (in the melt) Yb concentrated sample (2.3 mm thickness). The corresponding Yb concentration in the crystal is $C=3.2 \times 10^{20}$ ions/cm^3, so the cross-section of the main absorption peak at 981.2 nm in Fig.1 is calculated from the coefficient of absorption k to be $\sigma_a=k/C= 3.7\ 10^{-20}\ cm^2$. It is important to mention that this value is lower than the one of $KGd(WO_4)_2$:Yb ($3.7\ 10^{-20}\ cm^2$ [19]) but much higher than the one of the YAG :Yb : $0.6\ 10^{-20}\ cm^2$.

The energy levels of the ground and excited-states can be easily seen from both absorption and emission spectra in Fig .1 and 2. Three main peaks have been detected from the sublevel 1 in absorption (1-5 so-called zero-line, 1-6 and 1-7) and it is also possible to observe 2-5 by thermal population of level 2 at room temperature in good agreement with the emission spectrum (transition 5-2) so that other transitions 5-3 and 5-4 can be interpreted in addition of 5-1 0-line emission which is overlapping with the 1-5 absorption line.

The emission cross-sections σ_e were obtained from the polarized emission spectra, and were calibrated at 981.2 nm (5-1 zero-line transition, 10187 cm^{-1}) from the absorption cross-section using McCumber's relation [20] according to (2):

$$\sigma_e(981.2)=\sigma_a(981.2)\frac{N_{gs}}{N_{es}}\exp(-10187/kT) \quad (2)$$

where N_{gs}/N_{es} is the ratio of the number of Yb ions in the $^2F_{7/2}$ ground state to the number of Yb ions in the $^2F_{5/2}$ excited state at thermal equilibrium. This ratio is calculated knowing the energy sublevels of the Yb^{3+} ion : 0, 169, 407, 568 (cm^{-1}) for the $^2F_{7/2}$ ground state and 10187, 10476 and 10695 (cm^{-1}) for the $^2F_{5/2}$ level [18]. The result for the value of the emission cross section at the peak at 1025 nm, wavelength at which the laser emission is observed by 5-3 transition in Fig.3, is σ_e=3 x 10^{-20} cm^2, which is comparable than the one of YAG :Yb^{3+} (2 10^{-20} cm^2).

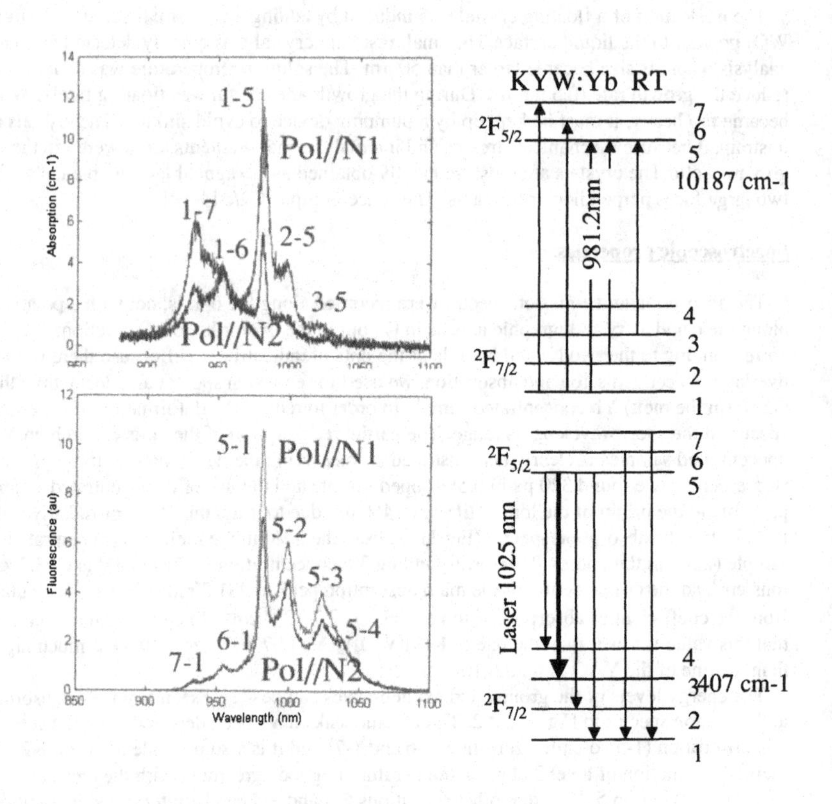

Figure 2. Absorption spectra of KYW:Yb^{3+} crystal at room temperature
Figure 3. Emission spectra of KYW:Yb^{3+} crystal at room temperature

(These spectra have been recorded along the b-axis in the two orthogonal directions of polarizations N_1 and N_2 of the optical indicatrix. The transions involved in both absorption and emission processes between the Stark levels of the $^2F_{7/2}$ and $^2F_{5/2}$ states have been drawn on the Yb^{3+} energy level diagram).

GROWTH BY LHPG AND SPECTROSCOPIC CHARACTERIZATION OF Yb^{3+}-DOPED Y_2O_3 CUBIC STRUCTURE REFRACTORY SESQUIOXIDES.

Growth of the fibers

Samples of Y2O3 fibers were elaborated by the LHPG (Laser Heated Pedestal Growth) floating zone technique in which the crystal is pulled from a ceramic rod melted with an annular focused CO_2 laser [5],[21].In the apparatus built at the Laboratory, a special mirror configuration creates an annular beam that heats homogeneously the melted zone without rotation.The floating zone motion is generated by the vertical displacement of the feed and the seed. During the translation, the feed progressively melts and behind the floating zone a crystallized rod is formed. The liquid zone remains in equilibrium between the feed and the crystallized rod due to the superficial tensions. Because of the lack of crucible and of the use of 200 watt power CO_2 laser, the LHPG technique is well adapted to grow refractory materials such as oxides under air atmosphere.Transparency of samples is good and no crack appearance have been checked by using optical microscopy.

Structural characterization

Rare earth sesquioxides are refractory matrices (they all melt above 2350°C) and they have been selected as hosts first because they present high and better thermomechanical properties than in YAG. As an example, the thermal conductivity at room temperature in Y_2O_3 is K= 27 W/mK versus K= 13 W/mK in YAG [5]. Moreover they are chemically related with the dopant that insures a very good substitution. Low phonon energy insures good laser efficiency by minimizing energy loss due to non-radiative processes; it is already low in YAG (700 cm^{-1}) but it is weaker in rare earth oxides with 377 cm^{-1} for Y_2O_3 [22].

Yttrium sesquioxide belongs to the bixbyite type $[^{VI}A_2][^{IV}O_3]$ which is body-centered cubic with space group Th^7 (or Ia3). The elementary cell contains 16 Y_2O_3 formula units with 32 cation (sixfold coordinated) sites for which the trivalent rare earth dopants can substitute: 8 centro-symmetric sites with symmetry C_{3i} (or S_6) and 24 non centro-symmetric sites with symmetry C_2. Lattice constants is 10.60 Å [23].

Previous neutron diffraction or spectroscopic studies [9, 10], have fairly established that the trivalent rare earth dopant ions substitutionnally occupy the two available crystallographic sites with no preferential site occupation. Although parity selection rules, time-resolved spectroscopy allows Aumüller et al. to establish without doubt luminescence of two different centres in praseodyme doped-Y_2O_3 [24].

Composition measurements have been achieved on a CAMECA SX-100 with a 20 kV-40nA accelerated beam and a LiF crystal receiving the response of K-lines of the RE atoms. In order to increase the resolution of these measurements we have calibrated the results using standards that are pure oxides synthesized in our laboratory

Spectroscopic properties

As can be seen in Fig.3, the absorption spectra of Y_2O_3 consists of the three expected lines assigned to transitions from the lowest level of the ground state to the three multiplets of the excited state. The 1-5 zero-line centered at 977 nm appears quite narrow compared with the two other lines,1-6 and 1-7, located around 950 and 903 nm respectively. These two lines are especially suitable to laser diode pumping because of their broad linewidth (11 and 7 nm respectively). Moreover the absorption and emission cross section values,1x10^{-20} cm2 and 1.2 x10 –20 cm2 respectively, sligthly less than those of YAG and KYW crystals are,however, quite high.

The emission spectra which has been drawn in the same Fig.3, exhibits broad lines of high intensity at 1033 nm that must contain the transition of the first excited multiplet to both the second and third Stark levels of the ground state because of the low energy splitting of these two levels (~100 cm^{-1} after [25]).A second line is located at 1079 nm characterized with a smaller intensity and the 5-1 resonant line scarcely appears at 977 nm. The high value of the ground state splitting, which appears to be the highest of Yb^{3+}-doped crystals fluorapatites and the low efficiency of the 0-line emission are favorable to a quasi 4-level scheme for laser application.

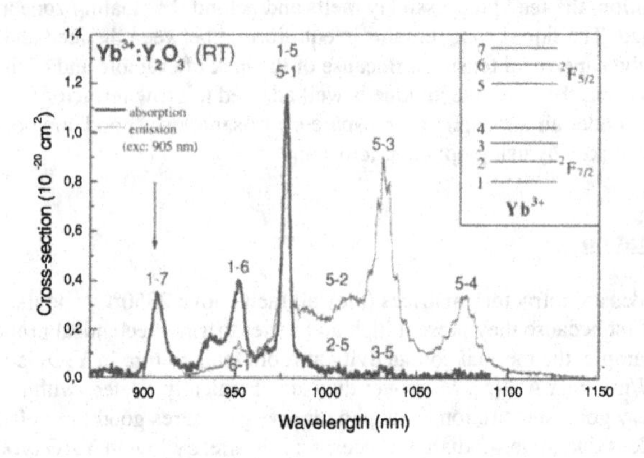

Figure 4. Absorption and emission spectra of 2.4% Yb^{3+}-doped Y_2O_3 under 905 nm pumping laser at room temperature.The transitions involved are mentioned by the energy level diagram.

Table 1: The manifold splitting of Yb^{3+} in some laser crystals

		Y_2O_3 [this work]	Y_2O_3 (calculated) [13]	YAG [16]
$^2F_{5/2}$	7	11025	11030	10902
	6	10504	10629	10679
	5	10225	10243	10327
$^2F_{7/2}$	4	931	1036	785
	3	526	590	612
	2	295-373	491	565
	1	0	0	0

The manifold splittings of Yb^{3+} in Y_2O_3 are given in Table 1 in which values can be compared to both, the usual YAG laser crystal and KGd(WO4)2 crystal studied in this research program.It can be seen that the 2F7/2 ground state splitting in Y2O3 is much more larger than in tungstate host which is more in favourable to get the 4-level laser system..

Our data are similar to those recently obtained by Petermann et al.[14] who are succefully growing bulky crystals with the help of the Bridgmann method associated with rhenium crucible

Another kind of fibers has been grown, so-called " concentration gradient samples", which are single crystals with dopant concentration varying regularly and continuously from one end to the other on a few cm length, between two well-defined compositions. The advantage of "concentration gradient samples" is that they allow a full and fast study of the system using only one sample. Actually every point of the sample can be considered as single crystal of well-known luminescent ion composition checked by EDS (Energy Dispersion Spectroscopy) microprobe. The first data on these new samples have been published in [5].

NEW EVALUATION OF Yb $^{3+}$-DOPED LASER CRYSTALS IN CW END-PUMPING CONFIGURATION

Approach of the evaluation

The purpose of this work is to present a new evaluation based on a quasi-three level laser model according equations (2,3,4) below, checked to be close to experimental laser data [2], [3]. The model deals with gaussian waves, takes into account,both,the saturation of the pump, which occurs for Yb ion because the $^2F_{5/2}$ level has a long lifetime up to 2.5 ms and then accumulates population, the stimulated emission at the pump wavelength, the variation of the pump and laser waists along propagation which are important for laser diode pumping, and the variation of the laser intensity along propagation. It is an extension of the model of Risk and Taira et al. [12-13].

The fractions of population of the initial laser level ($^2F_{5/2}$) and of the ground state ($^2F_{7/2}$), N and N_0 respectively, in a steady state, are such that :

$$N_0(r,z)+N(r,z)=1$$

(2)
$$0=-\frac{N}{\tau}+I_p(r,z)\left(\sigma_{ap}N_0-\sigma_{ep}N\right)+I_L(r,z)\left(\sigma_a N_0-\sigma_e N\right)$$

τ is the initial laser level lifetime, $I_p(r,z)$, $I_L(r,z)$ are respectively the densities (photons/(cm^2 s)) of the pump beam and of the two counter-propagating laser waves inside the crystal. σ_{ep}, σ_{ap} are respectively the emission and absorption cross-sections of the pump and σ_e, σ_a the emission and absorption cross-sections of the laser beam.

The equation of propagation for the $\psi_l^+(r,z)\exp(-k_l z)$ electric field of the laser wave stands in a nonparabolic approximation as :

(3)
$$\frac{\partial\psi_l^+}{\partial z}+\frac{i}{2k_l}\nabla_t^2\psi_l^+=[\sigma_e N(r,z)\frac{C}{2}-\sigma_a N_0(r,z)\frac{C}{2}-\frac{\alpha_l}{2}]\psi_l^+$$

C is the laser ion concentration. Equation (3) is valid for forward propagation; for backward propagation (ψ_l^- electric field), we have to change the signs of k_l, n_l and of the bracket. The equation of propagation for the $\psi_p(r,z)\exp(-k_p z)$ electric field of the pump wave stands as :

(4)
$$\frac{\partial\psi_p}{\partial z}+\frac{i}{2k_p}\nabla_t^2\psi_p=[\sigma_{ep}N(r,z)\frac{C}{2}-\sigma_{ap}N_0(r,z)\frac{C}{2}]\psi_p$$

Our evaluation is based on the laser output power and the differential slope extracted from the following crystals: $Sr_5(PO_4)_3F$ (S-FAP), $KY(WO_4)_2$ (KYW), $KGd(WO_4)_2$ (KGdW), $CaGd_4(BO_3)_3O$ (GdCOB), $CaY_4(BO_3)_3O$ (YCOB), $YAl_3(BO_3)_4$ (YAB), Y_2O_3, Lu_2O_3, Sc_2O_3, Y_2SiO_5 (YSO), $LiNbO_3$ (LNB), located inside the same laser cavity and calculated numerically with the model. The pump power has been fixed to 1 W. The laser waist was 22 µm and the pump waist 29 µm. The crystals concentration was chosen typical in laser experiments for each crystal. We used a rather high concentration : 8.97 10^{20} ions/cm^3 for crystals having a rare earth crystallographic site (YAG, YCOB, GdCOB, YAB, KGdW, KYW), and a lower concentration : 0.36x10^{20} ions/cm^3 [1] for crystals having a low crystals having a low segregation coefficient (C-FAP and S-FAP).

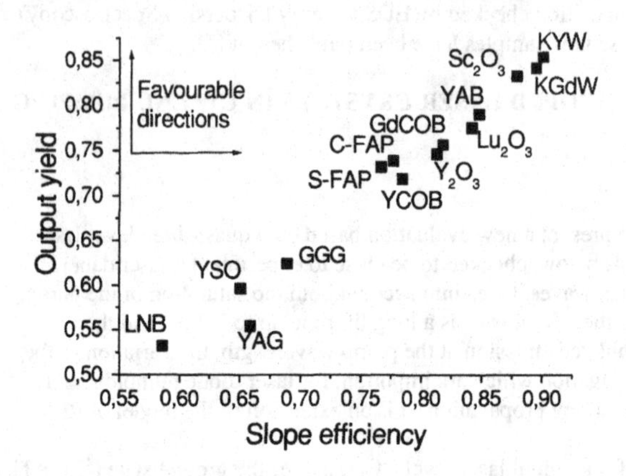

Figure 5. Output yield and slope efficiency predicted by the evaluation model for the most promising Yb 3+ -doped crystals located inside the same laser cavity which is described in the text [3].

Table 2. Spectroscopic data of some Yb3+-doped KYW tungstate and Y_2O_3 sesquioxide.

	Conc 10^{20}ions/cm^3	σ_{ap} 10^{-20}cm^2	σ_{ep} 10^{-20}cm^2	σ_e 10^{-20}cm^2	σ_a 10^{-20}cm^2	λ_p nm	λ_e nm	τ µs	L_{opt} mm	R_{opt} %
YAG	8.97	0.8	0.159	2.3	0.149	942	1029	951	2.9	93
S-FAP	0.36	8.6	0.076	7.3	0.407	899	1047.3	1260	9.5	92
KYW	8.97	13.3	16.0	3	0.299	981.2	1025.3	600	0.4	99.5
Y_2O_3	8.97	2.4	2.3	0.85	0.06	975	1031	850	1.6	99

Results

For each crystal we have determined numerically with the model the crystal length L_{opt} and the reflectivity R_{opt} of the output mirror leading to the maximum laser output power. All spectroscopic data of the two Yb-doped crystals which have been evaluated in this article are given in Table 2. YAG and S-FAP have been mentioned for comparison [2], [3]. The results of calculation of the output yield P_{out}/P_{pump} and the slope efficiency $dP_{out}/dPump$ are shown Fig.6. We can see that KGdW and KYW have the highest laser potentialities, it is surprising that YAG has the lowest. S-FAP and C-FAP appear less efficient than GdCOB and YAB borate crystals. Cubic sesquioxides (Sc_2O_3, Lu_2O_3, Y_2O_3) are also well positioned in this evaluation. The connection between this evaluation and DeLoach et al. [1] one has been discussed in [2]. First of all, σ_e is not an adequate parameter to be considered as figure-of-merit for CW end-pumping. It is clear that in this case I_{min} is a more meaningful figure-of-merit, but that helps only to evaluate the maximum output yield allowed to be extracted with a crystal. The calculated maximum output yield available from several Yb^{3+}-doped crystals versus I min has been shown in Fig.6. In this figure, we took into account, both, of P_{out} which comes from calculations and of the quantum defect by the term $\lambda p / \lambda e$. In order to have a more direct comparison, Fig.6 contains "fictive C-FAP" and "fictive S-FAP" meaning that it has been drawn using fictive high values of Yb concentration (8.97×10^{20} instead of 0.36×10^{20} ions/cm3) in C-FAP and in S-FAP, as in the 6 other crystals (YAG, YCOB, GdCOB, YAB, KGdW, KYW) which are, in fact, impossible to insert within the lattice of apatite structure. by using the real value of Yb 3+ concentration, the two points labelled "C-FAP" and "S-FAP" are shown, located below the tungstate crystals. Especially in this case, the quantum defect is unfavourable to apatite crystals since laser diode pumping wavelength is 905 nm, within the highest Stark component of the $^2F_{5/2}$ excited state. So, it was justified by this way that Fig.5 is a more adapted evaluation in CW end-pumping cpnfiguration than Fig.1 for Q-switch oscillators and then to nanosecond-pulse lasers.

In addition of these spectroscopic data, we can discuss on the choice of laser crystals by considering comparative values of the thermal conductivities of each type of hosts. Undoped

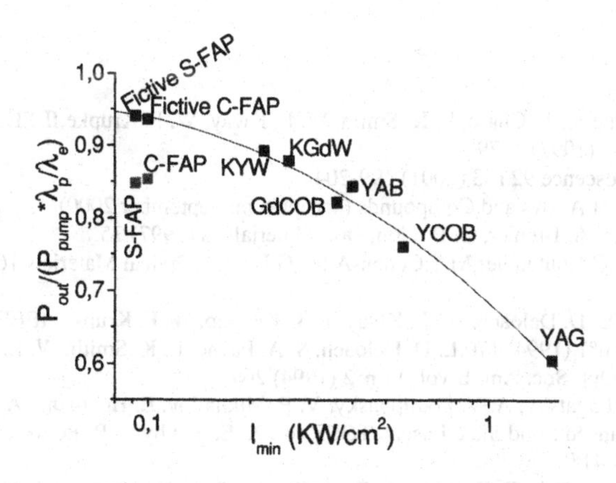

Figure 6. Calculated maximum output yield available from several Yb^{3+}-doped crystals versus I_{min} [2].

Y_2O_3 sesquioxide is characterized by the highest thermal conductivity (12.8 W/m°C) as compared with YAG (9.8 W/m°C) according last measurements in [26], excepted sapphire laser host (33 W/m°C) which is not involved by this study,due to the high difficulty to substitute enough Yb^{3+} ions within the octahedral Al^{3+} site of Al_2O_3. The other crystals are characrerized by smaller thermal conductivities: around 2 W/m°C for C-FAP, S-FAP, GdCOB, and around 3 W/m°C for YAB and KGW. It is very well known that thermal performances have to be considered in the final decision of the choice of new laser crystals in connection with the capacity of reproducible crystal growth of bulky samples. In this way, recent successful crystal growth of sesquioxide large crystals from rhenium crucibles by the Bridgmann method is an important advance [26].

CONCLUSION

We have used two kinds of crystal growth techniques to get Yb^{3+}-doped laser crystals which have up to a few mm thickness:double tungstate crystals are grown by the nucleated floating crystal method and sesquioxide crystals are grown by the LHPG method wich is very convenient for refractory oxide crystal growth since there is no need of crucible.The broad Yb^{3+}absorption lines in $KGd(WO_4)_2$ and $KY(WO_4)_2$ double tungstates as well as Y_2O_3 or Sc_2O_3 sequioxides are well suited for laser diode pumping between 900 nm and 980 nm. The Stokes shift under zero-phonon line pumping is relatively small:about 650 cm^{-1} in tungstates and 699 cm^{-1} in sesquioxides.This is quite well favourable to reduce the thermal loading of the material during laser operation. Such samples, are promising diode pumped laser crystals, according to a new model of evaluation of Yb^{3+}-doped crystals established for the CW regime in CW end-pumping configuration, which differs from DeLoach's model only valid for nanosecond pulse regime.Tungstates, sesquioxides and YAB have the highest laser potentialities in the CW regime, wheras fluorapatites, YAG and tungstates are the most efficient when considering the nanosecond pulse extraction regime. Moreover, another advantage especially of the sesquioxides is the high value of the thermal conductivities, the highest ones of the usual hosts, which gives to such crystals a very high potential by developping the crystal growth of bulky samples with the Bridgman method by using rhenium crucible.

REFERENCES

1.D. DeLoach, S. A. Payne, L. L. Chase, L. K. Smith, W. L. Kway, W. F. Krupke,IEEE J. Quant. Electr. Vol. 29 n°4 (1993) 1179
2.A.Brenier, J. of Luminescence 92 n°3 (2001) 199-204
3.A.Brenier,G.Boulon, J.of Alloys and Compounds (accepted on September 2000)
4.G. Métrat, N. Muhlstein, A. Brenier, G. Boulon, Opt. Materials, 8 (1997) 75.
5.L.Laversenne,Y.Guyot,C.Goutaudier,M.T.Cohen-Adad,G.Boulon,Optical Materials 16 n°4 (2001) 471
6.S. Payne, L. K. Smith, L. D. Deloach, W. L. Kway, J. B. Tassano, W. F. Krupke, IEEE J. Quant. Electron. Vol. 30 n°1 (1994) 170.L. D. Deloach, S. A. Payne, L. K. Smith, W. L. Kway, W. L. Krupke, J. Opt. Soc. Am. B vol. 11 n°2 (1994) 269.
7.N. V. Kuleshov, A. A. Lagatsky, A. V. Podlipensky, V. P. Mikhailov, E. Heumann, A. Diening, G. Huber, Advanced Solid State Lasers, TOPS vol. X, Edited by C. R. Pollock and W. R. Bosenberg, (1997) 415.
8.Augé, F. Mougel, F. Balembois, P. Georges, A. Brun, G. Aka, A. Kahn-Harari, Advanced Solid State Lasers, Topical Meeting of OSA (feb. 1999) Boston, paper TuC4-1/277.

9.A. A. Lagatsky, N. V. Kuleshov, V. P. Mikhailov, Advanced Solid State Lasers, Topical Meeting of OSA (feb. 1999) Boston, paper TuB12-1/247.
10.P. Wang, J. M. Dawes, P. Dekker, H. Zhang, X. Meng, Advanced Solid State Lasers, Topical Meeting of OSA (feb. 1999) Boston, paper ME14-1/151.
11.P. Wang, J. M. Dawes, P. Dekker, J. A. Piper, Advanced Solid State Lasers, Topical Meeting of OSA (feb. 1999) Boston, paper PD15-1.
12.P. Wang, J. M. Dawes, P. Dekker, D. S. Knowles, J. A. Piper, J. Opt. Soc. Am. B vol. 16 n°1 (1999) 63.
13.L. Fornasiero, E. Mix, V. Peters, K. Petermann, G. Huber, Cryst. Res. Technol. Vol. 34 n°2 (1999) 255.
14.K. Petermann, G. Huber, L. Fornasiero, S. Kuch, E. Mix, V. Peters, S. A. Basun, J. of Luminescence 87-89 (2000) 973
15.R. Gaumé, P. H. Haumesser, B. Viana, D. Vivien, G. Aka, B. Ferrand, « Photonic Materials for the 21st Century", May 28-31, 2000, Lyon (France).
16.W. P. Risk, J. Opt. Soc. Am. B vol. 5 n°7 (1988) 1412.
17.T. Taira, W. M. Tullier, R. L. Byer, Apple. Opt. Vol. 36 n°9 (1997) 1867.
18.G. Métrat, M. Boudeulle, N. Muhlstein, A. Brenier, G. Boulon, J. Cryst. Growth 197 (1999) 883.
19.A.Brenier, G.Métrat, N.Muhlsstein,F.Bourgeois, G.Boulon,Optical Materials 16 (2001) 189
20.D.E.McCumber,Phys.Rev. 136 (4A) (1964) 954
21.R. S. Feigelson, W.L. Kway, R.K. Route, *Proc. SPIE* 484 (1984) 133
22.G. Schaack, J. Koningtein, *J. Opt. Soc. Am.*, (1970), 60,1110
23.F. Hanic, M. Hartmova, G.G. Knab, A. A. Urusovskaya, K. S. Bagdasarov, *Acta Cryst.*, (1984), B40, 76
24.G.C. Aumüller, W. Köstler, B.C. Grabmaier, R. Frey, *J. Phys. Chem. Solids*, (1994), 55(8), 767
25.N.C. Chang, J. B. Gruber, R.P. Leavitt, C.A. Morrison, *J. Chem. Phys.*, (1982), 76(8), 3877
26.V.Peters, E.Mix, L.Fornasiero, K.Petermann, G.Huber, S.Basun, LPHYS 99, Book of abstracts (1999) p.147

Mat. Res. Soc. Symp. Proc. Vol. 667 © 2001 Materials Research Society

HIGH PRESSURE STUDIES OF Sm^{2+}-DOPED SOL-GEL GLASSES

Vilma C. Costa, Yongrong Shen, and Kevin L. Bray
Department of Chemistry, Washington State University, Pullman, WA 99164
Ana M. M. Santos
Center of Nuclear Technology Development / CDTN, CP 941, Pampulha,
Belo Horizonte, MG, Brazil, 30123-970

ABSTRACT

Glasses containing nominally 1, 2, 5, and 10 wt% Sm_2O_3 in Na_2O-Al_2O_3-SiO_2 and Al_2O_3-SiO_2 were prepared from metal alkoxide solution using the sol-gel process. After low temperature heat treatment in air, the glasses were heated up to 800 °C under a flowing H_2 atmosphere to reduce Sm^{3+} into Sm^{2+}. Samarium ions in the divalent and trivalent states were identified by fluorescence measurements. The fluorescence properties of Sm^{2+} ions are discussed in relation to concentration of Sm_2O_3 and the glass matrix composition. Preliminary results of pressure studies on the luminescence spectra and lifetime of Sm^{2+} in the glasses are presented as well.

INTRODUCTION

Samarium (II) doped crystals and glasses have attracted much attention for potential applications in lasers, fiber amplifiers and memory devices. Interest in materials containing Sm^{2+} has increased in recent years because of their hole burning properties and applications as optical data-storage materials. Macfarlane et al. [1] first observed persistent spectral hole burning spectra in Sm^{2+} in halide crystals at 2 K. Recently, persistent spectral hole burning has been observed at room temperature in fluoride crystals [2], borate glasses [3], and fluorohafnate glasses [4]. A new approach for the preparation of Sm^{2+}-doped aluminosilicate glasses using the sol-gel process has been reported [5] and room temperature persistent spectral hole burning has been observed in these systems. Glass systems are expected to be more desirable for high density memory devices than crystals because of their large inhomogeneous line widths and convenient sample preparation. The sol-gel technique is an especially attractive glass preparation method since lower processing temperatures are required relative to conventional melting techniques and the high porosity of sol-gel glass promotes reduction of Sm^{3+} to Sm^{2+} in presence of H_2.

Sm^{2+} has a ground $4f^6$ configuration and an excited $4f^55d$ configuration. The close proximity in energy of the two configuration leads to a mixing of the states from the two configurations and to an alteration of the optical properties of Sm^{2+}. With high pressure we can tune the extent of mixing of the two configurations by varying their energy separation. As a result, we can systematically investigate how the extent of mixing varies the static and dynamic emission properties of Sm^{2+} and correlate the emission properties with crystal field strength and coordination environment. There has been considerable progress in understanding the spectral properties of rare earth ions under pressure. The effects of pressure on spectral properties of Pr^{3+} [6], Nd^{3+} [7], Sm^{2+} [8] and Eu^{3+} [9] in crystals and Eu^{3+} [10] and Sm^{3+} [11] in glasses have been reported.

In this paper, we present preliminary results on fluorescence of Sm^{2+} in aluminosilicate derived sol-gel glass examined under pressure up to 190 kbar.

EXPERIMENTAL SECTION

Sample Preparation

Sol-gel glasses containing nominally 1, 2, 5, and 10 wt% Sm_2O_3 doped into the molar compositions $5Na_2O$-$10Al_2O_3$-$85SiO_2$ and $10Al_2O_3$-$90SiO_2$ were prepared from metal alkoxide solution following the procedure presented by Nogami et al [12]. Tetraethoxysilane (TEOS, $Si(OC_2H_9)_4$), $Al(OC_4H_4^{sec})_3$, $SmCl_3 \cdot 6H_2O$ or $Sm(NO_3)_3 \cdot 6H_2O$ and CH_3CO_2Na were used as starting reagents. TEOS was hydrolyzed for 1 h with a solution of H_2O, C_2H_5OH in the molar ratio of 1:1 per mol of TEOS. HCl was added to achieve an initial solution pH of 1.5 - 2. $Al(OC_4H_9^{sec})_3$ was then added. After stirring for 15 min, an ethanolic solution of either $SmCl_3 \cdot 6H_2O$ or $Sm(NO_3)_3 \cdot 6H_2O$ and CH_3CO_2Na was added and the solution was further stirred for 15 min. The resultant solution was hydrolyzed by adding the mixed solution of H_2O, C_2H_5OH and HCl in a 4:1:0.011 mol per mol of Al alkoxide.

After stirring 30 min, the sols were cast into plastic vials where they were allowed to gel at room temperature. After gelation, the samples were aged at 60 °C for 2 days and then dried at 90 °C for 2 days. After 2 h heat treatment in air at 500 °C and 700 °C, the glasses were heated up to 800 °C for 24 h, under a flowing mixed gas of 20%H_2-80%N_2 or at 750 °C for 10 h in H_2 atmosphere.

Spectroscopic Measurements

The fluorescence spectra were obtained upon $4f^55d$ excitation at 488.0 nm from an argon laser. Samarium luminescence was dispersed on a 1-m Spex 1704 monochromator and detected by a Hamamatsu R928 photomultiplier tube. All reported spectra have been corrected for instrument response.

For the luminescence decay measurements, a Nd:YAG laser (a second harmonic beam of 532 nm) was used to excite Sm^{2+} from the $^7F_0$ ($4f^6$) ground state to the $4f^55d$ excited state. The spectra were detected by a 1-m monochromator and decay data were recorded by a digital storage oscilloscope. Lifetimes were measured at room temperature. Fluorescence decays were monitored at the peak wavelength (683 nm) of the $^5D_0 \rightarrow {}^7F_0$ emission line of Sm^{2+}. Since the decays curves were nonexponential, average lifetimes were obtained by calculating the area under normalized curves.

High pressure was generated in a diamond anvil cell (DAC). The sample chamber consisted of a 250 μm hole drilled in a pre-indented inconel gasket. A small piece of glass sample was loaded in the sample chamber along with a ruby pressure calibrant. A spectroscopic oil (poly-dimetylsiloxane) served as the pressure transmitting medium.

RESULTS

Ambient Pressure Fluorescence

The gels were heated in air at 500 °C for 2 h before heating in H_2 atmosphere. The luminescence spectra of the samples after heating at 500 °C showed no signal from samarium ions. The absence of a luminescent signal is attributed to non-radiative quenching to residual water and organic fragments remaining in the sol-gel matrix after heating at 500 °C. Higher heat treatment temperatures are needed to remove residual water and organic fragments and are expected to promote the luminescence of Sm^{3+} and Sm^{2+}.

Figure 1a shows the luminescence spectrum $5Na_2O$-$10Al_2O_3$-$85SiO_2$ sample containing 2 wt% Sm_2O_3. The sample was heated in air at 700 °C for 2 h and was excited at λ_{exc}= 488 nm. The observed emission spectrum shows peaks around 565, 600, 650 and 710 nm attributed to $^4G_{5/2} \rightarrow {}^6H_{5/2}$, $^6H_{7/2}$, $^6H_{9/2}$, and $^6H_{11/2}$ transitions of Sm^{3+}. Figure 1b shows the luminescence of the same composition after heat treatment at 750 °C for 10 h in flowing H_2. The spectral changes are consistent with a partial reduction of Sm^{3+} to Sm^{2+} in the glass. The peaks at 683, 700, 725 and 763 nm are assigned to the $^5D_0 \rightarrow {}^7F_0$, $^7F_1$, $^7F_2$, and $^7F_3$ transitions of Sm^{2+} ions, respectively. Similar behavior was observed for the $10Al_2O_3$-$90SiO_2$ glass composition.

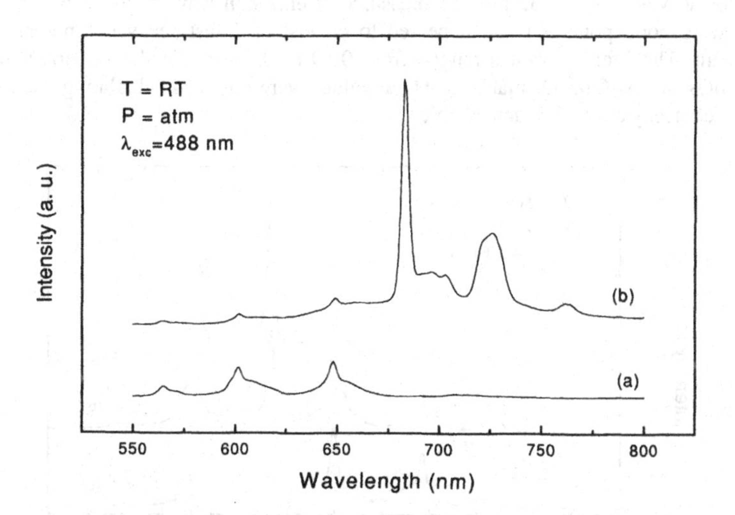

Figure 1. Emission spectra of a $5Na_2O$-$10Al_2O_3$-$85SiO_2$ glass composition containing 2 wt% Sm_2O_3: (a) glass heated at 700 °C in air, (b) glass heated at 750 °C in H_2 gas for 10 h.

The emission spectra of $10Al_2O_3$-$90SiO_2$ glasses containing 1, 2, and 10 wt% Sm_2O_3 after heating at 800 °C in H_2 for 24 h are shown in figure 2. The spectra indicate that the relative proportion of Sm^{2+} decreases with increasing Sm_2O_3 doping concentration. This result is consistent with a recent NMR and X-ray study of Sm_2O_3-doped Al_2O_3-SiO_2 glasses by Jin et al. [13]. Jin et al. propose that the presence of Al^{3+} significantly facilitates the reduction of Sm^{3+} to Sm^{2+}. They argue that glasses doped with Sm^{3+} contain Al^{3+} almost exclusively in the form of network forming AlO_4 tetrahedra. They further show that heat treatment in a reducing (H_2)

atmosphere leads to the conversion of AlO_4 tetrahedra into AlO_6 octahedra and that this coincides with the formation of Sm^{2+}. According to their $AlO_4 \leftrightarrow AlO_6$ model, we expect that the increasing non-bridging oxygen concentration resulting from the formation of AlO_6 octahedra satisfies the increased coordination needs of the larger Sm^{2+} ions. The data presented by Jin et al. showed for a given Al/Si ratio, that the relative amount of Sm^{2+} decreases with increasing Sm_2O_3 doping concentration. In other words, the fractional conversion of Sm^{3+} to Sm^{2+} decreases as the number of non-bridging oxygens available per Sm^{3+} decreases. Our results for the $5Na_2O$-$10Al_2O_3$-$85SiO_2$ glass composition are also consistent with this interpretation. Addition of Na_2O to Al_2O_3-SiO_2 glasses at fixed Al_2O_3 content leads to an increase in the non-bridging oxygen concentration. A comparison of the results for $5Na_2O$-$10Al_2O_3$-$85SiO_2$ (2 wt%, Fig 1b) with those for $10Al_2O_3$-$90SiO_2$ (2 wt%, Fig 2b) shows that the relative amount of Sm^{2+} is greater in the presence of Na_2O.

Ambient Pressure Decay

Fluorescence decay curves at room temperature for the $^5D_0$ level of Sm^{2+} were obtained at excitation wavelength of 532 nm and measured at emission wavelength of 683 nm. The decay curves were nonexponential and appeared to consist of a fast-decay component and a slow-component. The lifetime values ranged from 0.60 to 0.86 ms for the samarium doped Na_2O-Al_2O_3-SiO_2 and Al_2O_3-SiO_2 matrices. These values were obtained calculating the area under the normalized decay curve for each sample.

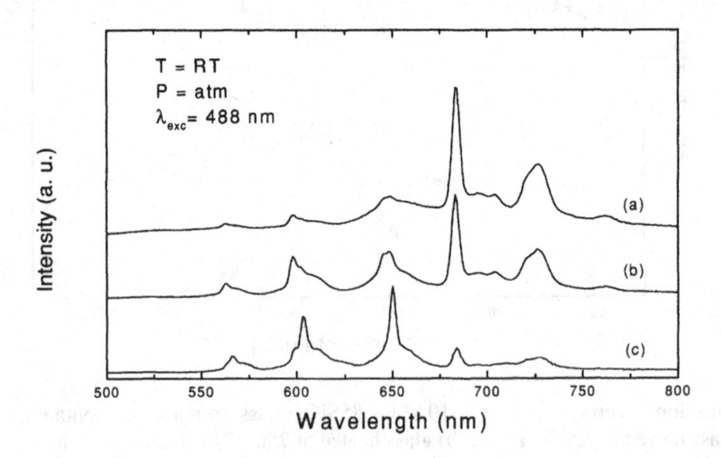

Figure 2. Normalized emission spectra of $10Al_2O_3$-$90SiO_2$ glasses containing: (a) 1 wt% Sm_2O_3, (b) 2 wt% Sm_2O_3, and (c) 10 wt% Sm_2O_3. Glasses heated in H_2 gas at 800 °C for 24 h.

High Pressure Fluorescence

Upon excitation into $4f^5 5d$ state (λ_{exc} = 488 nm), luminescence of Sm^{2+} is associated to the nonradiative $4f^5 5d \rightarrow {}^5D_0$ relaxation and the radiative transition from the ${}^5D_0$ level into the ground ${}^7F_J$ states.

Figure 3 presents representative high pressure emission spectra for the $10Al_2O_3$-$90SiO_2$ glass composition containing 2 wt% Sm_2O_3. The sample was heated to 750 °C in an H_2 atmosphere for 10 h. The ambient pressure spectrum (Fig. 2b) indicates the presence of emission from both Sm^{3+} and Sm^{2+}. A slight redshift (~0.03 nm/kbar) and pronounced intensity decrease were observed for the ${}^5D_0 \rightarrow {}^7F_0$ transition of Sm^{2+} as pressure was increased. By ~74 kbar, no Sm^{2+} emission intensity was observed. Upon release of pressure, the emission spectrum did not revert back to the original ambient pressure spectrum. An overall decrease in emission intensity and no Sm^{2+} emission was observed upon release. The released spectrum indicates that an irreversible transformation occurs in the glass with increasing pressure. The irreversible transformation may involve a permanent structure change in the glass or, possibly, a pressure-induced shift of the sol-gel reaction toward greater completion.

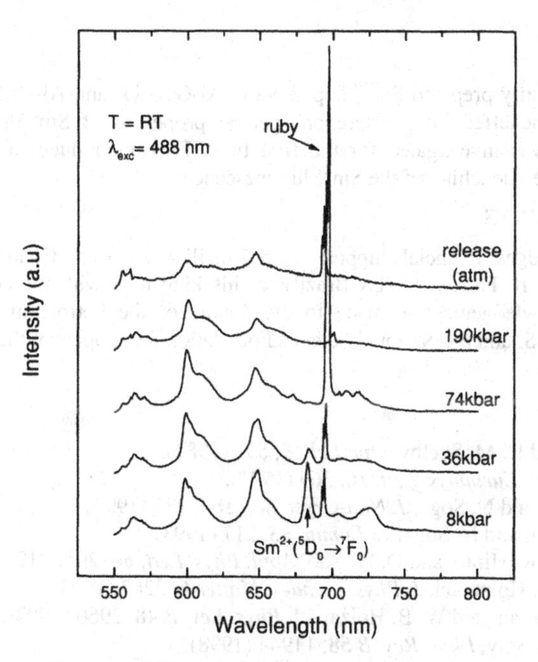

Figure 3. Representative high pressure emission spectra of $10Al_2O_3$-$90SiO_2$ glass containing 2 wt% Sm_2O_3. The sample was heated at 750 °C in H_2 gas for 10 h. The release spectrum is the result obtained upon returning the pressure to ambient pressure at the conclusion of the experiment. The sharp lines near 700 nm originate from the ruby pressure calibrant used in the experiment.

The loss of Sm^{2+} emission intensity with pressure could be due to an enhancement of the non-radiative decay of Sm^{2+} or a pressure-induced oxidation of Sm^{2+} to Sm^{3+}. High pressure studies of Sm^{2+}-doped crystals [8, 14] show that high pressure induces a large red shift of the excited $4f^5 5d$ state of Sm^{2+}. Since the $4f^5 5d$ state of Sm^{2+} is close in energy to the emitting $^5D_0$ sate of the ground $4f^6$ configuration of Sm^{2+}, it is conceivable that pressure induces a $4f^5 5d$-$4f^6$ ($^5D_0$) electronic crossover. Since the $4f^5 5d$ state interacts strongly with the glass matrix, a pressure-induced electronic crossover could greatly enhance the non-radiative decay rate of Sm^{2+}.

Alternatively, pressure may lead to an oxidation of Sm^{2+} to Sm^{3+}. Based on the work of Jin et al. [13], such an oxidation would likely be accompanied by a structural transformation of the glass involving conversion of AlO_6 octahedra to AlO_4 network forming tetrahedral. In other words, pressure could induce the reverse of the reduction reaction. This process is plausible because, thermodynamically, pressure stabilizes the lowest volume state of a system. In two out of three glasses, Jin et al. reported a decrease in sample density upon reduction of $Sm^{3+}$to Sm^{2+}. Thus, oxidation of Sm^{2+} is consistent with stabilization of the lowest volume state of the glass. Further work is necessary to more definitively establish the effect of pressure on the glass.

CONCLUSION

We have successfully prepared Sm^{2+}-doped Na_2O-Al_2O_3-SiO_2 and Al_2O_3-SiO_2 glasses via the sol-gel method. The effect of pressure on spectral properties of Sm^{2+}in sol-gel derived aluminosilicate glass was investigated for the first time. Pressure induces a decrease in the intensity and a complete quenching of the Sm^{2+} luminescence.

ACKNOWLEDGEMENTS

V.C.C. acknowledges financial support from Brazilian National Council of Research, CNPq and Dr. Wilmar B. Ferraz (CDTN-Brazil) for his kind help with the heat treatment of glasses. Further acknowledgement is made to the donors of the Petroleum Research Fund, administered by the ACS, and the National Science Foundation for support of this research.

REFERENCES

1. R. M. Macfarlane and R. M. Shelby, *Opt. Lett.* **9**, 533 (1984).
2. R. Jaaniso and H. Bill, *Europhys. Lett.* 16, **569** (1991).
3. D. H. Cho, K. Hirao, and N. Soga, *J. Non-Cryst. Sol.* **189**, 181 (1995).
4. K. Hirao, S. Todoroki, and N. Soga, *J. Lumin.* **55**, 217 (1993).
5. M. Nogami, Y. Abe, K. Hirao, and D. H. Cho, *Appl. Phys. Lett.* **66**, 2952 (1995).
6. Y. R. Shen and W. B. Holzapfel, *J. Phys: Condes. Matter* **6**, 2267 (1994)
7. Th. Tröster, T. Gregorian, and W. B. Holzapfel, *Phys. Rev. B* **48**, 2960 (1993).
8. Y. R. Shen and K. L. Bray, *Phys. Rev. B* **58**, 11944 (1998).
9. Y. R. Shen and W. B. Holzapfel, *Phys. Rev. B* **52**, 12618 (1995).
10. M. J. Lochhead and K. L. Bray, *Phys. Rev. B* **52**, 15763 (1995).
11. C. K. Jayasankar, P. Babu. Th. Tröster, and W. B. Holzapfel, *J. Lumin.* **91**, 33 (2000).
12. M. Nogami, N. Hayakawa, N. Sugioka, and Y. Abe, *J. Am. Ceram. Soc.* **79**, 1257 (1996).
13. J. Jin, S. Sakida, T. Yoko, and M. Nogami, *J. Non-Cryst. Sol.* **262**, 183 (2000).
14. C. S. Yoo, H. B. Radousky, N. C. Holmes and N. M. Edelstein, *Phys. Rev. B* **44**, 830 (1991).

Mat. Res. Soc. Symp. Proc. Vol. 667 © 2001 Materials Research Society

Near-field Scanning Optical Microscopy and Electron Microprobe Microscopy Investigations of Immiscibility Effects in Indium Gallium Phosphide Grown by Liquid Phase Epitaxy

C. A. Paulson, A. B. Ellis
Department of Chemistry, The University of Wisconsin, Madison, WI 53706
T. F. Kuech
Department of Chemical Engineering, The University of Wisconsin, Madison, WI 53706

Abstract

We have used Near-field Scanning Optical Microscopy (NSOM) and Electron Probe Microanalysis (EPMA) to study the topographic and microscopic optical properties of indium gallium phosphide ($In_{1-x}Ga_xP$) samples grown by Liquid Phase Epitaxy (LPE) on gallium arsenide substrates. Photoluminescence (PL) intensity images gathered using NSOM exhibit strong, highly localized variations in the optical properties of these samples that are seen to occur roughly in registry with the surface topography. Shifts in the PL peak position (by 27 meV) occur across highly mismatched samples with high In content, whereas no shifts were seen for $In_{1-x}Ga_xP$ films with a nearly lattice matched composition. Compositional fluctuations lead to these PL peak energy shifts, measured by NSOM with a resolution of 250 nm. These composition fluctuations arise from the known solid-solid miscibility gap in the $In_{1-x}Ga_xP$ system at temperatures used for the growth of these samples.

Introduction

Indium gallium phosphide ($In_{1-x}Ga_xP$) is an important alloy system that is used in many semiconductor electronic devices. Growth of $In_{1-x}Ga_xP$ is accomplished by a variety of methods, including Metal-Organic Vapor Phase Epitaxy (MOVPE) [1], Molecular Beam Epitaxy (MBE) [2], and Liquid Phase Epitaxy (LPE) [3]. Among these, LPE is recognized as having industrial appeal due to its high deposition rate. LPE deposition occurs under conditions that are close to thermodynamic equilibrium [3]. Under such conditions, immiscibility can be an important parameter for the LPE deposition of films [4]. Previous work has demonstrated that growth of an alloy composition within the thermodynamic miscibility gap can increase dislocation generation [5] and surface roughness [5], while decrease carrier mobility through alloy scattering [6]. Spatially-resolved investigations of immiscibility effects on LPE films have used Cathodoluminescence (CL) [7], Electron Microprobe Analysis (EPMA)[8], and Transmission Electron Microscopy (TEM) [8] among other techniques. Only a few studies report results on LPE films that have large mismatches to their substrates ($\Delta a/a_0 > 0.5\%$) [5][9][10]. We present results of Near-field Scanning Optical Microscopy (NSOM) investigations of LPE $In_{1-x}Ga_xP$ samples grown on gallium arsenide (GaAs) substrates over a range of lattice mismatch of 0 to 2%. The NSOM images reveal a high level of local variation in the photoluminescence (PL) for $In_{1-x}Ga_xP$ at compositions that are well removed from the lattice match condition of x = 51%. Since the NSOM experiment simultaneously determines surface topography, we also find that the PL fluctuations and the surface topography are correlated. Through comparison with EPMA results, the contrast in the NSOM images can be related to composition fluctuations in these films. This composition fluctuation results from thermodynamic immiscibility.

Experimental

Samples of $In_{1-x}Ga_xP$ were grown by LPE on <100> GaAs substrates. The average composition for these samples was confirmed by far-field PL spectroscopy [5]. Samples of $In_{0.49}Ga_{0.51}P$, $In_{0.64}Ga_{0.36}P$, and $In_{0.77}Ga_{0.23}P$ were investigated. These samples have approximately 0, 1, and 2% lattice mismatch with GaAs, respectively. All epitaxial layers were greater than 1 μm thick. Far-field collection mode NSOM was performed in an air ambient at room temperature. An argon ion laser was used for the excitation source, and the aperture of the NSOM probe was approximately 50 nm in diameter. The luminescence was gathered using a parabolic mirror and analyzed by a 1/4-meter spectrometer with a water-cooled PMT detector. The instrumental resolution was 12 meV at a photon energy of 1.7 eV. Separate measurement scans were performed with the spectrometer centered at a specific energy, allowing spectrally and spatially resolved images of the $In_{1-x}Ga_xP$ materials to be obtained. Spatial mapping of the alloy composition was performed using a Cameca SX51 Electron Microprobe instrument. For all the microprobe results, the beam current was 80 nA (Faraday current) at 1500 V, and the peak integration time was 30 s. The small local changes in alloy composition necessitated long measurement integration times. The electron beam excitation area was approximately 1 μm, and the pixel spacing was 1 μm. The x-ray mapping was performed using the indium Lα, gallium Lα, phosphorus Kα, and arsenic Lα x-ray lines.

Results

NSOM was used to gather 8 by 8 micron images of the topography and optical properties of the LPE samples. A topographic image gathered during NSOM of $In_{0.64}Ga_{0.36}P$ is shown in figure 1(a). Generally, the bright and dark areas of the image (representing high and low areas of the topography, respectively) occur along a diagonal connecting the upper right to the lower left corner of the image. The PL peak energy was found to vary between 1.676 and 1.703 eV for this sample, as will be discussed below. NSOM PL images of the same area as the topographic image, were acquired with the spectrometer centered at 1.703 eV, figure 1b, and with the spectrometer centered at 1.676 eV, figure 1c. The PL intensity image gathered at 1.703 eV (figure 1b) was obtained three times, from the same area, with all images being nearly identical. Images obtained with the spectrometer centered at 1.676 eV are different from those obtained at 1.703 eV. The topography for all the imaging remained unchanged, illustrating that the tip did not drift significantly between scans. The spatial variation in the luminescence images show bright and dark areas occurring along the diagonal connecting the upper right to the lower left corner of the image, corresponding to the spatial variation in the topography.

NSOM imaging was also performed on the $In_{0.77}Ga_{0.23}P$ and $In_{0.49}Ga_{0.51}P$ (nearly lattice-matched) LPE samples. The $In_{0.77}Ga_{0.23}P$ sample results were similar to those for $In_{0.64}Ga_{0.36}P$, with a rough morphology and strong contrast in the luminescence image. The nearly lattice-matched sample exhibited smooth morphology and very little luminescence contrast.

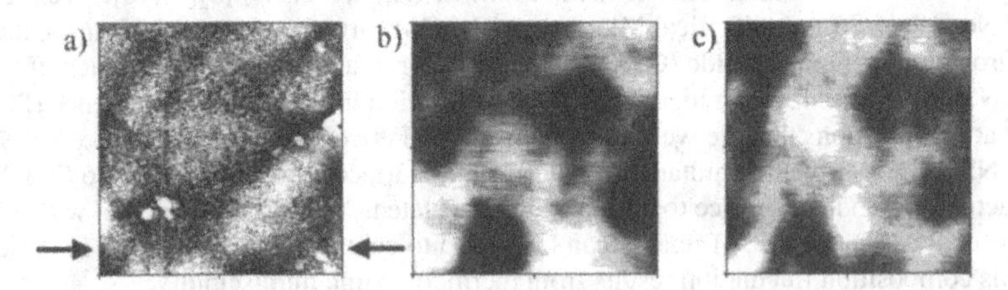

Figure 1. NSOM images from the $In_{0.64}Ga_{0.36}P$ sample, all from the same 8 by 8 micron region. (a) Topography. The gray scale ranges from 0 to 37 nm. Arrows next to this image indicate a line along which point spectroscopy was performed. (b) An image of PL intensity at 1.70 eV. The gray scale represents 12-fold changes in luminescence intensity. (c) An image of PL intensity at 1.676 eV with the gray scale representing 6-fold changes in luminescence intensity.

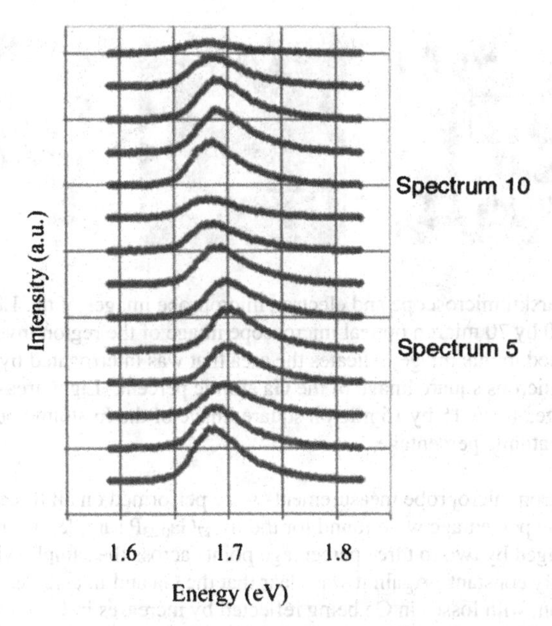

Figure 2. Fourteen PL point spectra gathered from the $In_{0.64}Ga_{0.36}P$ sample. These point spectra were evenly spaced along the line indicated in figure 1(a). The peak maximum ranges from 1.676 eV to 1.70 eV in these spectra.

Point spectroscopy was performed on the $In_{0.64}Ga_{0.36}P$ sample. During point spectroscopy, the NSOM tip was held in a fixed position over the sample while a PL spectrum was recorded. Fourteen point spectra are shown in figure 2, acquired from points roughly evenly separated by ~0.5 microns, along the line indicated in figure 1(a). The PL maxima ranged from a minimum energy value of 1.676 eV in spectrum 10, to a maximum of 1.703 eV in spectrum 5. Changes in the magnitude of the PL peak maximum are also seen in the PL point spectra.

These samples were also investigated by EPMA, the results of which are shown for the $In_{0.64}Ga_{0.36}P$ sample in figure 3. An optical microscope image of the sample is shown in figure 3a. This image has dimensions of 70 by 70 microns. Carbon, deposited by the electron beam during the EPMA measurement, was seen in the optical microscope after EPMA was performed and was used to determine the measurement area. A box (15 by 15 microns) in the optical microscope image (figure 3a) indicates the area probed by the electron beam. The distribution of Ga atomic percentage (within the box) determined by EPMA x-ray mapping is shown in 3b, and that of In is shown in 3c (each 15 by 15 microns). Light areas in these images indicate a higher atomic percentage of Ga (3b) or In (3c). By comparing the images, the Ga atomic percent is high where the In atomic percent is low, indicating their respective atomic percentages are inversely correlated. X-ray mapping was also performed for the elements phosphorus and arsenic, the only other elements present in the heterostructure, but no trends were noted for their distributions. The As signal arising from the substrate was very weak, since it arises from more than a micron below the surface of the film.

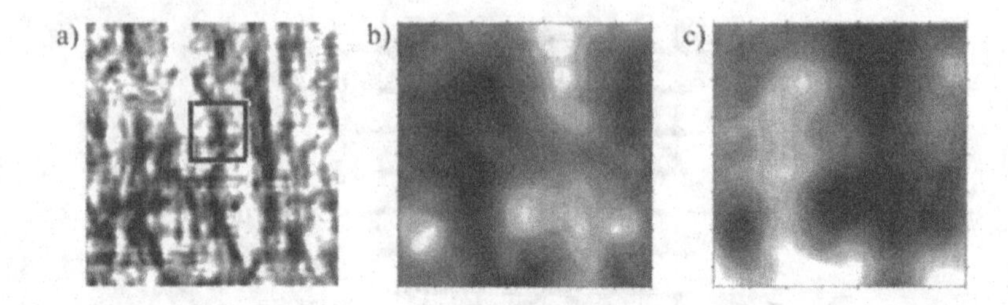

Figure 3. Nomarskii microscope and electron microprobe images of the LPE $In_{0.64}Ga_{0.36}P$ sample. (a) A 70 by 70 micron optical microscope image of the region investigated by EPMA. The square marked in this image indicates the area that was interrogated by the electron beam. (b) A 15 by 15 microns square image of the Ga atomic percent. Light areas have a higher Ga atomic percentage. (c) A 15 by 15 micron square image of the In atomic percentage. Light areas have a higher In atomic percentage.

The electron microprobe measurements were performed on all three samples. The largest changes in atomic percentage were found for the $In_{0.77}Ga_{0.23}P$ sample, where the amount of Ga and In each changed by two to three percentage points across the sample with their sum remaining roughly constant. Again it was clear that the Ga and In changes occurred in a reciprocal fashion, with losses in Ga being reflected by increases in In and vice-versa. The results found for the $In_{0.64}Ga_{0.36}P$ sample showed smaller changes in atomic percent of about 1.5 percentage points. For the $In_{0.49}Ga_{0.51}P$ sample, the changes in In and Ga were less than one percentage point, and they were uncorrelated.

Discussion

In all NSOM experiments, the source of contrast recorded by the images must be determined. In the present measurements, both the PL peak intensity and PL peak position spatially vary across the $In_{0.64}Ga_{0.36}P$ sample, as shown in figure 2. The change in PL peak position can change due to strain in the material or due to a change in the alloy composition [11], or both. Since the epitaxial layers were all greater than a micron in thickness, they are all totally or near totally relaxed [12]. Therefore, composition variations are the most likely source of the variation in PL peak energy position. The electron microprobe results corroborate this conclusion. Consider the NSOM point spectroscopy results of figure 2, where the highest PL peak energy value was 1.703 eV and the lowest energy was 1.676 eV. For $In_{1-x}Ga_xP$, the band edge emission energy varies as [5];

$$h\upsilon = 1.35 + 0.73x + 0.70x^2$$

where x is the percent Ga. Using this expression, the PL-derived composition variation is ~2 percentage points (from figure 2) which is in reasonable agreement with the corresponding microprobe result. Also, both the NSOM and the electron microprobe experiments showed little variation across the nearly lattice-matched sample. Therefore, the NSOM appears to measure spatial variation in alloy composition. Data in figure 1b can be interpreted to mean that the regions that are lighter, as compared to figure 1c, have a higher In atomic percentage in the alloy, which results in a smaller PL band gap energy nearer to 1.676 eV. The regions of 1c where the image is brighter than it is in 1b, are regions of higher Ga atomic percent in the alloy, exhibiting a larger PL band gap nearer to 1.703 eV.

The spatial variation in the In and Ga composition are observed to occur with roughly the same spatial directions as the surface topography in the NSOM results shown in figure 1. These effects are also noted in the EPMA data from the same sample shown in figure 3. Further, we find that for the near lattice-matched sample there is little fluctuation in the alloy composition, and the surface morphology was smooth by both NSOM and EPMA. Spatial variation in the alloy composition is correlated with local lattice mismatch and surface roughness. Therefore, a correlation between the surface topography and the variation in atomic percentages of Ga and In is indicated on both the several micron length scale probed by EPMA, and on the submicron scale probed in NSOM.

Only a few reports on the surface morphology exist for LPE heterostructures for films with compositions that are more than 1% lattice mismatched relative to their substrate. A previous study had also indicated that the morphology becomes rougher, as does dislocation density, as the lattice mismatch increased [5]. The regions of the NSOM optical images (figures 1 b and c) that remained dark in both images may contain dislocations. These dark regions were also found on the $In_{0.77}Ga_{0.23}P$ sample but not on the near lattice-matched sample. Our data indicates that the alloy fluctuation is correlated with a high dislocation density.

Immiscibility exists in the $In_{1-x}Ga_xP$ ternary system and provides a driving force for the alloy fluctuations seen in these samples. However, only a few studies have looked in detail at the spatial variation in alloy composition. Previous studies reported variations on either much shorter length scales - tens of nanometers as probed by TEM [8] - or longer length scales of several microns, probed by EPMA. For comparison, the smallest features in figure 1c are about 250 nm in their minimum spatial dimension, providing an estimate for the NSOM resolution. Higher resolution may be achievable by improvements in the PL signal to noise ratio.

Conclusion

NSOM and EPMA were used to spatially quantify composition fluctuations in $In_{1-x}Ga_xP$ heterostructures grown by LPE on GaAs substrates. Alloy composition fluctuation was correlated with changes in surface topography. Immiscibility provides a driving force for this alloy fluctuation under common growth conditions. NSOM provided higher resolution images of the composition fluctuations than EPMA, and provides an alternative method for probing the local composition changes in a broad class of metastable semiconductor materials.

References

1. O. Ueda, M. Takikawa, J. Komeno, I. Umebu, Jap. J. Appl. Phys. Lett. 26, L1824, (1987).
2. S. F. Yoon, K. W. Mah, H. Q. Zheng, J. Alloys and Compounds, 280, 299, (1998).
3. E. Bauser, "Atomic Mechanisms in Semiconductor Liquid Phase Epitaxy," *Handbook of Crystal Growth, Vol. 3*, ed. D. T. J. Hurle (Elsevier, 1994), pp.880-939.
4. T. W. Kim, H. J. Ko, H. L. Park, Solid State Comm., 110, 29, (1999).
5. G. B. Stringfellow, J. Appl. Phys. 43, 3455, (1972).
6. J. H. Marsh, Appl. Phys. Lett., 41, 732, (1982).
7. T. Kato, T. Matsumoto, T. Ishida, Jpn. J. Appl. Phys. Part 1, 28, 1513, (1989).
8. P Henoc, A. Izrael, M. Quillec, H. Launois, Appl. Phys. Lett., 40. 963, (1982).
9. T. F. Kuech, J. O. McCaldin, J. Vac. Sci. Tech., 17, 891, (1980).
10. S. Mukai, J. Appl. Phys. 54, 2635, (1983).
11. G. D. Gilliland, Mater. Sci. Eng. Rep., R18, (1997).
12. J. W. Matthews, A. E. Blakeslee, S. Mader, Thin Solid Films, 33, 253, (1976).

Acknowledgements

The financial support of the Army Research Office and Office of Naval Research is gratefully acknowledged, along with the UW-Materials Research Science and Engineering Center (MRSEC) supported by the National Science Foundation.

Mat. Res. Soc. Symp. Proc. Vol. 667 © 2001 Materials Research Society

Abnormal Spectral Behavior of Trivalent Neodymium Ions in Potassium Yttrium Fluoride Crystals

Chunlai Yang and Baldassare Di Bartolo
Department of Physics, Boston College, Chestnut Hill, MA 02467

ABSTRACT

Nd^{3+} ions in KYF_4 can occupy different sites. The energy transfer processes that take place between them are responsible for the unusual behavior of Nd^{3+}. From the spectroscopic measurements at different temperatures we found an anomalous spectroscopic behavior in the low temperature range (20K - 100K). Based on an energy transfer model, we proposed an explanation to this abnormal behavior. We derived the formula to describe the deexcitation process and make a data fitting on our dynamic decay measurements.

INTRODUCTION

The optical activity of lanthanides as impurities in various host materials is centered on the states of the 4f shell. These states are strongly screened from the influence of the environment by the $5s^2 5p^6$ shells and therefore are not greatly influenced by the nature of host materials. Thus it is generally believed that the luminescence of these materials may be simply interpreted. This , however, is not true in the case of KYF_4.

The laser transition of Nd^{3+} laser materials originates in an $^4F_{3/2}$ energy level. Between the $^4F_{3/2}$ state and the next lower state $^4I_{15/2}$ there is a large energy gap of about 6000cm^{-1}. Nonradiative transition between these levels are improbable because, in solid state hosts, the upper energy of the phonon spectrum is only a few hundred cm^{-1}. In the case of fluoride hosts it is about 400 cm^{-1}, one fifteenth of the energy gap. As a consequence the lifetime of $^4F_{3/2}$ state is expected to be temperature independent. This general feature is not present in KYF_4 crystals.

Our spectroscopic measurements on Nd^{3+} doped KYF_4 crystal at different temperatures show a complicated and anomalous behavior in a low temperature range (T < 100K): at ~ 50K, the lifetimes and the emission intensity undergoes dramatic changes with the temperatures.

EXPERIMENTS

Three Nd^{3+} doped KYF_4 crystals were used in this research with different Nd^{3+} concentrations (2%, 3%, and 4%). These samples were provided by CREOL (Center for Research in Electro-Optics and Laser) of the University of Central Florida. Related experiments for laser performance were also performed there [1]. The crystal structure of KYF_4 is reported by Y. L. Fur et al [2].

Luminescence spectra were carried out by using a Omnichrome 532 Argon laser to excited the sample and a SPEX 1800 3/4 - meter Czerny Turner Spectrometer as the optical analyzer. A RCA 7102 photomultiplier tube and a EG&G P.A.R. 5101 lock-in amplifier were used to detect the signal. We used a Leybold-Heraeus refrigerator and a Lake Shore Cryotronics 805 temperature controller to set the sample at different temperatures from 20K to 300K.

The excitation source for the lifetime measurements was a Molectron DL-12 pulsed tunable dye laser, which was pumped by a Molectron UV-12 Nitrogen laser. The excitation wavelength was chosen to correspond to the strong transition from $^4I_{9/2}$ to $^4G_{5/2}$, which can effectively absorb and transfer the energy to $^4F_{3/2}$ by nonradiative transition. The radiation was detected by a cooled RCA 7102 photomultiplier and the signal was analyzed using a Tektronix 2232 digital storage oscilloscope. We used a 1.06 μm interference filter to isolate the particular transition from $^4F_{3/2}$ to $^4I_{11/2}$, that gives the normal laser emission.

RESULTS AND DISCUSSION

The three Nd^{3+} doped KYF_4 crystals show similar emission spectra. Fig. 1 presents one of the spectral behavior (2% sample) in the 1030 nm - 1070 nm range at different temperatures. The emission peaks are due to the radiative transitions of Nd^{3+} ions from the $^4F_{3/2}$ to the $^4I_{11/2}$ manifolds. One of these transitions give us the usual laser emission. At room temperature, the position of the strongest emission peak in Nd^{3+} doped KYF_4 is shifted to shorter wavelength compared to most other Nd^{3+} laser crystals. In KYF_4 crystal, this emission peak appears at 1.042 μm, instead of the 1.064 μm in YAG, 1.060 μm in GSGG, 1.050 μm in $LiYF_4$, and 1.055 μm in $NaYF_4$ [3].

In the 1030 nm - 1070 nm spectral range, there are two main emission peaks. They come from the Nd^{3+} ions at two main different sites in KYF_4 crystal [1, 4]. The peak

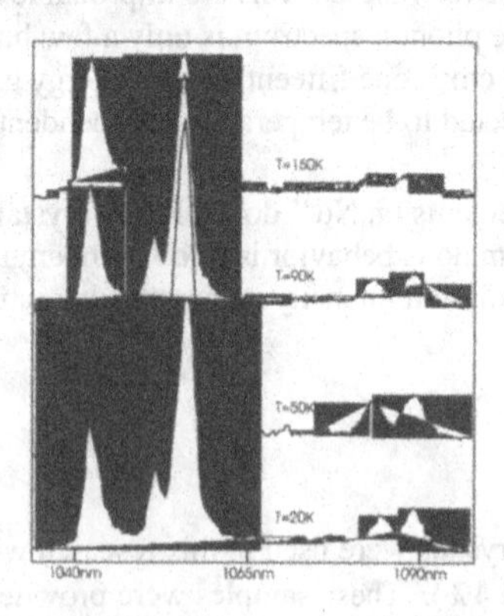

Fig. 1. Emission spectra from the $^4F_{3/2} \rightarrow ^4I_{11/2}$ transition of Nd^{3+} ions in KYF_4: Nd(2%) crystal at different temperatures.

at 1040 nm belongs to site 1 and the peak at 1055 nm comes from site 2. From the emission spectra, it can be seen that an unusual feature appears at low temperatures (in the vicinity of 50 K). The emission intensity of site 2 has a dramatic increase from 20K to 50K, and reaches its maximum value at 50K, then the intensity decreases with the increasing temperature after 50K. The temperature dependence of emission intensity of site 1 is just opposite to that of site 2, i.e., the emission intensity of site 1 has a minimum point at ~50K, where the intensity of site 2 shows a maximum. The lifetime measurements at different temperatures reveal this anomalous behavior again. The temperature dependence of Nd^{3+} lifetimes at site 2 is shown in Fig 2. Differently from most other Nd-doped crystals, the lifetime of Nd^{3+} ions in KYF_4 is strongly dependent on the temperature in the 20 K-100 K region; it shows a peak at about 50 K, similar to that of its emission intensity. We call this phenomena an "abnormal behavior".

To explain this abnormal spectral behavior we must consider the crystal structure of KYF_4. KYF_4 has a distorted fluoride structure [2]. In this crystal there are two different classes of Y^{3+} sites with significantly different environments. The Y^{3+} ion at site 1 is in a distorted fluoride environment; while the Y^{3+} ion at site 2 can only be described as being in a bypyramid coordination environment. When Nd^{3+} ions are doped into the KYF_4 crystal, they substitute for Y^{3+} ions without charge compensation. The Nd^{3+} energy levels at site 1 are somewhat higher than those at site 2.

The Nd^{3+} ions at two different sites are not independent of each other. An energy transfer can take place between them [4]. Considering the energy transfer between the two sites Nd^{3+}, the rate equation for Nd^{3+} ions at site 2 may be written as follows [5, 6]:

$$d\rho_2/dt = [-1/\tau_{2rad} - 1/\tau_{2nonrad} - w_{21} + (\rho_1/\rho_2)w_{12}]\, \rho_2 \qquad (1)$$

where ρ_i (i=1, 2) is the population of the excited Nd^{3+} ions ($^4F_{3/2}$ energy levels) at site i (i = 1 or 2), $1/\tau_{2rad}$ is the radiative transition probability from the $^4F_{3/2}$ level of Nd^{3+} ions at

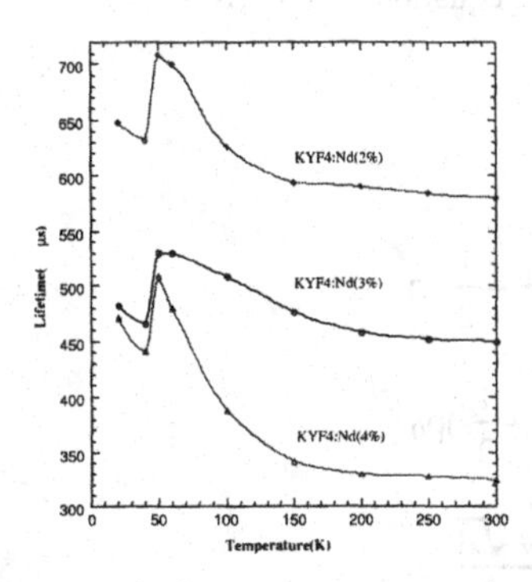

Fig. 2. Lifetimes measurements of $^4F_{3/2}$ energy level of Nd^{3+} ions at site 2 in three KYF_4:Nd crystals at different temperatures.

site 2, $1/\tau_{2nonrad}$ is the corresponding nonradiative transition probability, and w_{ij} (i, j=1,2) is the energy transfer rate from site i to site j. Due to the large energy gap below the $^4F_{3/2}$ energy level, the nonradiative transition probability can be neglected.

From this rate equation it can be seen that the deexcitation process of excited Nd^{3+} ions at site 2, and so its luminescence lifetime, is related not only to the radiative transition and nonradiative transition (≈0), but also to the energy transfer rates. The energy transfer from site 2 to site 1 is as though another path is opened to the deexcitation process. It will speed up the deexcitation process of the excited Nd^{3+} ions at site 2, and will shorten the luminescence lifetime of site 2 [5]. The energy transfer from site 1 to site 2, will slow down the deexcitation process of the excited Nd^{3+} ions at site 2, and will lengthen the luminescence lifetime of site 2 [7]. The radiative transition rate ($1/\tau_{2rad}$) is basically independent of the temperature, so, all the temperature dependence of luminescence lifetimes come from the energy transfer rates w_{21} and w_{12}. And it is known that the phonon assisted energy transfer rates are actually temperature dependent [7].

This sudden change of energy transfer rate will also have an influence on the emission spectra. The integrated emission intensity is proportional to the population of ions in excited states. At ~50K, the energy transfer rates undergo a large change with the increase of w_{12} being much larger than the increase of w_{21}. This will make the excited population of Nd^{3+} ions at site 2 undergo a large increase. Therefore the emission intensity of site 2 will have a large increase at 50K, as well, showing a behavior similar to that of the lifetime. At the same time, the population of excited Nd^{3+} ions at site 1 will have a large decrease, with a corresponding decrease of the emission intensity of site 1, a behavior that is just the opposite of that of site 2. These points are all consistent with our measurements (see Fig. 1, 2).

DATA FITTING OF DECAY PATTERNS

To make a detailed analysis of the decay patterns at different temperatures, the analytical solution of the rate equations was derived.

$$\rho_2 = ae^{-\alpha t} + be^{-\beta t} \tag{2}$$

where,

$$a = \left(-\frac{w_1 - w_2 + 3w_{12} - w_{21}}{2\sqrt{\Delta}} + \frac{1}{2}\right)\rho_0 \tag{3}$$

$$b = \left(\frac{w_1 - w_2 + 3w_{12} - w_{21}}{2\sqrt{\Delta}} + \frac{1}{2}\right)\rho_0 \tag{4}$$

$$\alpha = \frac{(w_1 + w_2 + w_{12} + w_{21}) + \sqrt{\Delta}}{2} \tag{5}$$

$$\beta = \frac{(w_1 + w_2 + w_{12} + w_{21}) - \sqrt{\Delta}}{2} \tag{6}$$

and Δ is given by:

$$\Delta = (w_1+w_2+w_{12}+w_{21})^2 - 4(w_1w_2+w_1w_{21}+w_2w_{12}) \tag{7}$$

Using these formulas, we performed a data fitting on the decay pattern. The fitting results are shown in Table 1, Fig. 3 gives the fitting results for w_{12} and w_{21} at different temperatures. It can be seen from these fitting results that w_{12} and w_{21} do have a significant increase at 50K, and the increase of w_{12} is greater than that of w_{21} as was assumed previously.

CONCLUSIONS

Spectroscopic measurements at different temperatures indicated that Nd^{3+} ions in KYF_4 have an abnormal spectral behavior in the low temperature range (20 – 100K). Based on the particular crystal structure of KYF_4, we proposed a justification for this abnormal behavior by considering the energy transfer between Nd^{3+} ions at different sites. This energy transfer model can explain the anomalous behavior on both emission spectra and lifetime measurement and make the explanation consistent with each other.

The sudden increase of energy transfer rates at ~50K may be an indication of a possible phase transition in KYF_4. This point may need confirmation by other related experiments on this material.

Table 1. Data fitting parameters of decay patterns at different temperatures.

T(K)	$w_{12}(s^{-1})$	$w_{21}(s^{-1})$	ρ_0(a.u.)	a	b	α	β
20	125	0	14.0	-4.23	18.2	1977	1563
40	218	129	13.8	-4.22	18.0	2134	1628
50	1682	1247	14.9	-2.75	17.6	4665	1679
70	1649	1262	13.9	-2.36	16.2	4645	1681
100	1368	1283	12.6	-1.02	13.6	4371	1695
150	1264	1268	11.9	-0.62	12.5	4248	1700
200	1296	1300	13.5	-0.69	14.2	4311	1700
250	1189	1242	12.9	-0.45	13.4	4144	1702
300	1108	1191	11.7	-0.29	12.0	4010	1704

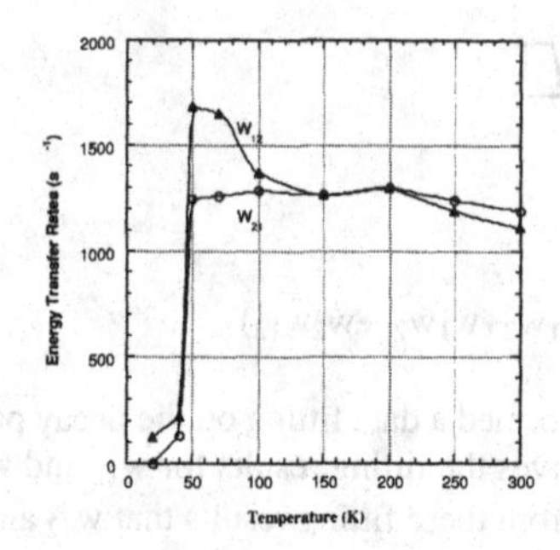

Fig. 3. Energy transfer rates (w_{12} and w_{21}) derived from the data fittings of decay patterns at different temperatures.

ACKNOWLEDGMENTS

We would like to thank Dr. Xinxiong Zhang for supplying us the interesting laser crystals used in this research and many useful information and references. We wish to acknowledge Dr. George Goldsmith and Dr. Charles W. Struck for their helpful discussions and suggestions.

REFERENCES

1. Toomas H. Allik, Larry D. Merkle and Richard A. Utano, Bruce H. T. Chai and J. L. V. Lefaucheur, Heika Voss and G. Jeffrey Dixon, "Crystal growth, spectroscopy, and laser performance of Nd^{3+}:KYF_4", J. Opt. Soc.Am. B 10, (1993) 633.
2. Y. L. Fur, N. M. Khaidukov, and S. Aleonard, "Structure of KYF4", Acta Cryst. C48, (1992) 978.
3. A. A. Kaminskii, *Laser Crystals*, Springer-Verlay, Berlin Heidelberg, New York, 1990.
4. J Sytsma, S J Kroest, G Blasse and N. M Klaidukov, "Spectroscopy of Gd^{3+} in KYF_4: A System with Several Luminescent Sites", J. Phys: Condens. Matter 3, (1991) 8959.
5. B. Di Bartolo, *Energy Transfer Processes in Condensed Matter*, Plenum Press, New York, 1984.
6. J. K. Neeland and V. Evtuhov, "Measurement of the Laser Transition Cross Section for Nd^{3+} in Yttrium Aluminum Garnet", Phys. Rev. 156, (1967) 244.
7. B. Di Bartolo, J. Danko and D. Pacheco, "Norradiative energy transfer without lifetime quenching in doped Mn-based crystals", Phys. Rev. B, 35, (1987) 6386.

Mat. Res. Soc. Symp. Proc. Vol. 667 © 2001 Materials Research Society

Influence of Pressure on 5d → 4f Emission Transitions of Ce^{3+}

Garry B. Cunningham, Yongrong Shen, Kevin L. Bray
Department of Chemistry, Washington State University, Pullman, WA 99164-4630

Ulisses R. Rodriguez Mendoza
Departmento de Fisica Fundamental y Experimental, Universidad de La Laguna,
S/C de Tenerife, Spain

ABSTRACT

High pressure is used to tune the emission and band structure of the phosphors Ce^{3+}:Lu_2S_3 and Ce^{3+}:Lu_2SiO_5. A significant red shift of the broad 5d → 4f emission of Ce^{3+} was observed in both phosphors. In Ce^{3+}:Lu_2S_3, we also observed a significant decrease in the emission intensity of Ce^{3+} and attribute the quenching to a pressure-induced electronic crossover of the Lu_2S_3 conduction bandedge with the emitting 5d state of Ce^{3+}. In Ce^{3+}:Lu_2SiO_5, two Ce^{3+} sites are present and we observed energy transfer from one site (Ce(2)) to the other ((Ce(1)) at low pressure. At high pressure, the energy transfer ceases and emission is no longer observed from the Ce(1) site. We propose an exciton recombination model of the energy transfer process.

INTRODUCTION

The efficiency of Ce^{3+}-activated luminescent materials varies widely with host lattice. In some host lattices, 5d → 4f emission of Ce^{3+} is highly efficient and in others, the emission is completely quenched. Several models of emission quenching in Ce^{3+} systems have been proposed in the literature, but a definitive understanding of the factors responsible for controlling the emission efficiency of Ce^{3+} has proven elusive. Evidence to date, however, suggests that the energy of the lowest 5d crystal field state of Ce^{3+} relative to the host lattice conduction band edge, charge transfer state, and/or local defect levels are important factors in determining its emission efficiency [1, 2, 3].

High pressure provides a versatile new method for perturbing the energy level structure and thermal activation barriers relevant to efficient luminescence in phosphors. We have recently shown, for example, that pressure provides us with the ability to alter the electronic states of luminescent centers as well as the band structure of host lattices [4, 5]. By using high pressure, we are able to systematically control the energies, intensities, linewidths, and time-resolved properties of luminescence transitions as well as the excitation and quenching mechanisms of luminescent centers. Pressure also influences coordination geometry and phase stability and can be used to gain insight into structure-property relationships of phosphor materials [4, 6].

In this paper, we present initial results of high pressure luminescence studies of Ce^{3+}-activated phosphors. Our objective is to use pressure to vary the energy of the emitting 5d state of Ce^{3+} relative to the host conduction band in an attempt to understand how host band structure crystal field strength (5d splitting) and covalency (5d barycenter energy) influence the energy and intensity of 5d → 4f emission of Ce^{3+}. We also consider simultaneous pressure-temperature variations to investigate thermal activation barriers to emission quenching and the energy of 5d states of Ce^{3+} relative to host lattice conduction bands. All reported luminescence spectra were corrected for the spectral response of the measurement system.

$Ce^{3+}:Lu_2S_3$

Van't Spijker et al. [7] first reported that $Ce^{3+}:Lu_2S_3$ is an efficient red scintillator. The luminescence spectra of $Ce^{3+}:Lu_2S_3$ as a function of pressure are shown in Figure 1. We observed two overlapping emission bands originating from transitions from the lowest $5d^1$ excited state to the two spin-orbit split components ($^2F_{5/2}$ and $^2F_{7/2}$) of the $4f^1$ ground state of Ce^{3+} (~596.0 nm and ~677.5 nm at ambient pressure). A large red shift of -30(3) cm^{-1}/kbar was observed for each of the two components with pressure up to ~100 kbar.

We also observed a rapid decrease of Ce^{3+} luminescence intensity at room temperature upon compression (Figure 2). At ~100 kbar, the intensity was a factor of ~20 weaker than at ambient pressure. With increasing pressure above ~100 kbar, the intensity continued to decrease until complete quenching occurred at ~120 kbar. Above 120 kbar, we considered two additional luminescence experiments in order to understand the origin of the quenching. First, we wanted to determine if the quenching was due to shifting of the 4f → 5d absorption band away from our initial excitation wavelength of 457.9 nm. Consequently, we considered several excitation wavelengths expected to excite the absorption band based on estimated positions of the red-shifted absorption band. Neither excitation at 488 nm and 514 nm nor dye laser excitation over the range 575 - 620 nm was able to produce luminescence between 120 and 300 kbar. Second, we completed 20 K experiments. No Ce^{3+} emission was observed at 20 K with any of the excitation wavelengths used in the study.

Upon decompression, the Ce^{3+} luminescence reappeared below ~100 kbar. At ~90 kbar, we completed a series of luminescence experiments between 300 K and 20 K. Strong temperature-dependent Ce^{3+} luminescence was observed at ~90 kbar (Figure 3). We also completed a variable temperature luminescence experiment upon full release of pressure and observed no temperature dependence of the Ce^{3+} emission intensity between 20 - 300 K.

Upon increasing pressure, we also observed a gradual change in the appearance of the sample from transparent at ambient pressure to orange-brown at high pressure. A preliminary high pressure absorption experiment at ~90 kbar was completed and showed a strong absorption

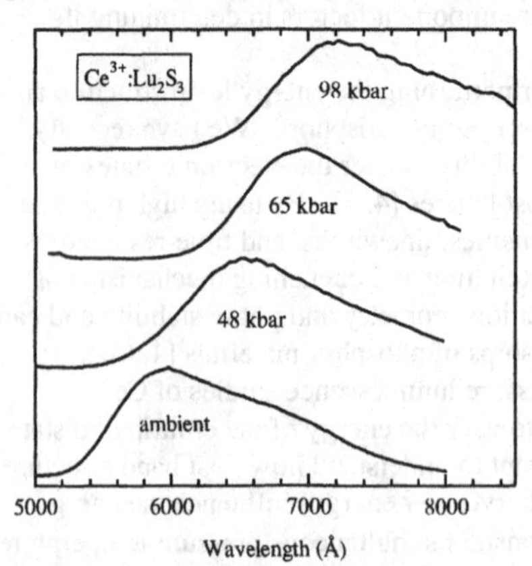

Figure 1. Normalized luminescence spectra of $Ce^{3+}:Lu_2S_3$ at 295 K upon excitation at 457.9 nm.

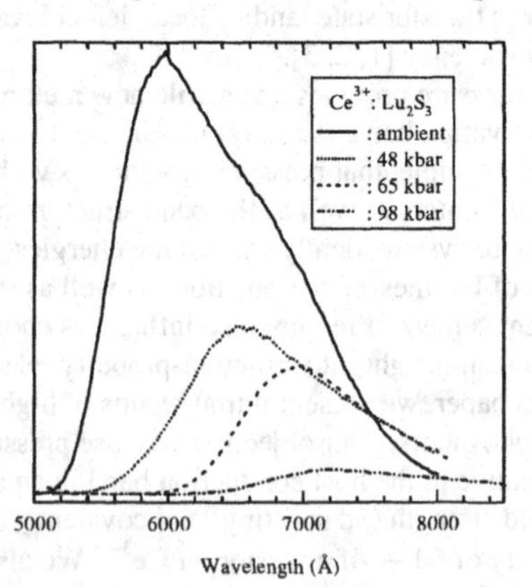

Figure 2. 295 K luminescence spectra of $Ce^{3+}:Lu_2S_3$ at several pressures upon excitation at 457.9 nm.

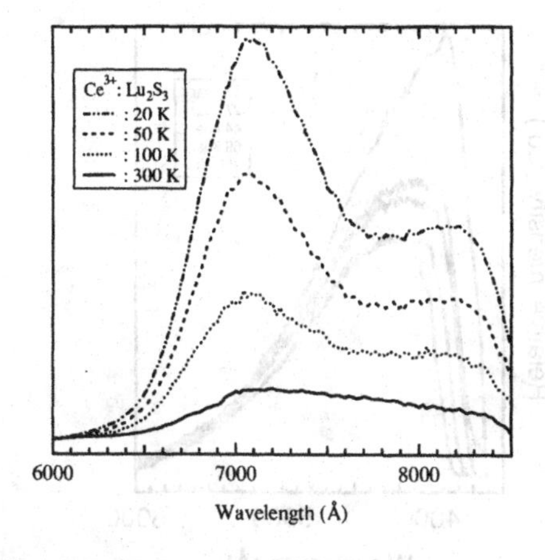

Figure 3. Luminescence spectra at ~90 kbar (on release) of Ce^{3+}:Lu_2S_3 (λ_{exc} = 354.7 nm).

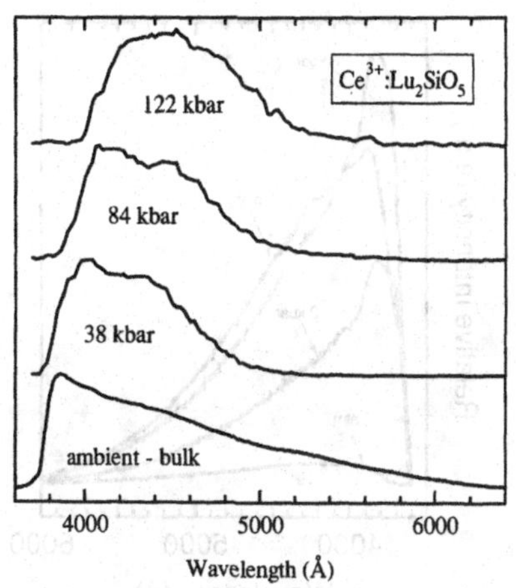

Figure 4. Normalized luminescence spectra of Ce^{3+}:Lu_2SiO_5 at 295 K (λ_{exc} = 354.7 nm).

beginning at ~450 nm (not shown). The absorption is attributed to the band edge absorption of the Lu_2S_3 host lattice. Based on the reported ambient pressure band edge absorption of ~330 nm for the host lattice [7], we see that the red shift rate of the absorption edge is much larger than that of the 5d state of Ce^{3+}. We consequently attribute the luminescence quenching to a pressure-induced crossing of the conduction bandedge of the host lattice with the emitting 5d state of Ce^{3+}. A quantitative description of the quenching process is in preparation.

Ce^{3+}:Lu_2SiO_5

Figure 4 shows room temperature luminescence spectra of Ce^{3+}:Lu_2SiO_5 upon pulsed excitation at 354.7 nm. At ambient pressure, a broad, asymmetric emission band, attributable to overlapping contributions from two crystallographically distinct Ce^{3+} sites, was observed. Suzuki et al. [8] and Naud et al. [9] identified a major 'blue' Ce^{3+} center, Ce(1), (emission peaks at ~393 and 427 nm) and a minor 'green' Ce^{3+} center, Ce(2), (emission peak at ~465 nm) and reported a 4:1 ratio of "blue" Ce^{3+} centers to "green" Ce^{3+} centers in Lu_2SiO_5. Upon increasing pressure, we observed a shift of –18(3) cm^{-1}/kbar for the Ce^{3+} emission in Lu_2SiO_5.

Unexpected intensity behavior was observed with pressure upon CW excitation at 325 nm (Figure 5). Between 1 atm and 6 kbar, a strong increase in the intensity of the Ce(1) emission relative to Ce(2) emission was observed. Between 6 and ~60 kbar, however, a significant quenching of emission from the Ce(1) site was observed. The emission intensity and lineshape remained unchanged between ~60 kbar and ~100 kbar. A gradual decrease in emission intensity, with no change in lineshape, was observed above ~100 kbar.

According to the ambient pressure excitation spectra presented by Raukas [10], 325 nm excitation is at the peak of an excitation band for Ce(2) and in a valley between excitation bands that peak at 296 and 356 nm for Ce(1). Consequently, 325 nm excitation is capable of directly exciting only the Ce(2) sites. Since the expected shift of the Ce(1) excitation band between 1 atm

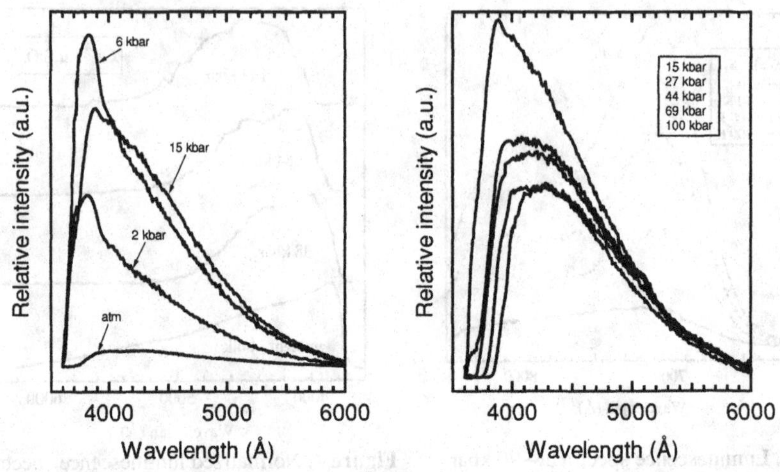

Figure 5. 295 K luminescence spectra of Ce^{3+}:Lu_2SiO_5 as a function of pressure (λ_{exc} = 325 nm).

and 6 kbar is not sufficient to permit direct excitation of the Ce(1) sites, we propose that the increase in the Ce(1) emission intensity observed between 1 atm and 6 kbar is a consequence of energy transfer from Ce(2) sites.

The energy transfer process is most conveniently described in the context of the energy level diagram for Ce^{3+}:Lu_2SiO_5 proposed by Raukas et al.[10] on the basis of photoconductivity experiments and reproduced in Figure 6. The excitation spectra presented by Raukas [10] for Ce(2) (not shown) revealed 4f → 5d excitation bands that peak at 376 nm (3.30 eV) and 326 nm (3.80 eV). (Note that the energies shown in Figure 6 correspond most closely with zero phonon energies. Depending on bandwidth, peak excitation energies are ~0.2 - 0.3 eV higher in energy than the zero phonon energies.) Consequently, 325 nm excitation induces a 4f → 5d(2) transition and places an electron into a resonant Ce^{3+} state. In the standard emission process, an electron in the 5d(2) level remains in the Ce^{3+} sub-system and decays non-radiatively to the 5d(1) state which subsequently emits. Since the 5d(2) state is resonant, however, ionization of an electron from the 5d(2) state to the host conduction band to produce an unbound electron becomes a competing process to the standard emission process. Once ionized, the electron decays

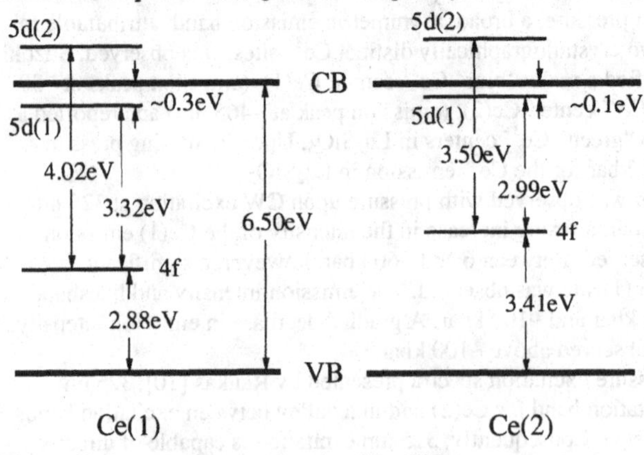

Figure 6. Proposed energy levels of Lu_2SiO_5.

non-radiatively to the bottom of the conduction band. Upon ionization, a hole is formally left behind at the 4f ground state of Ce and Ce^{3+} becomes Ce^{4+}.

The unbound ionized electron can decay via one of three processes. First, it can be recaptured by the 5d(1) state of Ce^{4+} to re-form Ce^{3+} and, as in the standard emission process, internally decay to the 4f ground state of Ce^{3+}. Second, the unbound electron can recombine directly (radiatively or non-radiatively) with the hole at the 4f ground state of Ce^{4+} to re-form Ce^{3+}. Given the spatially delocalized nature of the unbound electron in the conduction band of the host and the spatially localized (bound) nature of the hole at Ce^{4+}, radiative recombination is statistically unlikely. Non-radiative recombination of the unbound electron, however, from the conduction band to Ce^{4+} is expected to occur. Non-radiative transitions from the conduction band to deep defects are well known in semiconductors and are typically accompanied by a small thermal activation barrier. Finally, the unbound electron can recombine with or form an exciton with unbound holes in the valence band. Although the concentration of intrinsic holes in the valence band is expected to be low in Ce^{3+}:Lu_2SiO_5, photoexcitation can significantly increase the number of unbound holes in the valence band. In our experiment, the excitation wavelength is well below the bandgap energy of Lu_2SiO_5, so direct electron excitation from the valence band to the conduction band does not occur. Upon photoionization of an electron to the conduction band and creation of a hole at the Ce^{3+} site to form Ce^{4+}, it becomes possible to photoexcite an electron from the valence band to Ce^{4+} to re-form Ce^{3+} and produce an unbound hole in the valence band. The unbound hole is subsequently available to interact with the unbound photoionized electron. This process is in effect a two photon process for the formation of an unbound electron-hole pair via a Ce^{4+} intermediary. In the energy transfer model that we are developing, we propose that the unbound electron and unbound hole form an exciton that migrates through the lattice and binds to Ce(1) sites. Once bound, we further propose that the exciton recombines to excite Ce(1) sites to their 5d(1) or 5d(2) levels and that this excitation process leads to the Ce(1) emission observed upon 325 nm excitation in our experiment. Similar bound exciton recombination processes have recently been proposed to explain energy transfer from host lattice to lanthanide ion dopants in lanthanide doped III-V semiconductors [11].

The proposed excitonic energy transfer model provides a tentative explanation for the intensity changes observed with pressure. The intensity of Ce(2) emission depends on the number of photoionized electrons that are captured and trapped by the 5d(1) state. The trapping efficiency of the 5d(1) state is controlled primarily by the depth of the 5d(1) state in the bandgap. The observed redshift of 5d $\rightarrow$ 4f emission and expected increase in the bandgap of Lu_2SiO_5 indicate that the depth of the 5d(1) state increases with pressure. As a result, a greater fraction of captured electrons becomes trapped and an increase in Ce(2) emission intensity is expected with pressure. We further expect the increase in Ce(2) intensity to level off as the activation energy from the 5d(1) level to the conduction band becomes significantly larger than kT. These observations are consistent with the experimental intensity data for the Ce(2) sites.

In the proposed model, the intensity of Ce(1) emission depends on the concentrations of unbound electrons and holes available for exciton formation and energy transfer. During the pronounced increase in Ce(1) intensity between 1 atm and 6 kbar, we expect the number of unbound (photoionized) electrons to be significant because of the appreciable photocurrent observed by Raukas [1,10,12] upon 325 nm excitation at 1 atm. Consequently, we believe that the availability of unbound holes controls Ce(1) intensity between 1 atm and 6 kbar. We thus attribute the large increase in Ce(1) intensity between 1 atm and 6 kbar to an increase in the number of unbound holes. Since unbound holes are formed by promotion of electrons from the valence band to $Ce^{4+}(2)$ sites, the number of unbound holes is proportional to the number of $Ce^{4+}(2)$ ions. The number of $Ce^{4+}(2)$ ions, in turn, is determined by the efficiency of internal decay from the 5d(1) state and the efficiency of thermally activated non-radiative recombination of photoionized electrons from the conduction band to $Ce^{4+}(2)$. As discussed above, we expect the internal decay process to become more efficient with increasing pressure as the 5d(1) state shifts red and experiences less thermal depopulation to the conduction band. This effect accounts for the increase in $Ce^{3+}(2)$ intensity observed between 1 atm and 6 kbar (Fig. 5) and acts to decrease the number of $Ce^{4+}(2)$ ions. This effect alone, therefore, contributes to a reduction in $Ce^{3+}(1)$ intensity.

Temperature dependent intensity data (not shown) suggest that the increase in the number of $Ce^{4+}(2)$ ions needed to explain the increased $Ce^{3+}(1)$ intensity is due to a larger opposing decrease in the number of photoionized electrons that decay non-radiatively from the conduction band to the 4f ground state of $Ce^{4+}(2)$. The data support an increase in the thermal activation barrier associated with the non-radiative decay with pressure. As a result, despite the greater internal decay efficiency of $Ce^{3+}(2)$, a net increase in the number of $Ce^{4+}(2)$ ions occurs between 1 atm and 6 kbar. The larger number of $Ce^{4+}(2)$ ions increases the number of unbound holes produced by a second photon in the proposed model upon promotion of an electron from the valence band to $Ce^{4+}(2)$ to reform $Ce^{3+}(2)$ in its 4f ground state. The increased number of unbound holes leads to an increased number of excitons, more efficient excitation of $Ce^{3+}(1)$ via excitonic energy transfer, and increased $Ce^{3+}(1)$ intensity. We attribute the decrease in $Ce^{3+}(1)$ intensity between 6 and ~60 kbar to a reduction in the number of unbound photoionized electrons as the 5d(1) energy decreases and the 5d(1) state becomes a more effective trap. In contrast to the low pressure situation, the $Ce^{3+}(1)$ intensity is limited by the availability of unbound electrons as more electrons become trapped by the 5d(1) state.

ACKNOWLEDGEMENTS

Acknowledgment is made to the donors of the Petroleum Research Fund, administered by the ACS, for partial support of this research. We gratefully acknowledge additional support provided by the National Science Foundation. We wish to express our thanks to Prof. W. M. Yen and Dr. P. Dorenbos for generously supplying the Ce^{3+} crystals used in this study.

REFERENCES

1. W. M. Yen, et al., *J. Lumin.* **69**, 287 (1996).
2. G. Blasse, et al., *Inorg. Chim. Acta* **189**, 77 (1991).
3. M. V. Korzhik and W. P. Trower, *Appl. Phys. Lett.* **66**, 2327 (1995).
4. K. L. Bray, Topics in Current Chemistry, vol. 213, p. 1-94 (2001).
5. T. D. Culp and K. L. Bray, *J. Appl. Phys.* **82**, 368 (1997).
6. Y. R. Shen and K. L. Bray, *Phys. Rev. Lett.* **84**, 3990 (2000).
7. J. C. van't Spijker, et al., *Nucl. Instr. and Meth. B* **134**, 304 (1998).
8. H. Suzuki, et al., *Nucl. Instr. and Meth. A* **320**, 263 (1992).
9. J. D. Naud, et al., *IEEE Trans. Nucl. Sci.* **43**, 1324 (1996).
10. M. Raukas, PhD. Thesis, University of Georgia (1997).
11. K. Takahe and A. Taguchi, *J. Appl. Phys.* **78**, 5614 (1995).
12. M. Raukas, et al., *Appl. Phys. Lett.* **69**, 3300 (1996).

Poster Session

Mat. Res. Soc. Symp. Proc. Vol. 667 © 2001 Materials Research Society

ENHANCEMENT OF PHOTOLUMINESCENCE FROM ORGANIC AND INORGANIC SURFACE PASSIVATED ZnS QUANTUM DOTS

Hatim Mohamed El-Khair, Ling XU, Minghai LI, Yi MA , Xinfan HUANG and Kunji CHEN

National Laboratory of Solid State Microstructures and Department of Physics
Nanjing University, Nanjing 210093, China

ABSTRACT

ZnS quantum dots (QDs) chemically synthesized in PVP stabilizing medium have been coated with $Zn(OH)_2$, SiO_2 and polystyrene (PS) shells as inorganic and organic passivation agents. to synthesize ZnS/$Zn(OH)_2$, ZnS/SiO_2 and ZnS/PS QDs. PL properties of inorganically passivated ZnS/$Zn(OH)_2$ and ZnS/SiO_2 had reported band edge enhancement of 8-10 times, while organically passivated ZnS/PS QDs exhibit tremendous enhancement of band edge emission as much as 10-15 times,. Therefore inorganic and organic coating can passivate trap states of different energies on the surface of ZnS QDs.

INTRODUCTION

Quantum confinement phenomena have been observed in II-VI compound semiconductor quantum dots whose sizes are comparable to Bohr radii.[1] By adopting various preparatory and passivation techniques monodispersed QDs with high quantum yields (>50%) can be obtained. Organic surfactants such as gelatin,[2] polyvinyl pyrrolidone-K30 (PVP),[3] sodium dioctyl sulfosuccinate (AOT),[4] were used as stabilizing agents for QDs chemically grown in different solvents. Large surface to volume ratio in semiconductor QDs made their optical properties to be affected by physical properties of their surfaces.[1] Hence detectable surface trap states of different energies have been observed in luminescence properties from different bare QDs such as CdS[5] and CdSe.[6] Recently new synthesis techniques based on organo metallic interface such as emulsifier,[7] molecular adhesion[8] and zeolite architecture[9] can not only improve the structure and properties of semiconductor QDs but can also give rise to new material with photonic band gap phenomena.[10] Semiconductor QDs coated with inorganic shell such as CdSe/CdS[11] and CdSe/ZnSe[12] and organic shell (TOPO) have exhibited enhanced band edge emission of high quantum yield .[1]

In this work we are enabled to synthesize ZnS QDs in PVP/ethanol and PVP/aqueous stabilizing medium. $Zn(OH)_2$, SiO_2 and polystyrene (PS)were used as inorganic and organic surface passivation agents to chemically synthesized ZnS/$Zn(OH)_2$, ZnS/SiO_2 and ZnS/PS QDs, respectively. The band edge emission has been enhanced as much as 8-10 times for inorganic coating and 10-15 times for organic coating.

EXPERIMENT

Synthesis of ZnS QDs

We described in detail the synthesis and capping of monodispersed ZnS QDs using polyvinyl pyrrolidone-k30 (PVP) stabilizer in our previous paper [13]. 200 ml PVP/ethanol (5%) solution was prepared as QDs stabilizing agent. $Zn(NO_3)_2$ and Na_2S dispersed in water and ethanol solvent (5%) were used as precursors for Zn^{++} and S^{--} ions, which reacted according to the following equation:

$$Zn(NO_3)_2 \quad + Na_2S \quad \rightarrow ZnS + 2NaNO_3$$

The reactions were carried out at room temperature and in the air. By varying of the S^{--} ion concentration we can obtained ZnS QDs with different sizes. All ZnS QDs coated with organic and inorganic shell were prepared from ethanol dispersed precursors and reacted in PVP/ethanol

Synthesis of ZnS/Zn(OH)₂ core/shell QDs

ZnS QDs coated with $Zn(OH)_2$ were prepared in the same manner as described in our previous work.[13]

Synthesis of ZnS/SiO₂ QDs

We used monodispersed ZnS QDs grown in PVP/ethanol solvent *of size ranging from 2 to 3 nm estimated from the absorption spectra is used* as starting solution to prepare ZnS/SiO_2 QDs. The following reagents were used: 1.Polyvinyl pyrrolidone-k30 (PVP) (surfactant). 2.Zinc nitrate (Zn^{2+} precursor). 3.Sodium sulfide (S^{2-}precursor). 4.Tetra ethyl ortho silicate (TEOS). 5.Ammonia. 6.Ethanol (solvent).

After ZnS QDs have been synthesized in PVP/ethanol stabilizing medium, TEOS was added to QD colloidal solution with a rate of 0.1 ml /minute at room temperature under continuously stirring,.

Synthesis of ZnS/PS QDs:

The following reagents were used for synthesis of ZnS/PS QD: 1.Zinc nitrate (Zn^{2+} precursors). 2.Sodium nitride (S^{2-} precursors). 3.Styrene (Polystyrene precursors). 4.Potassium persulfate (Polymer initiator). The experiment procedures were carried out as follows: ZnS QDs were synthesized in PVP/ethanol stabilizing medium (5%) by reacting $Zn(NO_3)_2$ with Na_2S. Under continuously stirring, with a rate of 345 rpm at 70 ^{0}C styrene was added. After that the polymer initiator was added and the reaction has been carried out for 18 hours.

RESULTS AND DISCUSSION

Figure 1 shows the UV-Vis absorption and PL spectra of bare structure ZnS QDs *of sizes ranging from 2 to 3 nm as estimated from the absorption spectra* grown in PVP/water solution. The absorption spectra for ZnS QDs shown in fig. 1 (a) are slightly blue shifted with decreasing size (from d4 to d1), which is due to the quantum size effect. The PL spectra for ZnS QDs shown in fig. 1 (b) display two bands. The weak and narrow peaks near absorption edges come from the band recombination of carriers whereas the stronger and broader emission bands centered 430-450 nm is due to the surface states such as dangling bonds and S vacancies.[14] These sites could trap electrons and holes and caused radiative recombination.

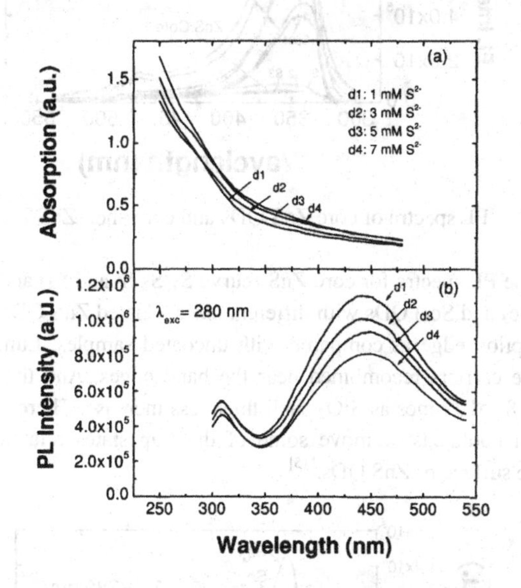

Figure 1. Absorption and PL spectra of bare ZnS QDs of different sizes synthesized from water dispersed precursors reacted in PVP/ethanol stabilizing medium

Figure 2 shows the PL spectra *(Hg lamb is used as excitation source)* of core ZnS (S1, S2 and S3) with different sizes and core/shell ZnS/$Zn(OH)_2$ QDs (Sc11, Sc12 were formed by coating S1 with different shell thickness while Sc21, Sc23 and Sc31; Sc32 formed by coating S2 and S3, respectively). Luminescence spectra of core ZnS QDs in figure 2 show one broad peak, which is different from figure 1. This is mainly due to the dispersed precursors in different solvents (water and ethanol) hence *bare ZnS QDs synthesized in PVP/ethanol do not have surface state*. The PL spectra for the ZnS/$Zn(OH)_2$ QDs(.Sc11, Sc12, Sc21, Sc23, Sc31 and Sc32) in figure 2, display narrowed emission peaks near the absorption edge (330- 350 nm) These peaks are much narrower than that of uncoated ZnS QDs. The emission intensity also increased approximately by order of magnitude compared with the core ZnS. Particularly interesting we found that, these emission intensities increased with increasing shell thickness of $ZnS/Zn(OH)_2$ QDs. These results are the direct evidence to show that $Zn(OH)_2$ coating has greatly removed surface trap states of different energies and enhanced the band edge emission of the ZnS QDs[2].

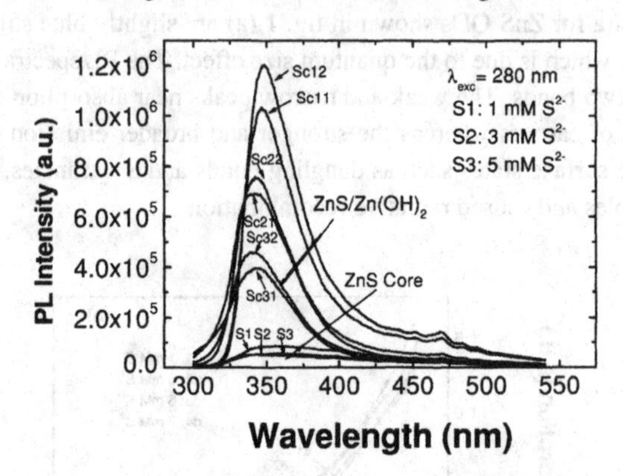

Figure 2. PL spectra of core ZnS QDs and core/shell $ZnS/Zn(OH)_2$ QDs.

Figure 3 shows the PL spectra for core ZnS (curve S_1, S_2, S_3 and S_4) and core/shell structured ZnS/SiO_2 (Sc_1; Sc_2; Sc_3 and Sc_4) QDs with different sizes. Coated ZnS QDs have sharp emission bands near the absorption edge as compared with uncoated samples. Luminescence properties prove that, the charge carriers recombined near the band edges. And the band edge emission enhanced as much as 8 to10 times as SiO_2 shell thickness increase. Therefore we suggested that amorphous SiO_2 shell could also remove some of the trap states related to the existence of dangling bonds on the surface of ZnS QDs.[15]

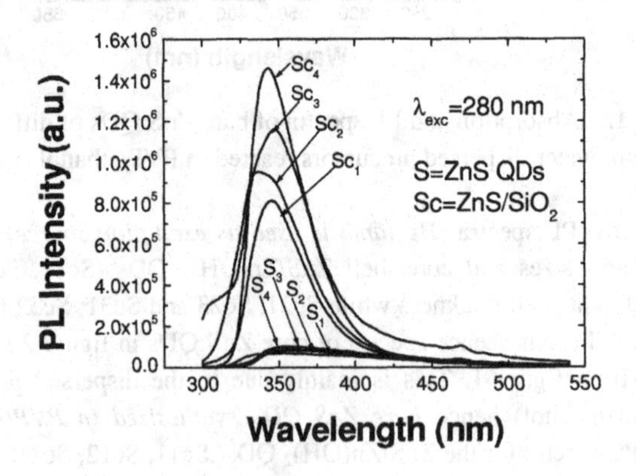

Figure 3. PL spectra for core ZnS (curve S_1, S_2, S_3 and S_4) and core/shell ZnS/SiO_2 (Sc_1, Sc_2, Sc_3 and Sc_4) QDs.

PL spectra of ZnS QDs and ZnS QDs coated with polystyrene (organic agent) were depicted in fig.4. Luminescence spectra of ZnS/PS QDs (curves Sc1, Sc2, and Sc3) had also exhibited sharp emission bands compared to that of bare ZnS QDs (S1, S2 and S3). *Bare ZnS QDs synthesized in PVP/ethanol do not have surface state.* We thought that PS has spherical structure and can easily attach to the Zn sites at the surface of ZnS QDs to form an organic shell. So defect states were removed from the surface of ZnS QDs after being coated with PS. And hence the band edge emission enhanced as much as 10-15 time, which is quite larger than enhancement by inorganic passivation (coated with $Zn(OH)_2$ and SiO_2 shell). These results have shown better surface passivation by coating ZnS QDs with organic PS shell. *Therefore organic passivation is rather comparable to inorganic passivation*

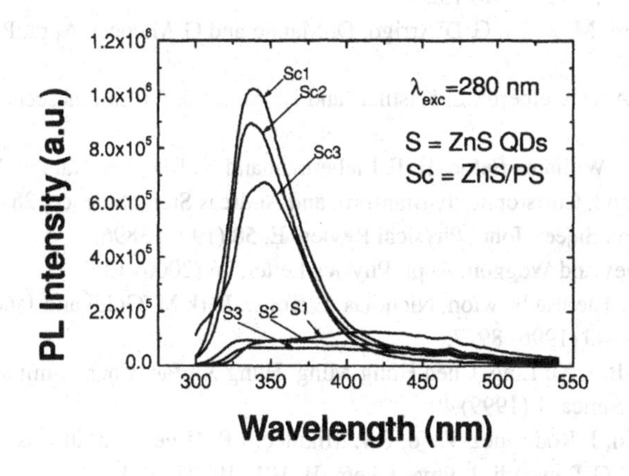

Figure 4. PL spectra of ZnS QDs (S1, S2 and S3) and ZnS QDs coated with polystyrene (Sc1, Sc2 and Sc3)

CONCLUSION

The PL properties of bare ZnS QDs synthesized in PVP/ethanol and PVP/aqueous stabilizing medium have shown wider emission bands, which are mainly attributed to the surface defects states. The enhancement of the band edge emission from $ZnS/Zn(OH)_2$ and ZnS/SiO_2 QDs (inorganic passivation) was observed due to the removal of surface states from the surface of ZnS QDs. Also the band edge emission from ZnS/PS (organic passivation) have been enhanced larger than that obtained by inorganic coating. *Also from the areas of recombination peaks we conclude that the organic passivation is rather comparable to the inorganic passivation.*

ACKNOWLEDGEMENT

This work is supported by National Natural Science Foundation of China under Grant No. 10074023 and No. 69890225.

REFERENCE

1. A. P. Alivisatos, J. Phys. Chem.100, (1996) 13226
2. Ling Xu, Xinfan Huang, Hongming Chen, Jun Xu and Kunji Chen, Jpn. J. Appl. Phys. 37, (1998) 3491
3. A. R. Kortan, R. L. Opila, M. G. Bawendi, M. L. Steigerwald, P. J. Carroll and L. E. Brus, J. Am. Chem. Soc., 112 (1990) 1327.
4. V. Turco Liveri, M. Rossi, G. D'Arrigo, D. Manno and G. Micocci, Appl. Phys. A, 69 (1999) 369.
5. A. Eychuler, A. Hasselberth, L. Kastikas and H. Weller. J. Of Luminescence, 48 & 49 (1991) 745.
6. R. J. Davey, L. Williams-Seton, H. F. Lieberman and N. Blagden, Nature, 402 (1999) 797.
7. Brian T. Holland, Christopher F. Blanford, and Andreas Stein, Science, 281 (1998) 538.
8. Kurt Busch and Sajeev John, Physical Review E, 58 (1998) 3896.
9. M. V. Artemyev and Woggon, Appl. Physics Letter, 76 (2000) 1353.
10. Yongchi Tian, Theresa Newton, Nicholas A. Kotov, Dirk M. Guldi and Janos H. Fendler, J. Phys. Chem., 100 (1996) 8927.
11. Huang Hong-Bin, Xu Ling, Chen Hong-Ming, Hung Xinfan, Chen Kunji and Feng Duan, Acta Physica Sinica, 1 (1999) 40.
12. B.O. Dabbousi, J. Rodriguez-Viejo, F.V. Mikulec, J.R. Hine, H. Mattoussi, R. Ober, K.F. Jensen, and M.G. Bawendi, J. Phys. Chem. B, 101 (1997) 9463.
13. H. M. El-Khair, XU Ling, HUANG Xin-Fan, LI Ming-Hai and CHEN Kun-Ji, Chin. Phys Lett., 18 (2001) 616.
14. W.G. Becker and A. J. Bard, J. Phys. Chem., 87 (1983) 4888.
15. L. Xu, X.F. Huang, J. Zhu, H.M. Chen, and K.J. Chen, J. Mater. Sci., 35 (2000) 1375.

Mat. Res. Soc. Symp. Proc. Vol. 667 © 2001 Materials Research Society

Photoluminescence and Raman Spectroscopy Studies of H^+ Ion Implanted SOI Structures Formed by Hydrogen Ion Slicing

Vladimir P. Popov, Ida E. Tyschenko, Konstantin S. Zhuravlev, Ivan I. Morosov[1]
Institute of Semiconductor Physics, Novosibirsk, 630090, Russia,
[1] Institute of Nuclear Physics, Novosibirsk, 630090, Russia,

ABSTRACT

H^+ ion implanted SOI structures formed by hydrogen ion slicing have been investigated by Raman spectroscopy and photoluminescence (PL). After implantation the wafers have been heat-treated by either furnace annealing (FA) or rapid thermal annealing (RTA). It has been found that implantation of 3×10^{17} H^+/cm^2 results in the formation of the amorphous Si layer (a-Si) inside silicon film on insulator. Structural transformations in a-Si depended on the annealing conditions. FA led to crystallization of a-Si and to the formation of monocrystalline silicon films. RTA results in the formation of the layers containing a high density of Si nanocrystals. A comparison of the Raman measurements with the PL data allows to conclude that PL bands obtained near 420 and 500 nm are not associated with the radiative recombination in Si nanocrystals.

INTRODUCTION

Recently, there has been an ever-increasing interest in the formation of nanocrystalline materials. Various methods of creating luminescent nanocrystalline materials have been reported [1-5]. An application area for nanocrystalline Si has been also rapidly growing. In particular, it is possible its using in the fabrication of Si based optoelectronics devices. Moreover, thin films containing a high density of the homogeneously oriented nanocrystals may be used as a "soft target" for the growth of the non-strained layers with different structure and the formation of Si based hetero-structures. In this paper, we report a series of experiments to study a behaviour of H^+ ion implanted SOI structures prepared by hydrogen ion slicing during heat treatments.

EXPERIMENT

SOI structures were formed by hydrogen slicing HS method. A thicknesses of the top silicon layer and buried SiO_2 were 500 nm and 280 nm, respectively. HS SOI structures have been implanted with H^+ ions at an energy of 24 keV to doses of 3.0×10^{17} cm^{-2}. Ion plasma source was used for implantation. The profiles of H atoms were controlled by SIMS. After implantation the samples were heat-treated by either FA or RTA annealing. The FA anneals was performed within the temperature range of 200 to 700°C in an N_2 ambient for 1 hour while the RTA anneals were carried out at temperatures ranging from 300 to 900°C in air for 10 s. Raman and PL measurements were used to investigate the structural and optical properties of the structures prepared. Both unimplanted and H^+ ion implanted HS SOI structures were investigated. PL spectra were excited by N_2 laser wavelength of 337 nm at room temperature and were recorded within the emission wavelength range of 340 – 850 nm. Raman spectra were measured in backscattering geometry with a DFS-52 spectrometer. The excitation wavelength was the 488-nm line of an Ar laser. Two geometry of the samples were used in this measurements: geometry of $x(y+z, y+z)\underline{x}$ (allowed geometry) and geometry of $x(y, y)\underline{x}$ (forbidden geometry), where x is (100) orientation, y is (010) orientation, and z is (001) orientation. Use of the forbidden geometry allowed us to reduce the 520 cm^{-1} line intensity from Si target. Raman experiments were performed also at room temperature.

RESULTS

Fig. 1 shows the depth distribution of hydrogen atoms measured by SIMS before and after RTA at T_a = 200 - 900° C. The inset to Fig. 1 displays the integral concentration of H atoms in the top silicon layer as a function of RTA annealing temperature. After implantation, practically all the implanted hydrogen is homogeneously distributed in the top Si layer of about 0.3 nm-thickness. No differences in H profiles were found after RTA up to T_a ~ 500° C. A reduction in the H concentration with increasing T_a was observed in samples annealed at T_a ¤ 500° C.

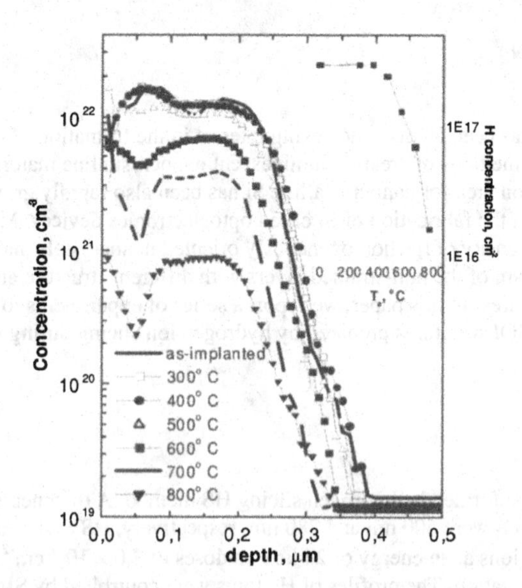

Figure 1. The depth distribution of H ion implanted SOI films measured by SIMS before and after RTA anneal. The inset displays integral concentration of H atoms as a function of RTA annealing temperature.

Fig. 2 shows PL spectra recorded both from the as-implanted samples and from those annealed by RTA at T_a = 300 - 700° C. The as-implanted samples exhibit broad PL band of the very low intensity in the wavelength range of about 400 - 700 nm. No differences in PL spectra were obtained after annealing up to 400° C. Annealing at T_a ¤ 500° C leads to the appearance of the PL bands in the violet and green-red spectral region. Their intensity increase with anneal temperature up to T_a = 600° C. PL spectra from the samples annealed at 600 and 700° C are identical. These broad PL spectra were fitted to three Gaussian profiles centered at 420, ~500 and ~600 nm. The inset to Fig. 2 displays PL spectrum and the curve fit results for the sample annealed at 600° C. In the case of FA (Fig. 3), violet PL peak of about 420 nm was obtained even after T_a = 200° C. Its intensity increases with T_a and reaches a maximum value after T_a = 600° C. A broad PL peak of the much less intensity appears also in the green region after annealing at 600° C. This PL spectrum was fitted to two Gaussian profiles centered at 420 and ~500 nm (see the inset to Fig. 3). Heat treatment at T_a = 700° C quenches progressively these PL bands. It should be noted, in the cases of both RTA and FA anneals, the PL observed was not too intensive.

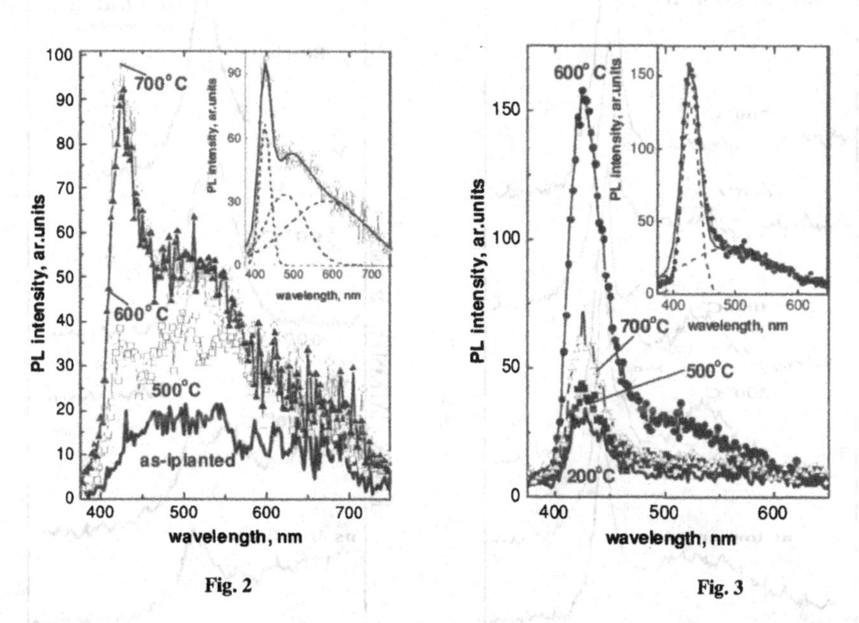

Figure 2. PL spectra from the H ion implanted SOI films before and after 10 s RTA at T_a = 500 - 700° C. The inset shows PL spectrum from the sample annealed at 600° C and the curve fit results.

Figure 3. PL spectra from the H ion implanted SOI films after 1 hour FA at T_a = 200 - 700° C. The inset shows PL spectrum from the sample annealed at 600° C and the curve fit results.

Fig. 4 shows an effect of RTA anneal on the Raman spectra of H^+ ion implanted SOI structures. Two distinct features from as-implanted sample were (1) a sharp peak at ~ 520 cm^{-1} due to the Si substrate, and (2) a broad peak centered around 480 cm^{-1}, which is connected with Si-Si binding in amorphous Si matrix [6] (Fig. 4a). After RTA at T_a = 300 – 400° C, a width at half maximum of 480 cm^{-1} peak decreases (Fig. 4a). The intensities of 480 cm^{-1} and 520 cm^{-1} peaks are almost invariable. At the same time, a shoulder of the low intensity forms in the low-frequency part of 520 cm^{-1} peak. As the annealing temperature is raised to 600° C, the intensity of 480 cm^{-1} peak drops. After T_a = 700° C the intensity of this peak is almost completely quenched. The intensity of 520 cm^{-1} peak is invariable inside the annealing temperature range studied. In detail a behavior of the low-frequency shoulder of 520 cm^{-1} peak as a function of RTA annealing temperature was investigated under conditions of the forbidden geometry for Si target. These results are shown in Fig. 4b. In this geometry, the intensity of 520 cm^{-1} peak is essentially reduced. At the same time, a low-frequency peak from Si-Si binding in crystalline matrix is seen clearly. After RTA at 300 – 400° C, this peak of low intensity is

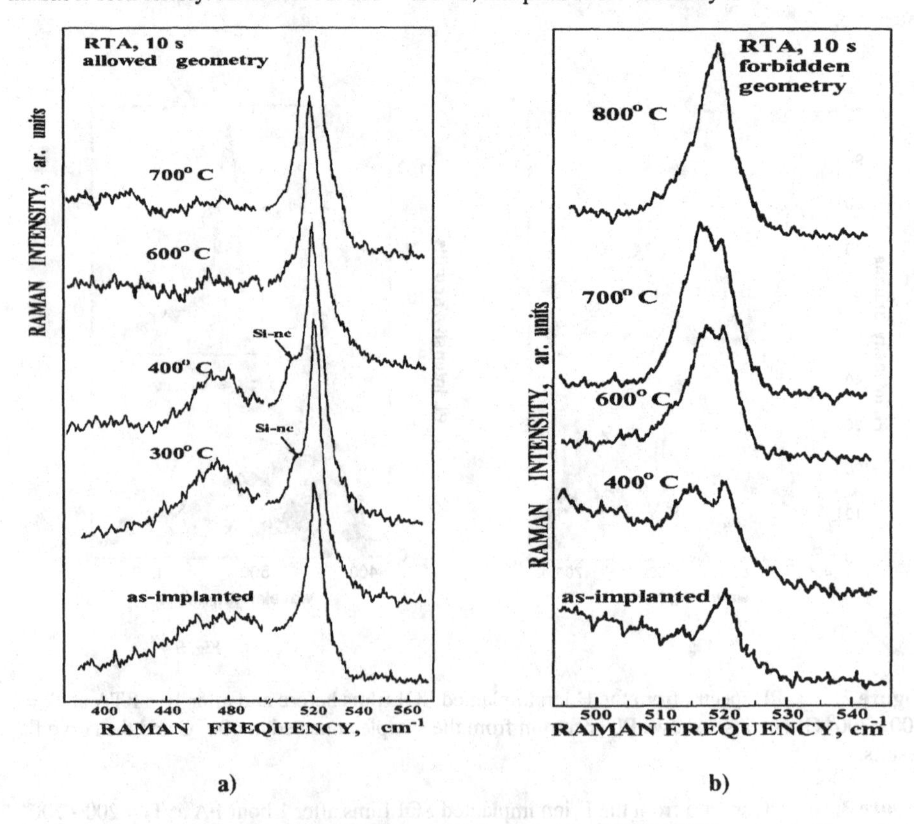

Figure 4. Raman spectra of H^+ ion implanted SOI structures after 10 s RTA measured in allowed (a) and forbidden (b) geometry.

centered around 514 cm^{-1}. The increase in T_a leads to its high-frequency shift and to rise in its intensity. After T_a = 600° C it was peaked at 517 cm^{-1}, after T_a = 700° C it was at about 518 cm^{-1}. The nature of this peak is connected with Si nanocrystals [7] formed inside amorphous Si layer containing the high concentration of H atoms. As T_a increases, dimensions of Si nanocrystals rise. It leads to the shift in this peak position toward a frequency of balk monocrystalline silicon 520 cm^{-1}. Because of different display of these two peaks under conditions of forbidden geometry, it may be suggested the different orientation of Si-nanocrystals formed during RTA inside the H enriched amorphous Si films and that of Si target. No Raman peak connected with Si nanocrystals was obtained from H^+ ion implanted SOI structures after FA (Fig. 5). Under conditions both of allowed and forbidden geometry for Si target, two Raman peaks at 480 cm^{-1} and 520 cm^{-1} were observed only. As the annealing temperature increased, the intensity of a peak connected with Si-Si binding in amorphous matrix reduced, but the intensity of a peak due to Si target raised. One reason for the absence of a Raman peak connected with Si nanoclusters could be the very low density of Si nanocrystals. Second reason of this effect could be the some orientation of Si nanocrystals and of Si target.

A comparison of the Raman results with PL data suggests that PL bands of about 420 and 500 nm are not associated with the radiative recombination in Si nanocrystals.

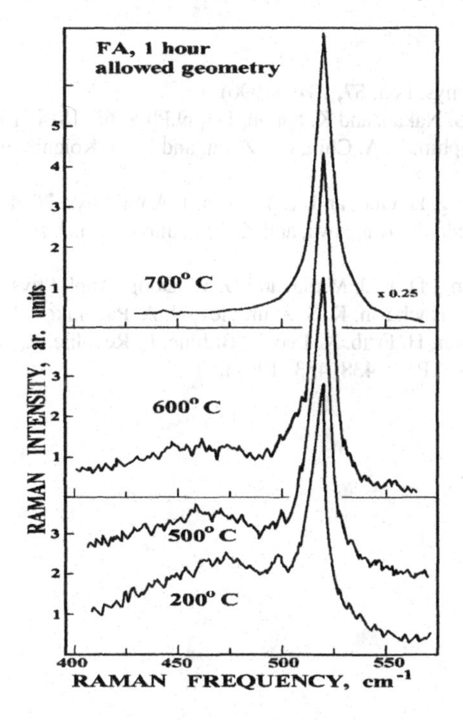

Figure 5. Raman spectra of H^+ ion implanted SOI structures after 1 hour FA measured in allowed geometry for Si target.

CONCLUSIONS

Hydrogen ion implantation into SOI structures and subsequent RTA annealing allow to form silicon nanocrystalline films with high concentration of nanocrystals. Nanocrystals formed during RTA have a different orientation regarding of Si target. A comparison of the Raman scattering results with PL data suggests that PL bands of about 420 and 500 nm from the structures formed after H ion implantation and subsequent FA and RTA anneals are not associated with the radiative recombination on Si nanocrystals.

ACKNOWLEGEMENTS

This work was done at the support of ISTC grant no.563.

REFERENCES

1. L. T. Canham, Appl. Phys. Lett. **57,** 1046 (1990).
2. T. Shimizu-Iwajama, S. Nakao, and K. Saitoh, J.Appl.Phys. **65**, 1814 (1994).
3. E. Werwa, A. A. Seraphin, L. A. Chin, Ch. Zhou, and K. D. Kolenbrander, Appl. Phys. Lett. **64**, 1821 (1994).
4. H. Morisaki, F. W. Ping, H. Ono, and K. Yazawa, J. Appl. Phys. **70**, 1869 (1991).
5. S. Hayashi, T. Nagareda, Y. Kanzawa, and K. Yamamoto, Jpn. J. Appl. Phys., Part 1 **32**, 3840 (1993).
6. A. Hartstein, J.C. Tsang, D. J. Di Maria, and D. W. Dong, Appl. Phys. Lett. **36**, 836 (1980).
7. I. E. Tyschenko, G. A. Kachurin, K. S. Zhuravlev, N. A. Pazdnikov, V. A. Volodin, A. K. Gutakovsky, A. F. Leier, H. Frцb, K. Leo, T. Bцhme, L. Rebohle, R. A. Yankov, and W. Skorupa, Mater.Res.Soc.Proc. **438**, 453 (1997).

Mat. Res. Soc. Symp. Proc. Vol. 667 © 2001 Materials Research Society

PL due to Discrete Gap Levels in Some Chalcogenide Glasses- A Configurational Coordinate Diagram Illustration

N. Asha Bhat, K.S. Sangunni and K.S.R.K. Rao
Department of Physics, Indian Institute of Science
Bangalore –560 012, INDIA

ABSTRACT

Photoluminescence (PL) studies were carried out on *a*-Se and a few $Ge_{20}Se_{80-x}Bi_x$ and $Ge_{20}Se_{70-x}Bi_xTe_{10}$ bulk glassy semiconductors at 4.2 K with Ar^+ laser as excitation source. While *a*-Se and samples with lesser at% of Bi show fine structured PL with a large Stokes shift, samples with higher at% of Bi did not show any detectable PL. The investigations show at least three radiative recombination transitions. Features extracted by deconvoluting the experimental spectra show that the discrete gap levels associated with the inherent coordination defects are involved in the PL transitions. Absence of PL in samples with higher Bi at% are explained on the basis of non-radiative transition mechanisms. Overall PL mechanism involving gap levels in chalcogenide glasses is illustrated with the help of a configurational coordinate diagram.

INTRODUCTION

Chalcogenide glasses show many interesting properties in addition to being promising materials in applications ranging from memory devices to x-ray imaging[1] to infrared optical fiber applications[2]. One such property is the appearance of n-type conduction with the addition of Bi and Pb in Ge-Se glasses[3,4]. Unlike in crystalline semiconductors and in *a*-Si and related materials, explanation for p or n type conduction in chalcogenide glasses is not very straight forward. This is so mainly because of the presence of large density of inherent defects with unique properties.

Among Bi and Pb doping, the former one was studied extensively and the earlier investigations have brought out two different descriptions for n-type conduction. The first one was based on the idea of phase separation at microscopic level in the form of *c*-Bi_2Se_3, a strong n-type semiconductor[5]. The second one, on the other hand, based on structural studies point out a coordination number of 3 for Bi, against c-Bi_2Se_3 phase separation and suggests the n-type conduction to be the consequence of reduction in inherent D^+ defects[6]. Though the two contradictory ideas were addressed recently in a consistent way[7], there were no direct and systematic studies on the role of defects in bringing out n-type conduction in Bi doped glasses.

Photoluminescence spectroscopy is one of the preferred techniques to study defect levels in the gap of semiconductors and ideal for chalcogenide glasses having a large number of defects in their forbidden energy gap. In the present paper we report our investigations on defects in *a*-Se and Bi doped Ge-Se and Ge-Se-Te glasses using PL spectroscopy. Multiple PL transitions observed in most of the samples except those with higher Bi concentrations are explained on the basis of phenomenological models and a configurational coordinate diagram representation.

EXPERIMENTAL DETAILS

Bulk glassy samples of interest were prepared by conventional melt-quenching technique from 5N pure elemental materials. The samples prepared in this way were confirmed to be x-ray amorphous. Differential scanning calorimetry scans were done on the samples to determine their glassy nature.

Samples having surface dimensions of 2X2 mm^2 were selected for PL studies. No grinding or polishing was done because samples already had shiny surfaces in their as-prepared form. To carry out photoluminescence studies a Fourier Transform Photoluminescence Spectrometer (MIDAC corp. USA) was used. Samples to be studied were suspended in a liquid helium cryostat and excited with an Ar^+ laser (λ = 514.5nm or E~2.41eV) with 2mm beam diameter. Experiment was carried out at 4.2 K with 150 mW/cm^2 excitation intensity. PL signals coming from the samples were analyzed using a Michelson Interferometer and detected by a cooled Ge-photodiode. The interferograms were recorded by averaging out for 10 coadded scans with a resolution of 2meV using a personal computer attached to the spectrometer. The excitation was turned on just before recording each spectrum.

RESULTS AND DISCUSSION

The PL spectra recorded at 4.2 K for *a*-Se is shown in figure1. It can be seen that the spectra falls in a broad band in 0.7 to 1.2 eV energy range with fine features indicating transitions at three separate energies. PL spectra taken similarly for different at% of Bi in $Ge_{20}Se_{80-x}Bi_x$ and $Ge_{20}Se_{70-x}Bi_xTe_{10}$ glasses are given in figures 2 and 3. Like in a-Se, at lower at% of Bi we see fine structured PL in broad energy range. However samples with higher Bi at% that show n-type conduction did not show any luminescence as can be seen in figures 2 and 3.

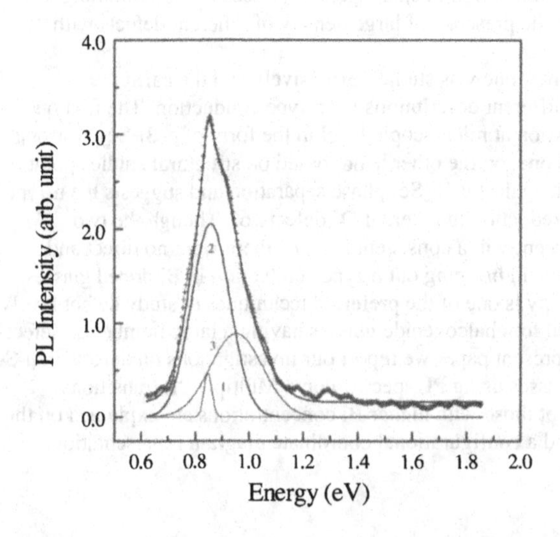

Figure 1 Experimental PL spectra for a-Se at 4.2 K (data points). The deconvoluted spectra are numbered 1-3 and their resultant fits well with experimental data.

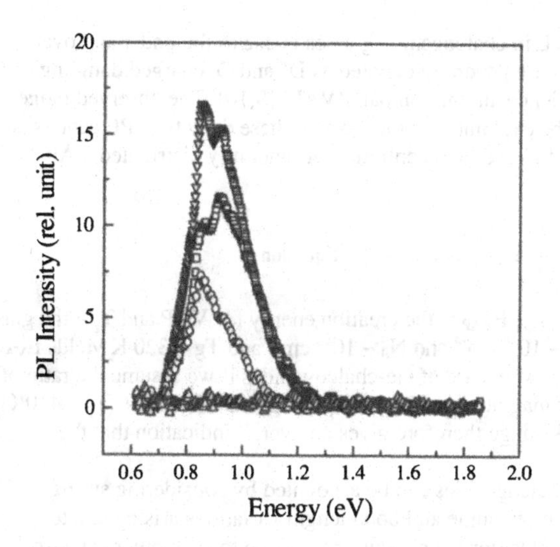

Figure 2 Experimental PL spectra in $Ge_{20}Se_{80-x}Bi_x$ glasses at 4.2 K for x=0(□), x=2 (∇), x=6 (o) and x=10(Δ).

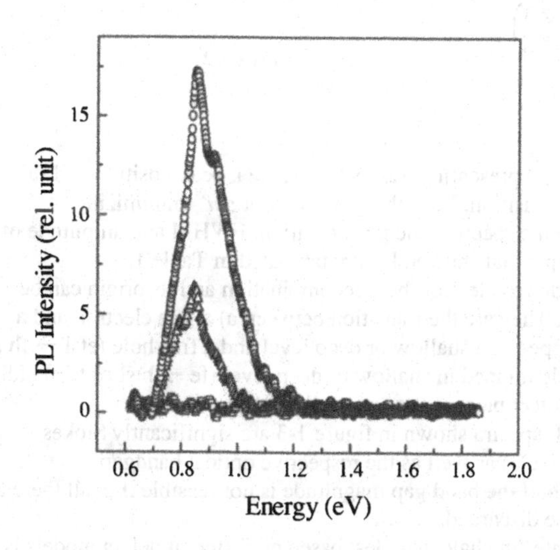

Figure 3 Experimental PL spectra in $Ge_{20}Se_{70-x}Bi_xTe_{10}$ at 4.2 K for x=0(o), x=5 (Δ) and x=11(□).

It has been commonly accepted that PL in chalcogenide glasses is due to the under and over coordinated centers [8]. These defects are generally denoted as D^+ and D^- charged dangling bonds or equivalently C_3^+ and C_1^- valence alternation pairs(VAP) [9,10]. The observed range of luminescence emission energies can be explained on the basis of these defects as PL centers as follows. Generally, for melt-quenched glasses, concentration of randomly distributed VAPs could be estimated using

$$N_{VAP} = N_0 \exp(-E_{VAP} / 2kT_g) \qquad \textbf{Equation 1}$$

where N_0 is the chalcogen atomic density, E_{VAP} is the creation energy for VAP and T_g – the glass transition temperature. For Se, $N_{VAP} \sim 10^{18}$ cm^{-3} and $N_0 \sim 10^{22}$ cm^{-3} and Tg ~ 320 K yields E_{VAP} ~ 0.8eV [10]. Proceeding in similar lines, in case of Ge-chalcogenides if we assume the ratio of N_{VAP} to N_0 to be of the same order of magnitude as that in Se, we get, $E_{VAP} \sim 1$ eV ($T_g \sim 410$K). Observation of PL in the 0.7 to 1.2 eV range therefore gives an overall indication that the recombination involves VAP centers.

The overall broadening of PL in chalcogenides can be accounted by considering strong electron-phonon coupling rather than bond angle and bond length variations arising due to disorder[8, 10]. Presence of multiple transitions at energies very close to each other can also result in an overall broadening of the spectra.

Luminescent transitions are generally described by Gaussian line broadening mechanism. Similar is the case proposed for chalcogenides too[8]. Accordingly, each experimental spectra were deconvoluted into a linear combination of three Gaussians (area version) given by

$$y = y_0 + \sum_{i=1}^{i=3} \frac{A_i}{w_i \sqrt{\pi/2}} \exp\left(-2\left(\frac{(x - xc_i)^2}{w_i^2}\right)\right) \qquad \textbf{Equation 2}$$

with y_0 is offset and A_i, xc_i and $1.177w_i$ representing, respectively, area, peak position and full width at half maxima(FWHM) of *i*th transition. Levenberg-Marquardt chi^2 minimizing algorithm is employed to deconvolute the spectra. The peak position, FWHM and amplitude of each transition extracted for each sample that showed PL are tabulated in Table 1.

PL in semiconductors is generally due to electron-hole recombination and its origin can be traced to six different transitions [11]. They are the transition between a) a free electron and a free hole (fe → fh), b) an electron trapped in a shallow or deep level and a free hole (etsl → fh or etdl → fh), d) a free elecrton and a hole trapped in shallow or deep level (fe → htsl or fe →htdl) and f) electron-hole pairs trapped at defect pairs in the gap (etdl → htdl).

Coming to the present results, the PL spectra shown in figure 1-3 are significantly Stokes-shifted from the excitation energy (2.41eV) as well as the respective optical bandgap energies(~2eV). Such a large shift of half the band gap magnitude is not feasible if at all there are any band-to-band PL and safely can be discarded.

An energy level scheme for gap states in chalcogenides, based on different defect models is sketched in figure 4 (a)[12]. Positioning of random D^- and D^+ are discussed elsewhre[13]. Greater configurational changes at nonrandom D^+ and D^- (also known as intimate VAPs)

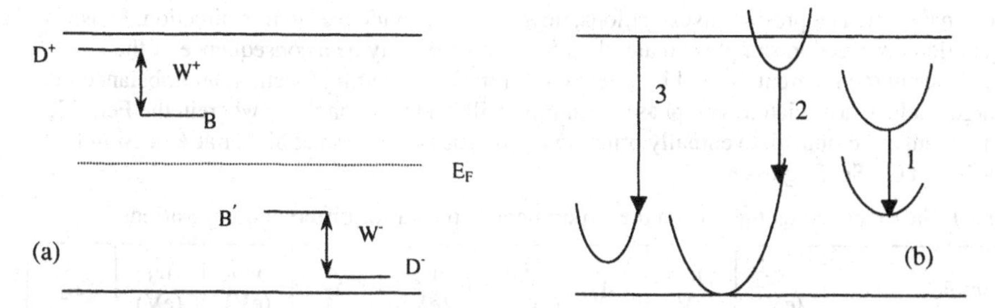

Figure 4 (a)Energy level scheme for chalcogenides and (b) Configurational coordinate diagram illustration of multiple PL transitions involving gap levels denoted by 1- the IVAP pair recombination. 2 - etsl → fh and 3- fe → htsl

arising due to distortion than at random D^+ and D^- can be expected due to reasons given elsewhere. Consequently, we anticipate them to lie deep in the gap[13].

Interstingly, out of the three, one deconvoluted spectra has almost same characteristics (peak position and FWHM) irrespective of the sample and therefore can be assigned to the recombination transition at IVAPs which are less sensitive to dopant atoms[12]. At the same time according to charged defect models, Stokes shift is greater for D^+ than for D^- and therefore the transition involving D^+ will be at lesser energy than that involving D^-. Utilizing these ideas, we can assign the second deconvoluted spectra in figure 1 to etsl → fh transition and the third one to fe →htsl. The multiple transitions discussed above can be illustrated with the help of a configurational coordinate diagram as shown in figure 4(b). The present findings highlight that in chalcogenides, be it elemental or compound, the origin of PL is solely due to the inherent defects.

Lastly, most important consequence of addition of Bi to Ge-Se and Ge-Se-Te glasses is a reduction in the width of the two broader transitions along with a shift in peak position to lower energy side. This is natural if we recall that the broader transitions are assigned to random VAPs which are more likely to get influenced by external dopants than the IVAPs. The scheme of identifying different transitions discussed above fits well for lower at% of Bi too. However, absence of any detectable PL for n-type conducting samples with higher at% of Bi can be due to two reasons. The first one is a reduction in the overall band gap with an increase in Bi concentration increase the probability of nonradiative processes such as photoconduction. In addition the characteristic PLE spectrum exhibited by chalcogenide gasses in general indicate that PL efficiency decreases drastically when the band gap of the sample under study is far less than the energy of exciting radiation. It could also be possible that the overall concentration of defects that give rise to PL decreases significantly as Bi replaces Se as a result of lesser and lesser number of coordination defects. From Table 1 we can see that at 6 and 5 at% of Bi respectively in Ge-Se and Ge-Se-Te systems, the relative area under the second deconvoluted spectra is less than that under third.

As the area under the curve depends on the density of defect centers giving rise to PL, it can be said that the concentration of D^+ defects decreases more rapidly than the concentration of D^- as Bi replaces Se. Such a decrease was assumed while explaining n-type conduction with 3-fold coordinated Bi. The present investigations, however, give evidences in this direction. Phase separation observed in samples greater than 5-6 at% of Bi may be a consequence of the imbalance in the inherent D^+ and D^- defect concentrations. Putting together, an imbalance in charged defects associated with phase separation will lead to a condition wherein the Fermi level can no longer be pinned, eventually bringing out n-type conduction at higher at% of Bi in both Ge-Se and Ge-Se-Te glasses.

Table 1 The extracted features from the experimental spectra for different compositions.

Sample	xc_1 (eV)	w_1 (eV)	A_1	xc_2 (eV)	w_2 (eV)	A_2	xc_3 (eV)	w_3 (eV)	A_3
a-Se	0.849	0.030	0.017	0.859	0.148	0.379	0.954	0.206	0.229
$Ge_{20}Se_{80}$	0.850	0.034	0.088	0.903	0.160	1.127	1.01	0.202	1.00
$Ge_{20}Se_{70}Te_{10}$	0.851	0.024	0.129	0.859	0.144	1.989	0.956	0.192	1.414
$Ge_{20}Se_{78}Bi_2$	0.851	0.028	0.122	0.877	0.148	1.944	0.986	0.204	1.852
$Ge_{20}Se_{74}Bi_6$	0.851	0.026	0.087	0.848	0.138	0.458	0.925	0.184	0.644
$Ge_{20}Se_{65}Bi_5Te_{10}$	0.847	0.028	0.090	0.784	0.072	0.220	0.873	0.140	0.628

CONCLUSIONS

PL studies carried out in a-Se and and Ge-Bi chalcogenide glasses show fine structured spectra. Multiple PL transitions involving gap levels are accounted on the basis of defect models and are illustrated using a configurational coordinate diagram representation. Present investigation shows that defects have a significant role to play in bringing out n-type conduction in Bi doped glasses.

Reference:

1. J. Rowlands and S. Kasap, *Physics today* **50** 24 (1997).
2. S. Ramachandran and S.G. Bishop, *Appl. Phys. Lett.* **73** 3196 (1998).
3. N. Tohge, T. Minami, Y. Yamamoto and M. Tanaka, *J. Appl. Phys.* **51** 1048 (1980).
4. N.Tohge, H. Matsuo and T. Minami, *J. Non-Cryst. Solids* **95 & 96** 809 (1987).
5. J.C. Phillips, *Phys. Rev.* B **36** 4265 (1987).
6. S.R. Elliot and A.T. Steel, *J. Phys. CL Solid State Phys.* **20** 4335 (1987).
7. N. Asha Bhat and K.S. Sangunni, *Solid State Commun.* **116** (2000) 297.
8. R.A. Stret, *Adv. Phys.*, **25** 395 (1976) and references therein.
9. R. A. Street and N.F. Mott, *Phys.Rev. Lett.* **35** 1293 (1975).
10. M. Kastner, D. Adler and H. Fritzche, *Phys. Rev. Lett.* **37** 1504 (1976).
11. G. Blasse and B.C. Grabmaier, *Luminescent Materials*, (Springer Verlag, 1994).
12. E.A. Davis, *Topics in applied Phys: Amorphous Semiconductors*, (Springer Verlag, Berlin, 1979) p 41.
13. N. Asha Bhat, K.S. Sangunni and K.S.R. K. Rao, *to be published*.

Mat. Res. Soc. Symp. Proc. Vol. 667 © 2001 Materials Research Society

Enhancement of Cathodoluminescent Characteristic from $CaTiO_3:Pr^{3+}$, by Ga^{3+} Addition

Jung-Woo Byun, Byung-Kyo Lee, Dong-Kuk Kim[1], Seong-Gu Kang[2], Seung-Youl Kang[3] and Kyung-Soo Suh[3]

Dep. of Inorg. Mat. Eng., Kyungpook Nat'l Uni., 1370, Sankyuk-Dong, Puk-Gu, Taegu, 702-701, Korea

[1]Dep. of Chem., Kyungpook Nat'l University, 1370, Sankyuk-Dong, Puk-Gu, Taegu, 702-701, Korea

[2]Dep.of Chem. Eng., Hoseo Uni., Asan, Chungnam, 336-795, Korea

[3]Micro-electronics Lab., ETRI, 161, Kajong-Dong, Yusong-Gu, Taejon, 305-350, Korea

ABSTRACT

$CaTiO_3:Pr^{3+}$ as an oxide phosphor is expected to be applied for a field emission display(FED) due to its relatively high conductivity. For the practical use, however, the CL intensity of $CaTiO_3:Pr^{3+}$ has to be enhanced. We introduced Ga^{3+} as a co-activator into the phosphor and investigated the CL characteristics with various Ga^{3+} concentrations. The CL intensity of $CaTiO_3:Pr^{3+}$ was remarkably increased when Ti^{4+} atom was replaced by the Ga^{3+}. When the Ga^{3+} concentration is 5 times of Pr^{3+} molar concentration, the emission intensity of the $CaTiO_3:Pr^{3+}$ phosphor with Ga^{3+} is about 5 times higher than Ga^{3+}-free samples. So, it was concluded that the addition of Ga^{3+} is essential to enhance CL property at low voltage. We proposed the following mechanism that excitation into the host lattice leads to the formation of electrons in the conduction band and holes in the valence band. The electrons in the conduction band recombine with the holes trapped at Ga^{3+} and this energy is effectively transferred to Pr^{3+} ion, which gives its own characteristic red emission.

INTRODUCTION

Recently, field emission displays (FEDs) have been developed as next-generation flat-panel displays, which have replaced cathode-ray tubes (CRTs) and liquid–crystal displays(LCDs). The low voltage operation and high emission source of FED lead to high performance for FED[1]. So the development of new phosphors for FEDs is necessary.

In 1994, a new red-emitting phosphor for FEDs, $CaTiO_3:Pr^{3+}$, was developed. This phosphor is highly regarded as a low-voltage red phosphor because of its good color quality and intrinsic conductivity. [2] In calcium titanates, one can expect efficient excitation processes through the conduction band (CB) of the host and/or the Pr^{3+} *4f5d* band, followed by strong luminescence intensities due to non-centrosymmetric cation sites.

Considering the ionic size, Pr^{3+} is very probable to replace the Ca^{2+}. Since the substitution of Pr^{3+} for Ca^{2+} will break the charge balance, it is necessary to add other ions with the oxidation state of +3 in the Ti^{4+} site. The Ga^{3+} is a good candidate for the co-dopant due to its similar ionic size (Ti^{4+}: 0.61 Å, Ga^{3+}: 0.62Å) and its oxidation state. The similar behavior was observed also for $SrTiO_3:Pr^{3+}$ of Al^{3+} addition [3-5]

In this work, we investigated the luminescence characteristics of $CaTiO_3:Pr^{3+}$, Ga^{3+} phosphors and suggested the effect of Ga^{3+} on the luminescence properties.

EXPERIMENTAL DETAILS

The $CaTiO_3:Pr^{3+}$, Ga^{3+} phosphor samples were prepared by solid solid reaction. The starting materials were $CaCO_3$(99.99%, Aldrich), TiO_2(99.99%, Aldrich), $Pr(NO_3)_3{\cdot}6H_2O$ (99.99%, Aldrich) and Ga_2O_3(99.99%, Aldrich). The concentration of Pr^{3+} and Ga^{3+} doped to the $CaTiO_3$ ranged 0 ~ 0.4mol% and 0 ~ 5mol%, respectively. The starting materials were mixed thoroughly with alcohol in an agate mortar and dried to 80 °C for about 2 hrs. These samples were fired at 1200 °C for 4 hrs in air.

The phase identifications of the prepared samples were carried out by x-ray powder diffraction analysis using Cu-Kα radiation(MAC-Science Co. Ltd.). Scanning electron microscopy(SEM) was performed by a Hitachi S800 microscope in order to investigate the morphology of the samples

For measurements of cathodoluminescence(CL) spectra, the synthesized phosphors were placed inside a vacuum chamber equipped with in-house assembled CL spectrophotometer(ISS PC1). The electron beam (Kimball Physics, EGPS-2X1) conditions during irradiation were 500V ~ 1kV, the beam current density was 10 or 20 $\mu A/cm^2$ and vacuum level was below 3×10^{-6} torr.

RESULTS AND DISCUSSION

According to the XRD analysis of Pr^{3+} and Ga^{3+} doped $CaTiO_3$, all the samples are found to be isostructural with $CaTiO_3$ as shown in Fig. 1.

Figure 2 shows the CL emission spectra of $CaTiO_3:Pr^{3+}$ with different Pr^{3+} concentration. The maximum intensity was observed at a Pr^{3+} molar concentration of 0.01%, which is very low than the previous reports [3, 4]. Above this value, concentration quenching rapidly occurs. This phenomenon is probably due to energy transfer to some unknown defect acting as a trap or due to strong cross-relaxation effects.

The luminescence lines of Pr^{3+} are mainly originated from the optical transitions of a higher excited state $^3P_0$ and a lower excited state $^1D_2$ to the ground state. Pr^{3+} shows some different emission spectra depending on the

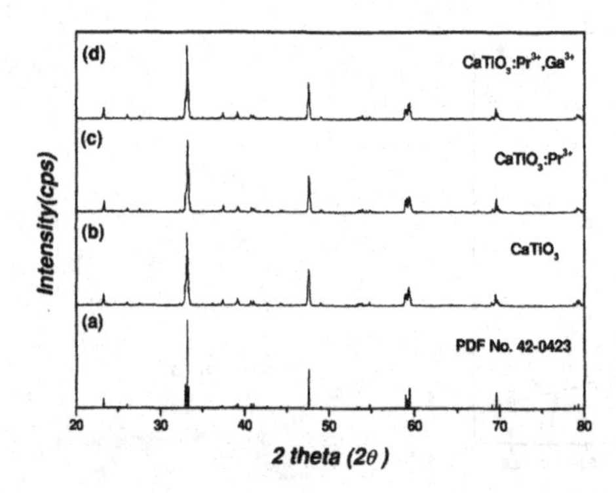

Figure 1. XRD patterns of (a) PDF No. 42-0423, (b) $CaTiO_3$, (c) $CaTiO_3:Pr^{3+}$ 0.01 mol % and (d) $CaTiO_3:Pr^{3+}$ 0.01mol%, Ga^{3+} 0.1mol%

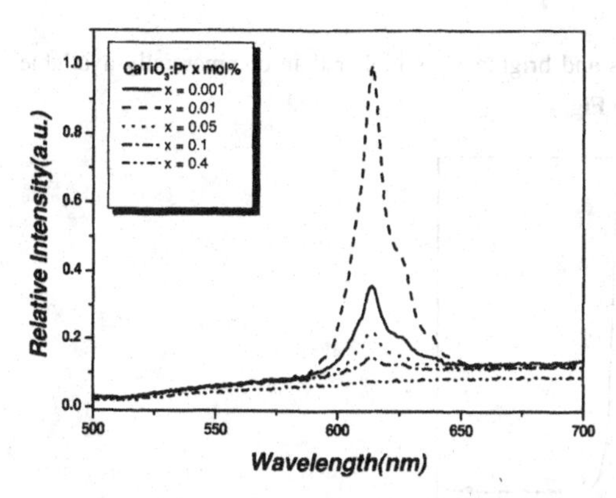

Figure 2. The CL emission spectra of $CaTiO_3:Pr^{3+}$ according to various Pr^{3+} mol%. (accelerating voltage: 500V, beam current: 20μA/cm^2).

host lattices [5, 6]. One of the relative intense.emission lines in the emission spectra is the transition of the red color from the $^1D_2$ level and the other is that of the green color from the $^3P_0$ level [7-9]. In the case of $CaTiO_3:Pr^{3+}$, in which the main emission peak is located in the well-defined wavelength region from $^1D_2 \rightarrow {}^3H_4$ transitions with the wavelength of 614nm, without the $^3P_0 \rightarrow {}^3H_4$ emission at 495nm.

Figure 3 shows the maximum CL intensity measured at 614 nm in $CaTiO_3:Pr^{3+}$ Ga^{3+} phosphor as a function of Ga^{3+} concentration. The optimum Ga^{3+} concentration is observed at 0.1 mol%. At this Ga^{3+} concentration with 0.01 mol% of Pr^{3+}, the CL emission intensity is

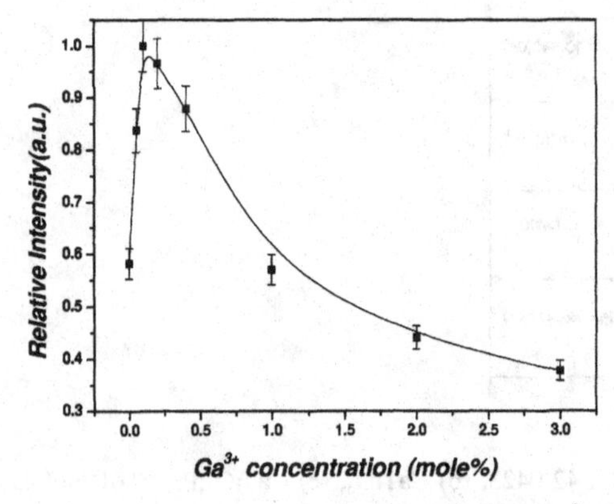

Figure 3. The CL maximum intensity of $CaTiO_3:Pr^{3+}$(0.01 mol%) phosphor as a function of Ga^{3+} concentration(accelerating voltage: 500V, beam current: 20μA/cm^2).

about twice then Ga^{3+}-free samples and brightness is higher than commercially available red phosphor, $Y_2O_2S:Eu^{3+}$ as shown in Fig. 4.

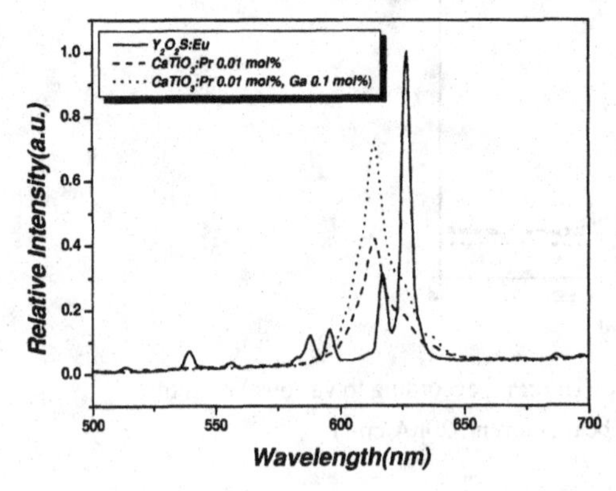

Figure 4. The CL spectra of $CaTiO_3:Pr^{3+}$ and $CaTiO_3:Pr^{3+}$, Ga^{3+} compared to commercially available red phosphor(accelerating voltage: 500V, beam current: 20μA/cm^2).

The molar concentrations and ratio of Pr^{3+} to Ga^{3+} in $SrTiO_3$ phosphor are very important. Although maximum CL intensity was observed in $SrTiO_3$: Pr^{3+}(0.01 mol%) doped with 0.1 mol% of Ga^{3+}, increasing rate is only 2 times higher than Ga^{3+}-free $SrTiO_3:Pr^{3+}$(0.01 mol%). When the molar concentration of Pr^{3+} is increased up to 0.2 mol%, the CL emission intensity is more intensified by Ga^{3+} addition, as shown in Fig. 5. In case of relatively high concentrations of Pr^{3+}, we obtained 5 times intensity in comparison with non-doped sample.

This enhancement ratio is greater than 2 in case of the Pr molar concentration of 0.01 %. The maximum brightness of the best $SrTiO_3$::Pr^{3+}(0.2 mol%), Ga^{3+}(1.0 mol%) sample obtained so far is more than 10 times of the brightness of Ga^{3+} -free sample.

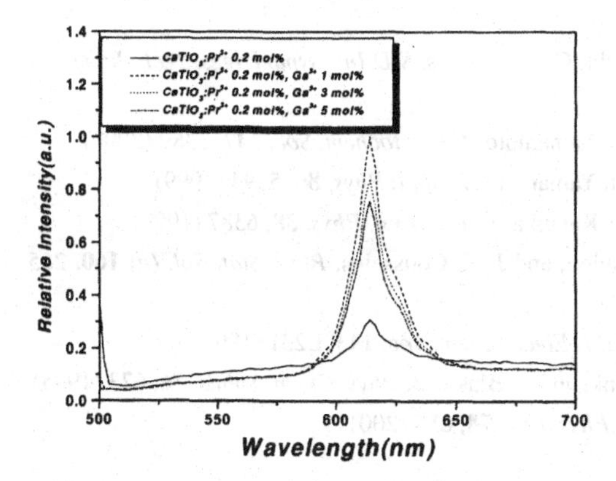

Figure 5. The CL spectra of $CaTiO_3:Pr^{3+}$ (0.2 mol%) and $CaTiO_3:Pr^{3+}$(0.2 mol%), Ga^{3+}(1, 3, and 5mol%) (accelerating voltage: 500V, beam current: 20μA/cm^2).

We have discussed the enhancement of CL property of $CaTiO_3:Pr^{3+}$ phosphor by Ga^{3+} addition. The added Ga^{3+} works as a charge compensator and hole trap center, that is, excitation into the host lattice leads to the formation of electrons in the conduction band and holes in the valence band. The electrons in the conduction band recombine with the holes trapped at Ga^{3+} and this energy is effectively transferred to Pr^{3+} ion, which gives its own characteristic red emission.

CONCLUSIONS

It was found that the Ga^{3+} addition substituted for Ti^{4+} in $CaTiO_3:Pr^{3+}$ phosphors greatly enhanced cathodoluminescence intensity. By the doping of 0.1 mol% Ga^{3+}, we obtained twice higher CL intensity than that without Ga^{3+}. The additive Ga^{3+} is expected to play a role of a hole trap center and so enhance the recombination rate of electron-hole pairs. This process results in the population of the excited Pr^{3+} 4*f* levels and we can obtain remarkable enhanced red emission. The green emission from $^3P_0$ is greatly suppressed by the non-radiative transition through a low-lying 4*f*5*d* level.

REFERENCES

1. A. Vecht, D. W. Smith, S. S. Chadha, C. S. Gibbons, J. Koh, and D. Morton, *J. Vac. Sci. Technol.* ***B*** **12**, 781 (1997)
2. S. S. Chadha, D. W. Smith, A. Vecht, C. S. Gibbons, *SID Intl Symp Digest Tech Papers* 53 (1994)
3. S. Okamoto, H. Kobayashi, and H. Yamamoto, *J. Electochem. Soc.* **147**, 2389 (2000)
4. S. Okamoto, H. Kobayashi, and H. Yamamoto, *J. Appl. Phys.* **86**, 5594 (1999)
5. S. Itoh, H. Toki, K. Tamura, and F. Kataoka, *Jpn. J. Appl. Phys.* **38**, 6387 (1999)
6. P. T. Diallo, P. Boutinaud, R. Mahiou, and J. C. Cousseins, *Phys. Stat. Sol.* ***(a)*** **160**, 255 (1997)
7. S. H. Cho, J. S. Yoo, and J. D. Lee, *J. Electrochem. Soc.* **143**, L231 (1996)
8. C. De Mello Dongega, A. Meijerink, and G. Blasse, *J. Phys. Chem. Solids.* **56**, 673 (1995)
9. S. Okamoto, H. Yamamoto, *Appl. Phys. Lett.* **78**, 655 (2001)

Mat. Res. Soc. Symp. Proc. Vol. 667 © 2001 Materials Research Society

MORPHOLOGY AND CATHODOLUMINESCENCE OF Li-DOPED $SrTiO_3:Pr^{3+},Ga^{3+}$, A RED PHOSPHOR OPERATING AT LOW VOLTAGES

Jin Young Kim, Duk Young Jeon, Seong-Gu Kang[1], Seung-Youl Kang[2], and Kyung-Soo Suh[2]

Dept. of Materials Sci. and Eng., Korea Advanced Institute of Sci. & Tech., 373-1, Kusong-dong, Yusung-gu, Taejon, 305-701, Korea

[1]Dept. of Chemical Eng., Hoseo Univ., Korea

[2]Electronics and Telecommunications Research Institute, Korea

Abstract

$SrTiO_3$:Pr,Ga phosphor using Li_2CO_3 as a flux has been investigated as a red phosphor for the application to fluorescent displays operated at low voltage. In $SrTiO_3$:Pr,Ga system, Pr^{3+} can substitute for Sr^{2+} because the ionic radius of Pr^{3+} almost coincides with that of Sr^{2+}. Previous work, it was found by XRF analysis of $SrTiO_3$:Pr,Ga single crystal that only a small fraction of Pr ions are incorporated in the $SrTiO_3$ lattice. In the present study, the effect of Li addition into $SrTiO_3$:Pr,Ga on the cathodoluminescence (CL) properties was examined at low acceleration voltage. Especially, thanks to the liquid phase of Li_2CO_3 during the sintering process, doped Li ions act as a lubricant for the efficient incorporation of Pr ions into $SrTiO_3$:Pr,Ga lattice. Furthermore, it is found that the Li addition could enhance the generation of the characteristic emission of Pr-activated $SrTiO_3$ phosphors.

Introduction

Recently, there has been a renewed interest in the development of more efficient cathodoluminescent (CL) phosphors for low voltage display applications such as field emission displays (FEDs) or vacuum fluorescent displays (VFDs) [1,2]. However, because of manufacturing and other practical constraints for low voltage display applications, the low voltage phosphors have serious concerns with respect to efficiency, spectral response, long-term reliability and stability, surface passivation, desorption, screening, and synthesis methods. Among these listed, the high CL efficiency at low voltage excitation of phosphors under the high current density of electron beam is of great technological importance. A new red-emitting low voltage phosphor, perovskite-structured $SrTiO_3$:Pr,Ga or Al has been investigated for its application to display devices because of high luminous efficiency and excellent color purity that it offers [3,4]. This study attempts, by adding Li_2CO_3 to $SrTiO_3$:Pr,Ga as a flux, to enhance the red Pr characteristic emission and not to deteriorate the color purity. The effect of the Li flux on the CL intensity of $SrTiO_3$:Pr,Ga fired in air are also investigated through analysis of the phosphors' morphologies.

Experiments

A single crystal of $SrTiO_3$ containing varying concentrations of Pr and Ga was grown by a standard floating zone method. The crystal boules were cut into wafers of about 1 mm thick and polished to an optical quality by successively finer diamond abrasives. The dopant concentrations in the crystal were determined by X-ray fluorescence spectroscopy(QuanX, KevexSpectrace).

The $SrTiO_3$:Pr,Ga powder phosphor sample using Li_2CO_3 as the flux agent was prepared with solid state reaction among $Sr(CO_3)_2$, TiO_2, $Pr(NO_3)_3$, Ga_2O_3, Li_2CO_3. Appropriate portions of the raw materials were weighed, mixed under alcohol in an agate mortar, and allowed to dry in air. The dried mixture were placed in a alumina crucible and fired at 1200°C for 8 hours. The CL(Cathodoluminesccence) emission spectra were obtained with in-house assembled CL spectrophotometer(ISS PC1). For CL measurements, electron acceleration voltage from 300 to 1000V, excitation current density of 53 $\mu A/cm^2$, and vacuum level of 1×10^{-7} Torr were used.

Results and Discussion

The amounts of Pr and Ga added to the starting materials and their concentrations in the crystals determined by XRF are summarized in Table 1. This result clearly shows that the concentration of Pr ions incorporated in the $SrTiO_3$ lattice is much lower than the added amount and it also indicates that when Ga ions are co-doped, the greater amount of Pr ions are incorporated into the crystal. Comparing the ionic sizes of the hosts and dopants it seems that Pr^{3+} and Ga^{3+} ions enter the Sr and Ti site, respectively, and the increase in solubility of Pr^{3+} is due to charge compensation achieved by the additional Ga^{3+}.

Table 1. Results of XRF analysis of the single crystal.

Specimen #	Concentration in starting materials (mole fraction)		Concentration in crystal (mole fraction)	
	Pr	Ga	Pr	Ga
1	1.0×10^{-3}	—	9.0×10^{-5}	—
2	1.0×10^{-3}	1.0×10^{-3}	2.4×10^{-4}	8.7×10^{-4}
3	3.0×10^{-3}	—	4.5×10^{-4}	—
4	6.0×10^{-3}	—	9.5×10^{-4}	—
5	6.0×10^{-3}	6.0×10^{-3}	1.56×10^{-3}	5.26×10^{-3}

Figure 1 shows the effects of Li_2CO_3 concentration on the CL spectra. When the Li_2CO_3 concentration is low, the intensity of the Pr characteristic emission increases as the Li_2CO_3 concentration increases. Owing to its small radius, the Li^+ ion can readily enter the $SrTiO_3$:Pr,Ga lattices interstitially and contribute the energy transfer from the $SrTiO_3$ host to the Pr^{3+} centers, thus increasing the emission spectra intensity. It can thus be appreciated that lithium has the effect of aiding electron transfer. Upon reaching 0.5 mol% of Li_2CO_3 concentration, maximum CL intensity which is more than 2 times greater than without Li-doped was obtained. However, when the Li_2CO_3 concentration is further increased, the intensity of the characteristic emission drops. This might be related to the Pr compound being melted by Li_2CO_3, coagulating and then being unable to disperse uniformly by itself.

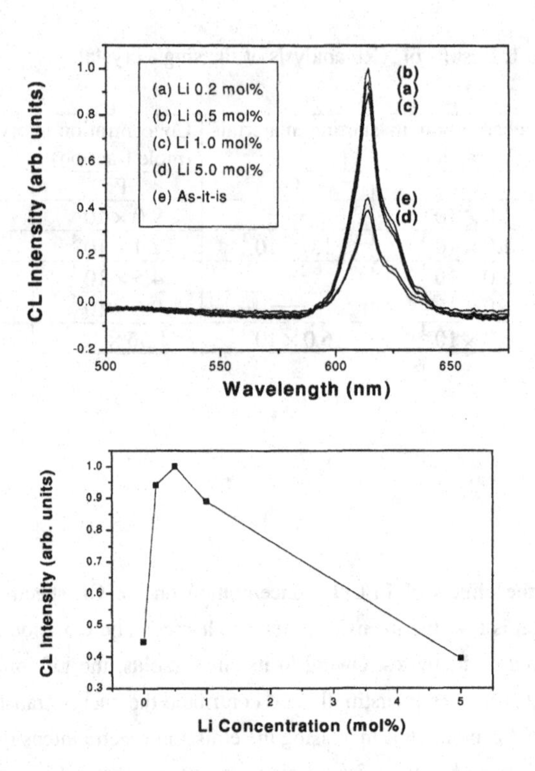

Fig. 1. Effect of Li_2CO_3 flux concentration (0.2 ~ 5.0 mol%) on emission spectra of $SrTiO_3$:Pr,Ga phosphors.

Figure 2 shows SEM micrographs of $SrTiO_3$:Pr,Ga phosphor particles without and with the addition of Li_2CO_3 flux agent. As shown in Fig. 2, the particle size of powder sample obviously increases with the addition of Li_2CO_3 flux agent. And also, thanks to the liquid phase of Li_2CO_3 during the sintering process, the dissolution of sharp edges of particulate solids makes the particle surfaces smoother, followed by the formation of larger grains with a rounded shape through the flux effect of Li. The optimization of chemical composition, particle size distribution, and morphology is required for the more efficient luminescence of phosphor materials. Among them, the spherical morphology is an important factor for lower light scattering at the surfaces as well as higher packing densities [5]. Consequently, it is understood that such a significant change of morphology may play a role in enhancing CL of $SrTiO_3$:Pr,Ga phosphors.

(a)

(b)

Figure 2. SEM micrographs of $SrTiO_3$:Pr,Ga (a) without and (b) with the addition of Li_2CO_3 flux.

Conclusions

Compositional analysis of $SrTiO_3$:Pr,Ga single crystals was investigated. It was found that only a small fraction of Pr ions added are incorporated in the lattice. So, this study attempted by adding Li_2CO_3 to $SrTiO_3$:Pr,Ga as an agent to do the efficient incorporation of Pr ions into $SrTiO_3$:Pr,Ga lattice. From the present work, it is understood that the addition of Li_2CO_3 flux favors electron transfer from the host to the Pr^{3+}, thus enhancing the intensity of the main characteristic peaks.

References

1. P. H. Holloway, T. A. Trottier, B. Abrams, C. Kondoleon, S. L. Jones, J. S. Sebastian, and W. J. Thomes, J. Vac. Sci. Technol. B,17, 758 (1999).
2. A. Vecht, C. Gibbons, D. Davies, X. Jing, P. Marsh, T. Ireland, J. Silver, A. Newport, and D. Barber, J. Vac. Sci. Technol. B,17, 750 (1999).
3. S. Itoh, H. Toki, K. Tamura, and F. Kataoka, Jpn. J. Appl. Phys., 38, 6387 (1999).
4. S. Okamoto, H. Kobayashi, and H. Yamamoto, J. Appl. Phys., 86, 5594 (1999).
5. Y. C. Kang, S. B. Park, I. W. Lenggoro, and K. Okuyama, J. Phys. Chem. Solids 60, 379 (1999).

Mat. Res. Soc. Symp. Proc. Vol. 667 © 2001 Materials Research Society

Amorphous Nitride Alloys as Hosts for Rare-Earth Luminescent Ions.

M. L. Caldwell, M. E. Little, C. M. Spalding, M. E. Kordesch[1] and H. H. Richardson[2]
Condensed Matter and Surface Science Program
Department of Physics and Astronomy[1]
Department of Chemistry and Biochemistry[2]
Ohio University, Athens, OH 45701, U.S.A.
[1]kordesch@helios.phy.ohiou.edu
[2]richards@helios.phy.ohiou.edu

ABSTRACT

Amorphous alloys of aluminum-gallium nitride doped with erbium (Er) were deposited at 300 K. The compositions ranged from 19% Al to 86% Al with optical band gaps varying linearly with composition from 3.4 eV (GaN) to 6.2 eV (AlN). The films were deposited on p-doped silicon (111) by a dc/rf dual gun system in a nitrogen/argon atmosphere at a pressure of 4.8 milli-Torr. After growth the films were thermally "activated" at 1070 K for 10 minutes in a nitrogen atmosphere. The cathodoluminescence emission intensities decreased linearly with Ga composition. This dependence suggests that the higher energy transitions in the Er ion are quenched by transitions to the conduction band of the alloys.

INTRODUCTION

Thin films (~200 nm) of amorphous AlN doped with Er, Tb, Cu, Mn and Cr have been grown by RF magnetron reactive sputtering at 77 and 300 K [1-9]. The doped amorphous AlN films were "activated" at temperatures up to 1250 K in a nitrogen atmosphere and showed luminescence in the visible spectrum when excited with 2.85 kV electrons. Colors ranging from blue to red (420 nm – 690 nm) have been observed from the luminescence of Er, Tb, Cu, Mn and Cr [1-9]. Electroluminescent devices have been fabricated using a glass substrate with an indium tin oxide transparent bottom electrode and an aluminum top electrode[10]. In these devices the phosphor showed luminescent efficiencies on the order of 0.03 lm/W.

Amorphous $Al_xGa_{1-x}N$ doped with rare-earth metals are promising phosphor materials because the bandgap of the host material can be tuned. The band-gap values range from 3.4 eV for GaN to 6.2 eV for AlN. Tuning the bandgap may alter the luminescent efficiency of the phosphor and could lead to more effective devices.

Because amorphous films can be deposited on a variety of substrates, the luminescence in these materials may be technologically important. Infrared microscopy, time-resolved photoluminescence spectroscopy and x-ray diffraction have been used to characterize the phosphor activation process. Luminescent properties of amorphous nitride alloy films doped with rare-earth metal Er will be discussed.

EXPERIMENTAL

The films were grown by a dual-source rf/dc sputtering system at 300 K on substrates of p-doped silicon. Prior to the growth, the silicon substrates were cleaned by ultrasonic agitation in isopropyl alcohol, rinsed with distilled water, a 5-minute dip in 25% HF solution and final rinse in de-ionized water to remove all native oxides. After placing the samples in the chamber, the dc sputtering was performed with an Al target of 99.999% purity with a slug (3/16 in diameter) of Er inserted in the target. The rf sputtering was performed with a GaN pressed powder target. The growth of the $Al_xGa_{1-x}N$ samples was performed in a 50:50 argon-nitrogen atmosphere. The growth pressure was around 4.8 milli-Torr during sputtering and the relative composition was determined by regulating the power to each sputtering source. The as-deposited films were in the thickness range of 220 to 270 nm. No additional diffraction peaks were observed in x-ray diffraction (XRD), and the films are inferred to be amorphous. The luminescence activation was performed in a tube furnace having a nitrogen/oxygen atmosphere at 750 C for up to 10 minutes. As documented previously, this is the optimum time length for thin-films doped with Er[10]. After activation, XRD showed that the alloy films were still amorphous.

The cathodoluminescence data was collected by using a 2.85 kV electron beam. The spot size of the electron beam was ~ 5mm in diameter. The sample was placed ~ 55 degrees relative to the surface normal direction. The light was collected through a CaF_2 window with an external lens assembly. The light was chopped and focused onto the entrance slit of a SPEX Industries double monochromator using a slit width corresponding to a resolution of 1 nm (0.5 mm slit width). The monochromator was not under vacuum and the total path length in air is around 6 feet.

RESULTS

We prepared three $Al_xGa_{1-x}N$:Er samples with the Al fraction *x* being 19%, 45% and 86%. The CL spectrum for each of these samples is shown in figure 1. The dominant peaks in the CL spectrum are a set of sharp bands centered on 535 nm and 559 nm respectively, and appear bright green when observed by eye. As the Ga content in the alloy approaches pure GaN, the intensity of the green emission dramatically decreases. There is a 50 time decrease in emission intensity from 14% Ga (curve C) to 55% Ga (Curve B). The band gap corresponding to this change in Ga composition is 5.5 to 4.8 eV.

The results of characterization by XRD of the alloy films are shown in figure 2. The figure displays typical XRD theta-2 theta scans of the Er doped $Al_xGa_{1-x}N$ thin films on Si (111) substrates before and after heating to 1020 K. The dominant feature in each spectrum are the Si (111) substrate peaks at 28° and 56°. No AlN peaks or GaN peaks are observed. We conclude that the films retain their amorphous structure during growth and luminescent activation.

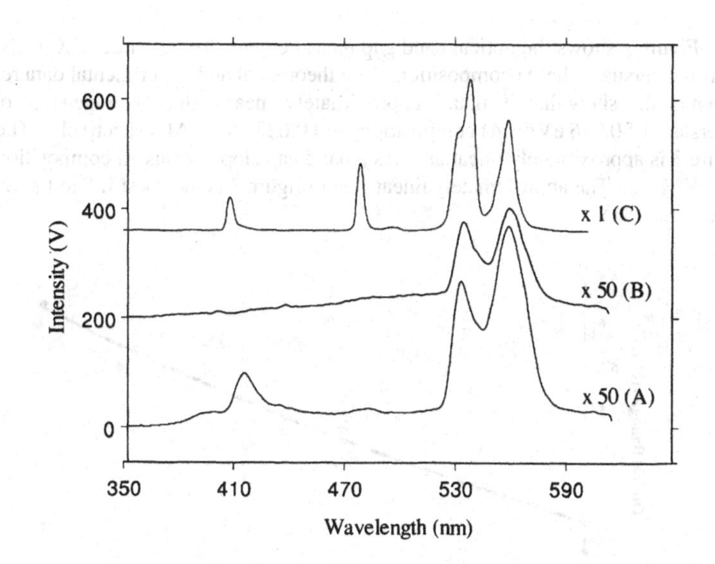

Figure 1. Cathodoluminescence of $Al_xGa_{1-x}N$:Er samples A, B, and C. Sample (A) contains 19% AlN, sample (B) contains 45% AlN, and sample (C) contains 86% AlN.

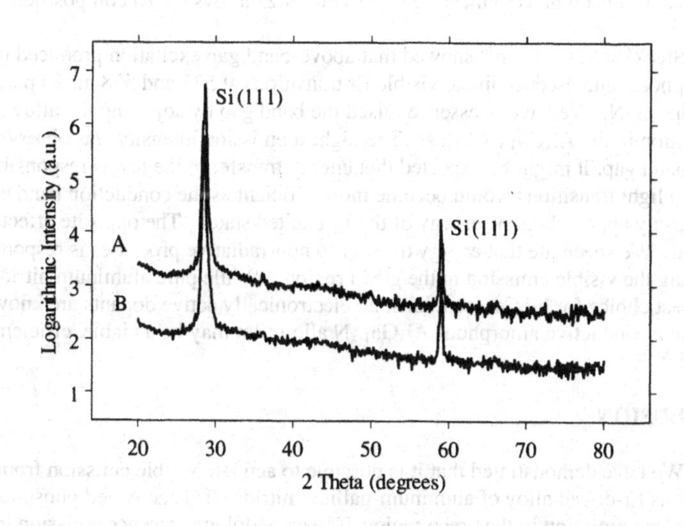

Figure 2. XRD profiles of amorphous $Al_xGa_{1-x}N$:Er before heating (B) and after heating (A) to 1070K.

Figure 3 shows the optical band gap measurements for the three $Al_xGa_{1-x}N$:Er alloy films versus x, the Al composition. The theoretical and experimental data reported by Chen et al.[11] show that the data is approximately linear with an average slope of band gap versus x of 0.026 eV/% Al composition, and 0.027 eV/% Al respectively. The data in figure 3 is approximately linear and has a band gap slope versus Al composition of 0.024 eV/% Al. The approximately linear line in figure 3 is the "best fit" to the experimental data.

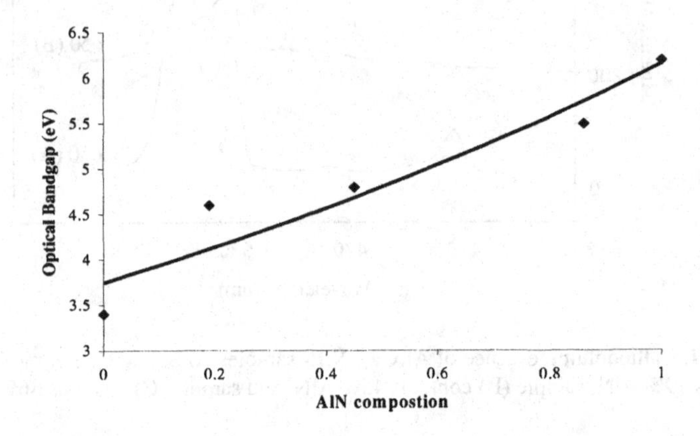

Figure 3. The band gap engineering of $Al_xGa_{1-x}N$:Er alloys vs. Al composition x.

Steckl and Birkhahn[12] showed that above-band gap excitation produced more intense photoluminescence in the visible Er transitions at 537 and 558 nm in pure crystalline GaN. We have in essence raised the band gap by adjusting the alloy composition in the $Al_xGa_{1-x}N$ films. The highest emission intensity was observed for the largest band gap. It might be expected that energy transfer to the levels responsible for the green light transitions would become more efficient as the conduction band edge more closely approached the energy of the Er excited states. The opposite effect was observed. We speculate that energy transfer to non-radiative processes is responsible for quenching the visible emission in the green region. While pure aluminum nitride might be the best choice for luminescent ions, no electronically active dopants are known for AlN, and a conductive amorphous $Al_xGa_{1-x}N$ alloy film may be a viable, efficient alternative.

CONCLUSION

We have demonstrated that it is possible to activate visible emission from an amorphous Er-doped alloy of aluminum-gallium nitride. This activated phosphor gives rise to strong emission in the green region. The cathodoluminescence emission intensities decreased linearly with Ga composition. This dependence suggests that the higher energy transitions in the Er ion are quenched by transitions to the conduction band of the alloys.

ACKNOWLEDGEMENT

This work is supported by BMDO URISP grant N00014-96-1782 entitled "Growth, Doping and Contacts from Wide Band Gap Semiconductors" and grant N00014-99-1-0975 entitled "Band-Gap Engineering of the Amorphous In-Ga-Al Nitride Semiconductor Alloys for Luminescent Devices from the Ultraviolet to the Infrared". The authors would also like to thank the Material Research Society for partial funding this project under the 2000-2001 Undergraduate Materials Research Initiative.

REFERENCES

1. H. J. Lozykowski, W. M. Jadwisienczak and I. Brown, Appl. Phys. Lett. **74**, 1129 (1999).
2. R. Birkhahn, M. Garter and A. J. Steckl, Appl. Phys. Lett. **74**, 2161 (1999).
3. H. J. Lozykowski, W. M. Jadwisienczak and I. Brown, Appl. Phys. Lett. **76**, 861 (2000).
4. M. L. Caldwell, H. H. Richardson and M. E. Kordesch, MRS Internet Journal Nitride Semiconductor Research, **5S1**, W3.26 pp. 1-7 (1999).
5. M. L. Caldwell, A. L. Martin, V. I. Dimitrova, C. M. Spalding, P. G. VanPatten, M. E. Kordesch, H. H. Richardson, J. Vac. Sci. Technol. **19**, (2001).
6. M. L. Caldwell, A. L. Martin, C. M. Spalding, P. G. VanPatten, M. E. Kordesch, H. H. Richardson, *Material Research Symposium Proceedings,* G6.6 (2000) *in press.*
7. M. L. Caldwell, A. L. Martin, V. I. Dimitrova, P. G. Van Patten, M. E. Kordesch, H. H. Richardson, Appl. Phys. Lett. **78**, 1246, (2001).
8. C. M. Spalding, M. L. Caldwell, V. I. Dimitrova, A. L. Martin, M. E. Kordesch, H. H. Richardson, P. G. Van Patten, *Material Research Symposium Proceedings*, G6.28 (2000) *in press.*
9. A. L. Martin, M. L. Caldwell, M. E. Kordesch, C. M. Spalding, P. G. Van Patten, H. H. Richardson, *Material Research Symposium Proceedings*, G6.5 (2000) *in press.*
10. V. I. Dimitrova, P. G. Van Patten, H. H. Richardson, M. E. Kordesch, Appl. Phys. Lett. **77**(4), 478 (2000).
11. H. Chen, K. Chen, D. A. Drabold, M. E. Kordesch, Appl. Phys. Lett. **77**(8), 1117 (2000).
12. A. J. Steckl, R. Birkhahn, Appl. Phys. Lett. **73**(12), 1700, (1998).

Mat. Res. Soc. Symp. Proc. Vol. 667 © 2001 Materials Research Society

Divalent and Trivalent Europium Doped Alumina Waveguides Elaborated by Pulsed Laser Deposition

Anne Minardi, Claudine Garapon, Jacques Mugnier, Corinne Champeaux[1]
Laboratoire de Physico-Chimie des Matériaux Luminescents, CNRS-Université Lyon I
10 rue Ampère, 69622 Villeurbanne Cedex, France
[1]Laboratoire de Science des Procédés Céramiques et Traitements de Surface,
CNRS-Université de Limoges, 123 avenue Albert Thomas, 87060 Limoges Cedex, France

ABSTRACT

Europium doped alumina Al_2O_3 optical waveguides were prepared by pulsed laser deposition (PLD) using a KrF laser. The targets were obtained by sintering doped powders synthesized by a sol-gel method. Depending on the oxygen pressure used during the deposition, Eu^{3+} (for 0.1 mbar) or Eu^{2+} (for 10^{-5} mbar) are obtained in the films. Two kinds of Eu^{2+} ions are present, with a 4f-5d broad excitation band peaking at 330 nm and emission bands located at 490 nm or 585 nm respectively. For Eu^{3+} doped films, the usual $^5D_0$ to the ^7F_J multiplets emission spectra were observed. The emission lines are strongly inhomogeneously broadened. Low temperature site selective fluorescence measurements were achieved in order to correlate the different Eu^{3+} sites observed with the structure of the films (amorphous or γ crystallized).

INTRODUCTION

Transition metal ions may easily dope α-Al_2O_3 and Cr:Al_2O_3 or Ti:Al_2O_3 are well-known examples of high performance laser materials. Rare earth ions, which have interesting fluorescence properties too, cannot be introduced into α-alumina by conventional crystal growth methods, due to their large size relative to that of Al^{3+} ions. Up to now, only rare earth doped transition alumina, such as γ-alumina, have been reported: thin films were prepared by implantation or PLD and monolithes by a sol-gel method[1-3].

Our aim is to elaborate by pulsed laser deposition optical waveguides of rare earth doped alumina and to compare their structural, optical and fluorescence properties with those of samples obtained by the sol-gel method, as powders or waveguiding thin films. Rare earth fluorescence properties may be thus considered either as the objective of this study or as a tool enabling to understand how these ions may be incorporated into alumina matrix. It is the reason why Eu^{3+}, which is a simple structural probe, was first chosen. In a preliminary study, we showed that, in case of europium doping, Eu^{2+} may be obtained instead of Eu^{3+}, when the film deposition is achieved under vacuum rather than under an oxygen pressure [4]. The aim of this article is to give further details about the spectral and dynamic properties of europium, divalent or trivalent and to try to draw conclusions from the structural point of view.

RESULTS AND DISCUSSION

Europium doped alumina films were obtained by pulsed laser deposition using a KrF excimer laser (λ=248 nm) with a fluence of 3 J/cm^2 [5]. The pressure in the ultra high vacuum deposition chamber was adjusted from 10^{-6} to 10^{-1} mbar by oxygen introduction. The silica substrates could be heated by an halogen lamp up to about 790°C. The targets were made from sintered powders, which were prepared by the sol-gel method from aluminium butoxide [6], the sol being doped by Eu^{3+} ions with a 0.5 or 1% atomic ratio relative to aluminium.

The influence of the oxygen pressure, used during the deposition, on the film structure and composition has been studied previously [6]. The composition of the films was checked using Rutherford Back Scattering (RBS) measurements: the stoichiometry of Al_2O_3 is maintained and does not vary with the oxygen pressure. The europium doping is homogeneous through the film thickness and the doping rate is equal to that of the target. From X ray diffraction, the film structure was shown to be γ for deposition pressure less than 10^{-2} mbar, the increase in the oxygen pressure inducing a decrease in the crystallinity, which is correlated to a decrease in the refractive index. Moreover the substrate temperature may favor the crystallization. For 0.1 mbar oxygen pressure, the films remain amorphous beyond about 700°C but γ phase crystallites are detected from 700 to 790°C, the maximum temperature available with the deposition set-up. Our study of europium sites is thus restricted to γ-alumina, as a first step to further extension to α-alumina.

Fluorescence of Eu^{2+} doped alumina waveguides

Time-resolved emission spectra obtained using 308 nm excitation by an XeCl excimer laser are shown in Figure 1 for an Eu^{2+} doped alumina film deposited on silica at a pressure of 1.2×10^{-5} mbar. Depending on the time scale, two broad bands, in the blue-green and in the orange, are successively observed. As europium is indeed present in the film according to RBS, this bands are attributed to the $4f^6$-5d to $4f^7(^8S_{7/2})$ transition of Eu^{2+}.

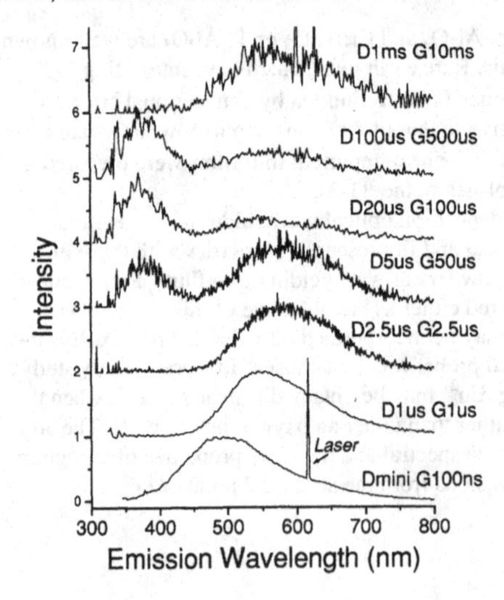

Figure 1. Time resolved emission spectra (delay D, gate G) of a 1% Eu^{2+} doped alumina film deposited at 790°C for a 1.2×10^{-5} mbar pressure (λ_{exc}=308 nm). The band located at 390 nm has a strictly exponential decay with a 100µs lifetime and has been checked to be due to the silica substrate. Two $4f^6$-5d to $4f^7$ bands at 490 nm and 585 nm are attributed to two Eu^{2+} sites

As a further argument in favor of the attribution of these bands to Eu^{2+} emission is the observation of the Eu^{3+} emission spectrum instead of the broad bands, after annealing of the film in an oxygen atmosphere at 800°C. At short times, less than 100 ns, the spectrum is constituted of a band peaking at 490 nm with a 6350 cm^{-1} width. The fluorescence decay, recorded at 420 nm, is slightly non-exponential with a mean time constant of 0.6 μs. For times longer than about 2 μs, a broad band with a maximum at 585 nm and a width of 4500 cm^{-1} is observed. The fluorescence decay, recorded at 570 nm, is highly non-exponential, the instantaneous time constant ranging from less than 1 μs to about 4 ms. Furthermore, a shift of the emission maximum with time indicates that this band corresponds to Eu^{2+} ions distributed in sites with slightly different environments.

The excitation spectrum, recorded using a C.W. xenon lamp and monitoring the 550 nm emission (Figure 2), is constituted by the $4f^7$ to $4f^6$-5d band, which is peaking at 330 nm and has a 10100cm^{-1} width. This very broad excitation band is due to the superposition of the unresolved components of the 5d level, split by the crystal field and indicates that the Eu^{2+} ions are located in low-symmetry sites with a broad crystal field distribution [7]. Time resolved excitation spectra of the short lived emission at 490 nm are underway.

We think that these results show the existence of two Eu^{2+} ions families. The first one, emitting at 490 nm, could be considered as regular ions with the μs range lifetime, usually observed for the $4f^6$-5d to $4f^7(^8S_{7/2})$ transition, which is allowed. Due to its large size relative to Al^{3+}, Eu^{2+} experiences a strong crystal field, explaining that emission takes place in the blue-green rather than in the blue. The second kinds of Eu^{2+} ions whose emission is red-shifted, are more perturbed ions. They experience a stronger crystal field and are probably close to trap centers, due to close charge compensation. As a matter of fact, life times longer than the usual μs range value, have been reported and are due to trapping and de-trapping from defects energy levels with various depths, as in the case of $SrAl_2O_4$ [8].

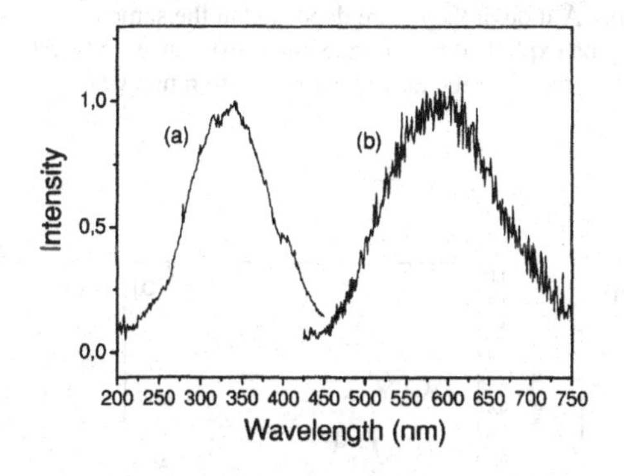

Figure 2. CW excitation spectrum (λ_{em}=550 nm) and TRS emission spectrum (D:2.5μs, G:2.5μs) (λ_{exc}=308 nm) of a 1% Eu^{2+} doped alumina film deposited at 790°C for a $1.2x10^{-5}$ mbar pressure.

Fluorescence of Eu^{3+} doped alumina waveguides

For low pressure (from 10^{-6} to 10^{-4} mbar) only Eu^{2+} is observed. As the oxygen pressure increases further up to 10^{-2} mbar, in addition to the decreasing contribution of the Eu^{2+} ions , a contribution due to the Eu^{3+} ions increases, each of them being easily distinguished at short times (μs range) and long times (ms range) respectively.

For pressure above 10^{-1} mbar, Eu^{3+} is obtained alone. The emission spectra of two films grown at 0.1 mbar at the same temperature (790°C) are presented in Figure 3, for a 465 nm excitation into the $^5D_2$ multiplet. The excitation was achieved using a pulsed dye laser with tangential excitation. In addition, Figure 3 gives the spectrum of a 0.5% Eu^{3+} doped alumina film, recorded in waveguided configuration after launching the laser beam into the waveguide by prism coupling.

The spectra are constituted by the $^5D_0$ to the ^7F_J transitions (only J=0 to 4 are detected). The multiplet splittings cannot be resolved due to the strong inhomogeneous broadening, which could be expected for amorphous films as well as γ-crystallized films due to the highly disordered structure of γ-alumina [9]. Furthermore, the high $^5D_0$ to $^7F_0$ and $^5D_0$ to $^7F_2$ intensities relative to that of $^5D_0$ to $^7F_1$ indicate that the Eu^{3+} ions are located in low-symmetry sites, as in the case of glasses [10]. The fluorescence decay, recorded at 620 nm is approximately exponential with a lifetime of 1.9 ms. On the excitation spectrum (Figure 3), recorded for the 620 nm $^5D_0$ to $^7F_2$ emission, the transitions to the $^5L_6$ and $^5G_{2,4,6}$ are observed between 375 and 400 nm but the poor resolution and sensitivity of the experimental set-up did not enable to detect the ^5D_J levels. The band observed at 260 nm is the charge transfer band Eu-O.

The three spectra show small differences indicative of changes in the ions environment. In order to get more information, site selective experiments were achieved. We restrict our attention here on the comparison of the AE-γ and AE-a films. These two films differ in their structure: the film labelled AE-γ is deposited on SiO_2/Si and has the γ-alumina structure, according to X ray diffraction with crystallites of about 15 nm as seen by transmission Electron Microscopy [11]. The film labelled AE-a is deposited on SiO_2 (Herasil®) and appears as amorphous by X ray diffraction. Although they were deposited at the same temperature, this structure difference may be explained by a temperature difference, induced by the nature of the substrate, sufficient to cross-over the narrow temperature range where crystallization takes place.

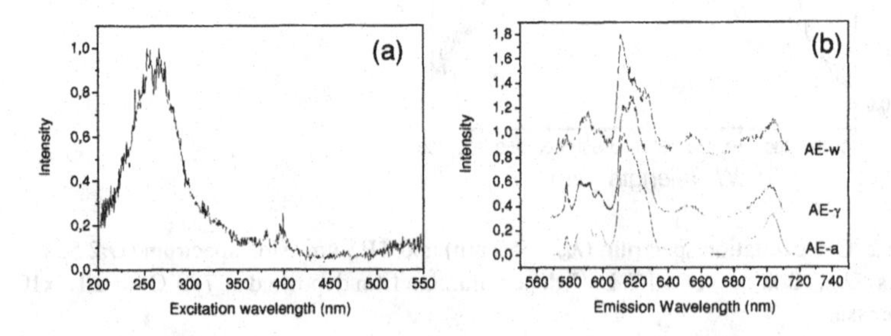

Figure 3. Eu^{3+} doped alumina films deposited at 790°C for a 0.1 mbar oxygen pressure
(a) Excitation spectrum (λ_{em}=620 nm) with the charge transfer band at 260 nm
(b) $^5D_0$ to ^7F_J emission spectra (λ_{exc}=465 nm). AE-γ has the γ-alumina structure, AE-a is amorphous, AE-w is 0.5% (instead of 1%) doped and is excited in waveguided configuration.

Site selective excitation into the $^5D_0$ level at 10K leads to a different evolution of the non-resonant emission spectra recorded in the $^5D_0$ to $^7F_1$ region, where the three $^7F_1$ levels get resolved (Figure 4).

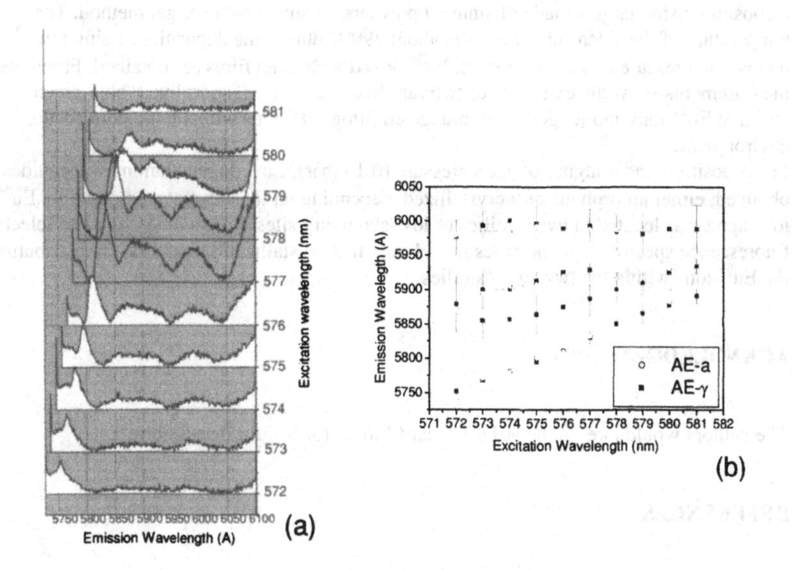

Figure 4. (**a**) FLN emission spectra in the $^5D_0$ to $^7F_1$ region for successive excitations into the $^5D_0$ level of 1% Eu^{3+} doped γ crystalized AE-γ alumina film deposited at about 790°C for 0.1 mbar oxygen pressure (**b**) FLN diagram of the emission wavelength versus excitation wavelength for the 1% Eu3+ doped alumina films, amorphous or γ crystallized.

Fluorescence line narrowing is observed, at least for the two high energy lines, the third line remains at the same wavelength (598.5 nm) with the same width (200 cm^{-1}) as for the inhomogeneous spectrum. The position of the emission lines as a function of the excitation wavelength within the inhomogeneously broaden $^5D_0$ level are given in Figure 5. A decrease in the crystal field strength is observed when the ions are excited in the low energy wing of the $^7F_0$ to $^5D_0$ excitation line, as often observed for Eu^{3+} doped disordered materials [12]. The two films exhibit different evolution: for the crystallized AE-γ film, the first line is clearly broad and asymmetric and the second line consist in two components. This shows that two sites families are present. For the amorphous AE-a film, the contribution of the two sites families are different: the first line is narrower and the contribution of the low-energy component of the second line is clearly preponderant. It appears thus that the crystallization induces a redistribution of the Eu^{3+} ions within the two families of sites. The reason of this behavior is not known at the present time and needs to be determined. It could be related to the influence of crystallites size, as we know that Al_2O_3 films, amorphous according to X ray diffraction, may contain nanocrystallites detectable by transmission electron microscopy, as observed with sol-gel films [6].

CONCLUSION

Optical waveguides of europium doped alumina were prepared by Pulsed Laser Deposition using targets made of sintered powders, obtained by a sol-gel method. The temperature of the silica substrates was about 790°C during the deposition. Using low deposition pressure (about 10^{-5} mbar), Eu^{2+} doped γ-alumina films are obtained. Fluorescence measurements show the existence of two families of Eu^{2+} ions: short lived, blue-green emitting Eu^{2+} ions and longer lived, orange emitting Eu^{2+} ions with a more perturbed environment.
For deposition under higher oxygen pressure (0.1 mbar), Eu^{3+} doped alumina waveguides are obtained, either amorphous or γ-crystallized, depending on the substrate temperature. Eu^{3+} ions appear as located in two families of low-symmetry sites. Low temperature site selective fluorescence spectroscopy indicates in addition, that crystallization induces a redistribution of the Eu^{3+} ions within the two sites families.

ACKNOLEDGEMENTS

The authors would like to thank Dr Bernard Moine for fruitful discussions.

REFERENCES

1. G. N. van den Hoven, R. J. I. M. Koper, A. Polman, C. van Dam, J. W. M. van Uffelen, M. K. Smit, *Appl. Phys. Lett.* **68** 1886 (1996)
2. R. Serna, C. N. Afonso, *Appl. Phys. Lett.* **69** 1541 (1996)
3. S. P. Feofilov, A. A. Kapianskii, R. I. Zakharchenya, *J. Lum.* 72-74 41 (1997)
4. A.Pillonnet, C. Garapon, C. Champeaux, C. Bovier, R. Brenier, L. Lou, A. Catherinot, B. Jacquier, J. Mugnier, *J. Phys. IV* **9** Pr 5 (1999)
5. A. Pillonnet, C. Garapon, C. Champeaux, C. Bovier, R. Brenier, H. Jaaffrezic, J. Mugnier, *Appl. Phys. A* **69** S735 (1999)
6. A.Pillonnet-Minardi, O. Marty, C. Bovier, C. Garapo, J. Mugnier, Opt. Mater. **16** 9 (2001)
7. G. Blasse, W. L. Wanmaker, J. W. ter Vrugt, A. Bril, *Philips res. Rep.* **23** 189 (1968)
8. W. Jia, H. Yuan, L. Lu, H. Liu, W. M. Yen, J. Lum. **76-77** 424 (1998)
9. R. S. Zhou, R. L. Snyder, *Acta Cryst. B* **47** 617 (1991)
10. G. Pucker, K. Gatterer, H. P. Fritzer, M. Bettinelli, M. Ferrari, *Phys. Rev. B* **53** 6225 (1996)
11. A.Minardi, O. Marty, C. Champeaux, C. Garapon, (to be published)
12. Th. Schmidt, R. M. Macfarlane, S. Völker, *Phys. Rev. B* **50** 15707 (1994)

Mat. Res. Soc. Symp. Proc. Vol. 667 © 2001 Materials Research Society

Electrical and Optical Studies of the Organic Thin Film Devices Produced by Cluster Beam Deposition Methods

J.Y. Kim, E.S. Kim and J.-H. Choi*
Department of Chemistry and Center for Electro- and Photo-Responsive Molecules, Korea University, 1, Anam-dong, Seoul 136-701, Korea

ABSTRACT

The neutral and ionized cluster beam deposition (NCBD and ICBD) methods have been applied to fabricate the new double-layer organic light emitting devices (OLEDs) with the structure of indium-tin-oxide (ITO)-coated glass/spin-coated poly[2-(N-carbazolyl)-5-(2-ethyl-hexyloxy)-1,4-phenylenevinylene] (CzEH-PPV)/8-hydroxyquinoline aluminum (Alq_3)/Li:Al. The surface morphology profiles measured by atomic force microscopy (AFM) showed that the cluster beam deposition methods are efficient in producing uniform and smooth film surfaces. Photo- and electro-luminescence (PL, EL) spectroscopies demonstrated that while the new hole transporting medium CzEH-PPV is susceptible to the bombardment of energetic ionized beam, the introduction of the neutral buffer layer significantly improves the device characteristics, i.e., lower threshold and turn-on voltages and higher external quantum efficiency (EQE). In addition, the effect of doping of highly fluorescent dye (DCM) into Alq_3 layer showed a complete energy transfer, color-tuning capability and higher EQE compared to the undoped devices.

INTRODUCTION

The recent progress made in the fabrication of efficient light emitting devices utilizing organic and polymeric materials is of extreme importance in developing future flat-panel display applications [1,2]. In preparing such high-quality functional devices, the acquisition of smooth and uniform thin films to form well-defined heterojunctions is one of the important prerequisites. For low molecular weight molecules the most common technique is the simple physical vapour deposition(PVD) method. Another less popular but promising scheme is applying cluster beam deposition methods [3,4]. Neutral and ionized cluster beams have been widely used in gas-phase reactions to understand the effect of solvation and intermolecular interactions at the microscopic level. The advantages of employing such cluster beam are the high directionality and translational kinetic energy obtained when the gas molecules undergo adiabatic expansion in high vacuum. Since the clusters are composed of weakly bound molecules, the collision of cluster beam with the substrate induces facile fragmentation into individual molecules followed by active surface migration resulting in more uniform film surfaces. In particular, the electric charge of the ionized cluster beam can provide more adjustable deposition parameters such as ionizing current, ionization and acceleration voltages, ion current density, etc.

In this paper, we describe the fabrication and characterization of the two-layer OLEDs utilizing CzEH-PPV, a newly synthesized derivative of a prototypal conjugated polymer PPV, as a hole transport layer and Alq_3 as an electron transport layer which is deposited by either NCBD or ICBD methods. The molecular and device structures are shown in Figure 1. In addition, to increase the EQE and tune the emitting colour the highly fluorescent dye dopant DCM has been co-deposited with Alq_3 and the dopant effect was investigated.

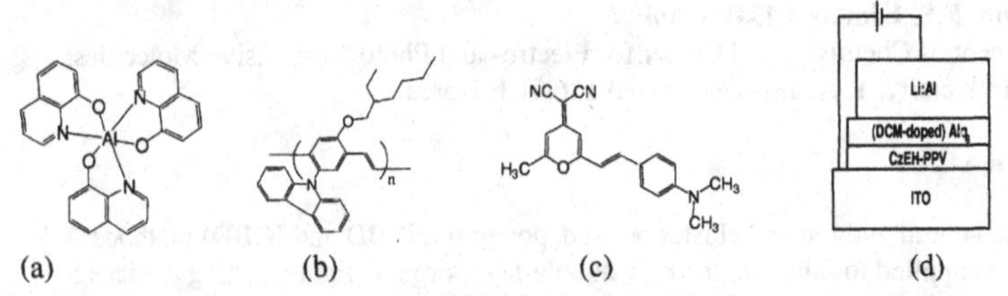

Figure 1. Structures of (a) Tris(8-hydroxy)quinoline aluminum (Alq_3), (b) Poly[2-(N-cabazolyl)-5-(2-ethylhexyloxy)-1,4-phenylenevinylene] (CzEH-PPV) and (c) 4 (dicyanomethylene)-2-methyl-6-(*p*-dimethyl aminostyryl)-4H-pyran (DCM), (d) structure of device: ITO (1,500 Å)/CzEH-PPV (600 Å)/(DCM-doped) Alq_3 (700 Å)/Li:Al (1,500 Å).

EXPERIMENT

CzEH-PPV was synthesized by the procedure described elsewhere [5] and was dissolved in 1,1,2,2-tetrachloroethane(15.38 mg/mL) and spin-coated on the ITO substrate to the typical thickness of about 600 Å. Commercial Alq_3 was deposited by the cluster beam deposition apparatus, which consists of the evaporation crucible cells, the ionization and extraction electrodes, the drift region, and the substrate. The chamber was pumped by a 10-inch baffled diffusion pump and the average pressure was maintained below 1×10^{-6} Torr. The source material Alq_3 placed inside the enclosed cylindrical crucible cell with a nozzle (1.0-mm diameter, 1.0-mm long) was evaporated by the resistive-heating of the cell at 620 K and expanded through the nozzle into the electrode region in high vacuum. As the gas molecules expanded from a high-pressure cell to the vacuum chamber, the gas condensation led to the formation of weakly-bound neutral molecular clusters.

In the NCBD scheme, after travelling through the drift region the neutral clusters were directly deposited onto the bare ITO- or CzEH-PPV spin-coated ITO-glass substrates. Normally seven substrates were simultaneously deposited to the thickness of about 700 Å and the temperature was kept at room temperature throughout the deposition process. The growth rate of the film was typically about 6.0 nm/min. Finally, Li:Al alloy as a low work function cathode was deposited by a vapor deposition. In the ICBD scheme, some fraction of the neutral clusters underwent ionization by the electron impact source. The source consisted of a cylinder-shaped grid anode (65.0-mm diameter, 20.0-mm high) surrounded by a filament cathode (0.20 mm diameter tungsten wire). The extent of ionization was adjusted by changing the cathode emission current (the cathode typically operated at 60 V with an emission current of 16 mA). The partially ionized cluster beam passed through an extraction electrode, where the beam was accelerated to travel the drift region and deposited onto the substrate described above.

For DCM dye (Exciton Co.) doping, DCM and Alq_3 were loaded in two separate crucibles, heated, and co-deposited onto the substrates by the same procedures described above. The mole fraction of DCM dopant (~ 0.5 mol %) could be determined from the previously known PL spectra, in which the maximum emission wavelength was found to be a sensitive measure of the doping level [2]. The PL and EL spectra were obtained using Amino-Bowman II spectrometer. The thickness and surface morphology of the films were measured by an alpha step surface profile monitor (Tencor Co.) and atomic force microscopy (PSI Co.), respectively. Device characteristics such as current density vs. voltage (J-V), EL intensity vs. voltage (L-V) and external quantum efficiency vs. current (EQE-I) curves were also obtained. The highest occupied and lowest unoccupied molecular orbitals (HOMO and LUMO) of CzEH-PPV were determined by UV photoelectron (ESCALab 220i spectrometer) and optical spectroscopies.

DISCUSSION

1. Surface morphology

Characterization studies of the Alq_3 films prepared by the NCBD, ICBD and PVD methods have been performed by recording AFM images and PL spectra. The AFM images and the root-mean-square roughness estimates (R_{rms}) of the films provided insight into the overall morphology profiles. Typically the R_{rms} for the ITO-glass substrate and spin-coated CzEH-PPV film were measured to be ~31 and ~17 Å, respectively. The measurements made for several different sections of the Alq_3 films showed that compared to the films prepared by NCBD (~21 Å) and PVD (~29 Å), the most uniform flat surface was provided by the ICBD (~14 Å at the acceleration voltage V_a of 250 V and ion current I_s of 30 nA). For the ICBD films the surface roughness has been found to strongly depend on the deposition parameters such as V_a and I_s, and decreases gradually as V_a and I_s increase. The lower roughness for NCBD and especially ICBD films suggests that after colliding with the substrate the weakly bound and highly guided cluster beam is quite efficient in fragmentation into individual energized molecules followed by immediate transforming of the energy into the surface migration energy leading to smooth and uniform thin films.

2. Characterization of devices without dopant molecules

Figure 2 (a) shows the typical normalized PL and EL spectra of undoped devices. The PL spectra with the excitation wavelength (λ_{ex}) of 400 nm did not depend on the deposition schemes and demonstrated that the fluorescence emission occurred in both CzEH-PPV (peak wavelengths λ_{pk} of 525 and 570 nm) and Alq_3 (λ_{pk} = 535 nm) layers. The profiles of EL were identical to those of PL spectra, indicating that the exciton formation and radiative process also occurred in both layers. Figure 2 (c) shows the schematic energy diagram for the devices. The HOMO and LUMO levels of CzEH-PPV determined by UV photoelectron and optical spectroscopies were found to be located very close to those of Alq_3. Therefore, it was expected that the recombination fabricated with the Alq_3 layer prepared through the ICBD scheme, however, the overall EL intensity was observed to fall off significantly, as can be seen in the poor

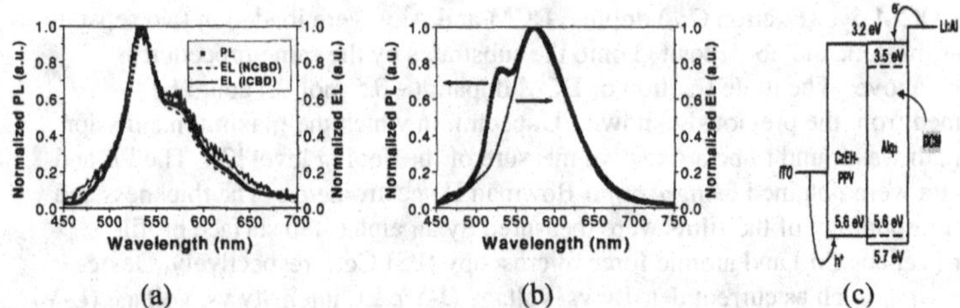

(a) (b) (c)

Figure 2. Comparison of the normalized PL and EL spectra of (a) undoped and (b) DCM-doped devices, and (c) schematic diagram of the energy level.

signal-to-noise ratio in Figure 2 (a). This can be attributed to the damage in the interface region caused by the bombardment of energetic ionic particles, which resulted in the formation of trap states promoting non-radiative processes. Similar phenomena were also observed in the recent OLED experiments by Usui and co-workers, in which the polyvinylcarbazole layer was susceptible to the radiation damage due to the accelerated ion beams [6].

To perform the systematic investigations of the effect of the neutral and ionic cluster beams on the polymeric layer and the device characteristics, the Alq_3 layer was deposited using a two-step scheme. In our scheme the first layer on the CzEH-PPV layer was deposited as a buffer layer using the NCBD method to avoid the direct ion bombardment and afterward the ICBD layer was added at V_a of 70 and 250 V, respectively. The relative ratio of the NCBD and ICBD layers for the total 700 Å thick film was changed as follows; NCBD(100%), NCBD:ICBD=6:1, 2:1, 1:6, and ICBD (100%). As the fraction of the ICBD layer and V_a increased, lower luminance and higher threshold voltage (defined at 10^{-4} $\mu W/cm^2$: Table I) were observed. The phenomena might be understood that the neutral buffer layer improved the luminance characteristics of the devices through protecting the interface region from the direct ion radiation. Furthermore, since the more energetic ion beam was expected to have a deeper propagation depth, the thicker buffer layer was required for higher V_a. The effect of ionic species on the device characteristics is also reflected in the EQE plot in a similar manner. Figure 3 represents the EQE-I characteristics of the undoped devices prepared at V_a of 70 V. As the fraction of the neutral buffer layer increased and V_a decreased, the device quality significantly improved. In particular the EQE prepared solely by the NCBD scheme has been improved by about a factor of 3 to 4 times compared to that prepared solely by the ICBD method.

Table I. Threshold voltages of undoped and DCM-doped devices.

Condition	Undoped devices		DCM-doped devices	
	V_a = 250 V	V_a = 70 V	V_a = 250 V	V_a = 70 V
NCBD (100%)	7.0 V	7.0 V	6.0 V	6.0 V
NCBD:ICBD=6:1	8.7 V	4.5 V	10.0 V	7.5 V
NCBD:ICBD=2:1	8.3 V	6.5 V	11.6 V	8.5 V
NCBD:ICBD=1:6	12.8 V	12.0 V	12.0 V	13.5 V
ICBD (100%)	14.0 V	13.0 V	12.5 V	13.0 V

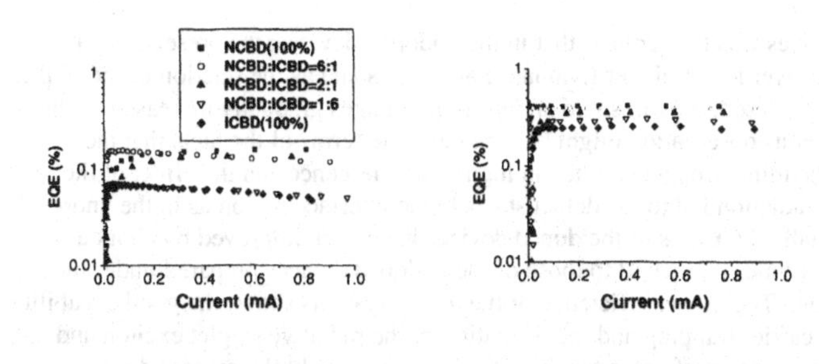

(a) (b)

Figure 3. EQE-I curves of (a) undoped and (b) DCM-doped devices as varying the ratio of NCBD/ICBD thickness of Alq_3 layer at V_a=70V.

3. Characterization of devices with dopant molecules

To increase the external quantum efficiency and tune the emission color of the EL devices, highly fluorescent organic dopant molecules such as DCMs were co-deposited in the Alq_3 layer. Due to the interdopant quenching in the dye-doped thin films, the typical optimum DCM concentration was found to be in the range of 0.25 to 1.0 mol% [2]. In our experiment 0.5 mol% was chosen and the strong fluorescence was observed in the region of 575-590 nm. Figure 2(b) shows the typical strong PL (λ_{ex} = 400 nm) and EL spectra of the DCM-doped OLED systems. The PL spectra did not show any dependency on the deposition schemes again as in the undoped cases and displayed the similar profile in terms of the luminescence emission occurring in both polymeric and organic layers. One conspicuous spectral feature observed in the PL spectra was the enormous increase of the DCM photoluminescence at about 580 nm, indicating that the efficient energy transfer process occurred from the Alq_3 matrix to the dopant molecules. More dramatic dopant effect was demonstrated in the EL spectra. Strong EL was observed only from the dye dopant molecules for all types of the doped devices. Such characteristic processes can be understood in terms of the carrier trapping and the rapid energy transfer. Figure 2(c) shows the energy diagrams of the DCM-doped devices. Since the HOMO and LUMO levels of the dopant (5.6, 3.5 eV) are located inside those of Alq_3 and CzEH-PPV, the dopant molecules are expected to act as both charge carrier traps and efficient energy acceptors from the Alq_3 matrix [7]. Therefore, in the PL spectra most of the emission in the organic layer resulted from the dye dopant with strong fluorescence intensity at around 580 nm. In addition for the EL processes the energy diagram suggests that the electrons are favourably trapped by the dopant in the organic layer, which leads to the formation of exciton followed by rapid radiative process. The significantly improved signal-to-noise ratio in the EL spectrum clearly shows the high efficiency of such processes.

The same type of the systematic investigations of the effect of the cluster beams on the polymeric layer and the device characteristics were performed as those in the undoped systems at V_a of 70 and 250 V, respectively. The overall trend observed in the doped devices was the same as that in the undoped devices; the presence of the NCBD buffer layer led to higher luminance and EQEs and as the fraction of the ICBD layer and V_a increased, the threshold and turn-on voltages gradually increased (Table I). The analogous observation might be understood in terms of the facts that the presence of the minor dopant resulted in higher luminescence and the effect of the energetic ion radiation led to the defect states in the interface region as in the undoped cases. Especially the EQEs of the doped devices have been improved by about a factor of 2 to 4 times compared to those of the undoped devices prepared under the same condition. The higher efficiency of the doped systems shows the good capability of the charge carrier trapping and the formation of the radiative singlet exciton and suggests the fabrication of color-tunable OLEDs using highly fluorescent dopants.

CONCLUSION

We have fabricated the new double-layer OLEDs with the structure of ITO-glass /CzEH-PPV/(DCM-doped)Alq_3/Li:Al by applying NCBD and ICBD methods. The investigations of AFM, PL, EL and device characteristics demonstrated that while the smoother surface morphology is obtained from the ICBD scheme, the new hole transporting medium CzEH-PPV is susceptible to the ion radiation damage and therefore, the NCBD–based OLEDs show better device performance, i.e., lower threshold and turn-on voltages and higher EQEs. In addition, highly fluorescent DCM dye was doped into Alq_3 layer and the doped devices showed a complete energy transfer, colour-tuning capability and higher EQE compared to the undoped devices.

ACKNOWLEDGEMENTS

We would like to thank Prof. J.I. Jin for providing the polymer samples. J.Y. Kim and E.S. Kim wish to acknowledge support of BK21 fellowships. This work was financially supported by the CRM-KOSEF (2001).

REFERENCES

1. R. H. Friend, et al., *Nature*, **397**, 121 (1999).
2. C.W. Tang, S.A. VanSlyke, and C.H. Chen, *J. Appl. Phys.*, **65(9)**, 3610 (1989).
3. T. Takagi, *Ionized-cluster Beam Deposition and Epitaxy*, Nyes Publication, Park Ridge, N.Y. (1988).
4. H. Usui, H. Kameda, K. Tanaka, and H. Orito, *Thin Solid Films*, **288**, 229 (1996).
5. K. Kim and J.I. Jin, manuscript in preparation.
6. H. Usui, K. K. Tanaka, H. Orito, S. Sugiyama, *Jpn. J.Appl.. Phys.*, **37**, 987 (1998).
7. M. Uchida, C. Adachi, T. Koyama, Y. Taniguchi, *J. Appl. Phys.*, **86**, 1680 (1999).

Mat. Res. Soc. Symp. Proc. Vol. 667 © 2001 Materials Research Society

Polygermyne: Germanium sheet polymers with efficient near-infrared luminescence

Günther Vogg[1], Martin S. Brandt[1], Lex J.-P. Meyer[1], Martin Stutzmann[1], Zoltán Hajnal[2], Bernadett Szücs[2], Thomas Frauenheim[2]
[1]Walter Schottky Institut, Technische Universität München, Am Coulombwall, D-85748 Garching, Germany.
[2]Theoretische Physik, Universität Paderborn, Warburger Str. 100, D-33095 Paderborn, Germany.

ABSTRACT

The structural, optical and electronic properties of the Ge sheet polymer poylgermyne are summarized. Prepared via topotactic transformation of Zintl-phase $CaGe_2$, $(GeH)_n$ forms a layered crystal in a tr6 stacking sequence with a distance of 5.65 Å between adjacent layers. The photoluminescence at 1.3 eV is excited nearly resonantly with a Stokes shift of 0.2 eV. Together with band structure calculations this shows that polygermyne has a direct band gap.

INTRODUCTION

The search for materials with a high luminescence efficiency on the basis of silicon has renewed the interest in the crystalline sheet polymer siloxene [1] and its derivatives. The layered crystal structure of this material based on the 3-fold coordination of Si backbone atoms leads to an electronic bandstructure with a direct bandgap and an intense luminescence with external quantum efficiencies of up to 10% [2-4]. Siloxene $(Si_2HOH)_n$ as well as the oxygen-free counterpart polysilyne $(SiH)_n$ [5] are formed in a topotactic reaction from the layered Zintl-phase $CaSi_2$ which contains extended puckered Si layers, similar to the {111} layers of crystalline Si and separated from each other by planar monolayers of calcium. Whereas Ca is removed during the deintercalation reaction in concentrated hydrochloric acid, the Si backbone remains unchanged since the fourth sp^3-valence of each Si atom pointing out of the layers is stabilized by H- or partly by OH-groups, leading to the formation of polysilyne or siloxene, respectively.

Similar sheet polymers based on germanium have only been described in the literature recently [6]. In this report we summarize the preparation and characterization of polygermyne $(GeH)_n$ which is obtained as a layered 2-dimensional Ge sheet polymer from epitaxially grown $CaGe_2$. Polygermyne is found to be the direct counterpart of polysilyne exhibiting semiconducting properties and strong infrared photoluminescence.

GROWTH AND STRUCTURAL PROPERTIES

Epitaxial $CaGe_2$ films were grown by reactive deposition epitaxy (RDE) of calcium on crystalline Ge(111) substrates as described in [7]. The film thickness was about 1μm. For the preparation of polygermyne, the $CaGe_2$ films were immersed into concentrated aqueous HCl for 12 hours at a temperature of -30°C, and then rinsed in acetone. X-ray diffraction (XRD) shows that the RDE growth of $CaGe_2$ leads to epitaxial thin films with an even higher crystalline quality than similar $CaSi_2$ films [7,8] which is also confirmed by scanning electron microscopy (SEM).

Figure 1. Scanning electron microscope side and top view of epitaxial $CaGe_2$ films on Ge(111).

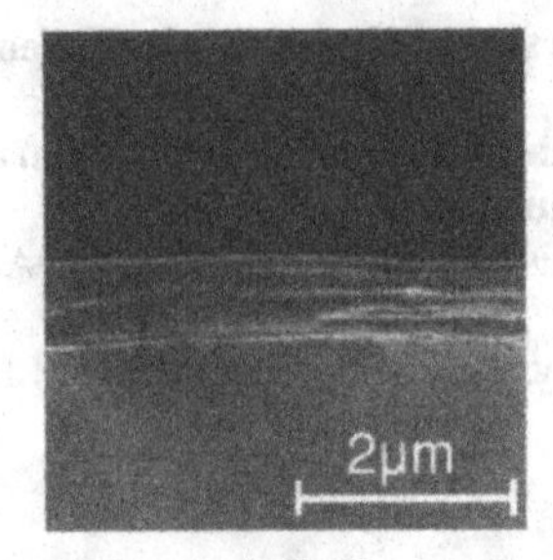

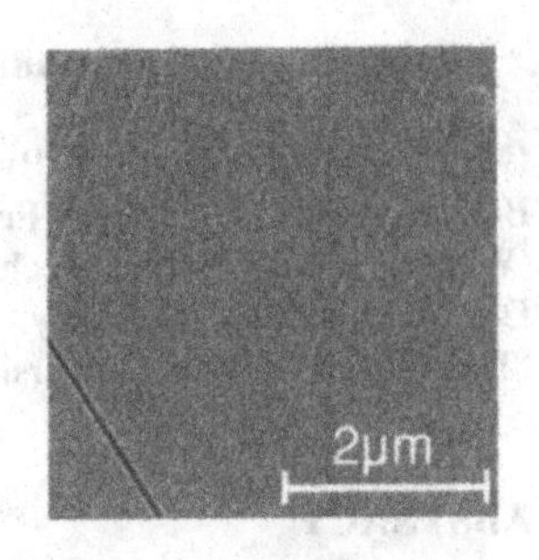

Figure 1 shows micrographs of the cleaving edge as well as the surface of a 1µm thick $CaGe_2$ film on Ge(111), respectively. $CaGe_2$ crystallizes in a trigonal-rhombohedral structure with a 6-fold stacking sequence of the layers (spacegroup $D_{3d}^5 - R\bar{3}m$ (no. 166), $c = 30.6$Å, $a = 4.01$Å), as shown in the left part of figure 2. The stacking as well as the spacegroup are preserved during the topotactic transformation into polygermyne (right part of figure 2). The c lattice constant in the sheet polymer increases to 33.9 ± 0.1 Å, the lattice constant a slightly decreases to 3.98 ± 0.02 Å [6]. The distance of the Ge layers is 5.65 ± 0.02 Å, which is in good agreement with the respective value observed in polysilyne (5.6Å [4]). This low c value indicates that only spurious quantities of OH groups are bonded to the Ge backbone. Energy dispersive X-ray analysis (EDX) indeed shows that less than 5 at% oxygen is present in polygermyne. This is further confirmed by the missing of strong oxygen-related modes in the respective infrared (IR) absorption spectra [6]. The Hydrogen effusion measurements in figure 3 show that polygermyne contains a similar amount of hydrogen as siloxene. The hydrogen is desorbed from polygermyne at significantly lower temperatures due to the lower binding energy of Ge-H compared to Si-H. The position of the desorption peak at 250°C is characteristic for hydrogen bonded to internal Ge surfaces [9] as expected for the layered structure of $(GeH)_n$.

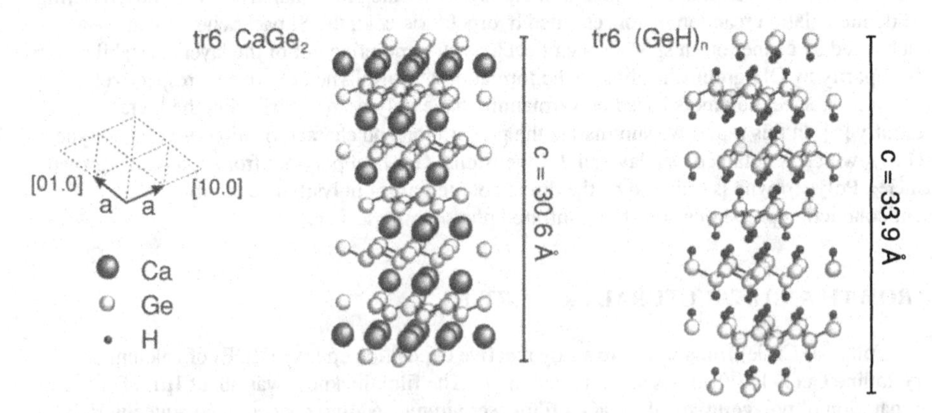

Figure 2. Crystal structure and stacking sequence of $CaGe_2$ and polygermyne. The structure is trigonal rhombohedral (spacegroup $D_{3d}^5 - R\bar{3}m$ (no. 166)). The position of the H atoms has not been refined. Instead, a typical Ge-H bond length of 1.58 Å has been used.

Figure 3. Thermally induced desorption of hydrogen from 1μm thick epitaxial polygermyne and siloxene films. The temperature gradient was 10°C/min.

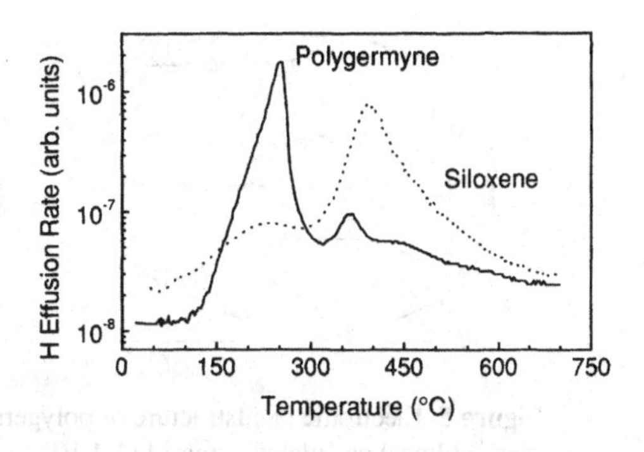

ELECTRONIC BANDSTRUCTURE AND OPTICAL PROPERTIES

A comparison of the absorption, photoluminescence (PL) and photoluminescence excitation (PLE) properties of polygermyne and siloxene is shown in figure 4. As a measure of the absorption, (1-Reflection) as determined from measurements of the diffuse reflection with an Ulbricht sphere in a UV-VIS spectrometer is used. The PL of siloxene and polygermyne was excited by an Ar^{+}-ion laser with 363 and 457 nm wavelength, respectively. For the determination of the efficiency of the PL excitation, the PL was excited with light from a tungsten lamp filtered by a monochromator. While the PL maximum is observed at 2.4 eV for siloxene, polygermyne has its PL maximum in the near infrared at 1.3 eV. In both Si and Ge sheet polymers, a nearly resonant excitation of the PL is found, with a Stokes shift between the PL and PLE maxima of 0.4 and 0.2 eV, respectively. Together with a radiative lifetime of about 10 ns and a polarisation memory of the PL on the same timescale, the resonant PLE has been taken as clear evidence that siloxene has an electronic bandstructure with a direct bandgap [10]. The strong similarity of the

Figure 4. Optical properties of epitaxial polygermyne and siloxene films as determined by photoluminescence (PL), photoluminescence excitation (PLE) as well as diffusive reflection measurements.

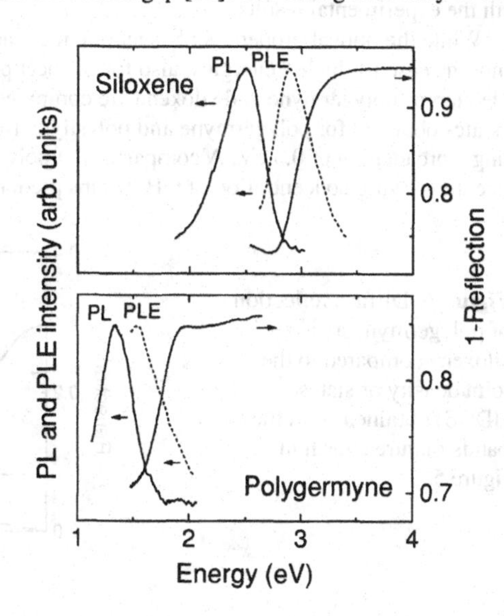

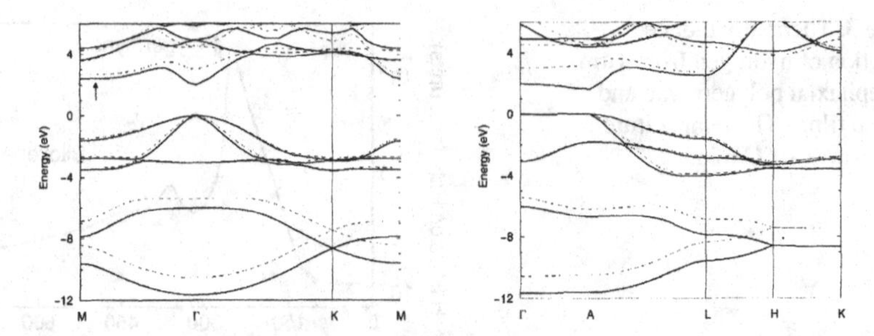

Figure 5. Electronic bandstructure of polygermnye (solid lines) and polysilyne (dashed lines) calculated using LDA-DFT.

optical properties of siloxene and polygermyne shown in figure 4 with an even reduced Stokes shift in the case of polygermyne therefore indicates that the Ge sheet polymer also has a direct bandgap, however at about 1.5 eV.

This conclusion is confirmed by bandstructure calculations using the local density approximation (LDA) of the density functional theory (FHI98MD code) applying a scissors operator of 0.7 eV to Ge and 0.6 eV to Si polymers [11]. Figure 5 shows a comparison of the electronic bandstructures obtained for polysilyne and polygermyne in a simple h1 stacking sequence. For polysilyne, a bandstructure with an indirect band gap of 2.7 eV (minimum in the conduction band indicated by the arrow) is predicted, in agreement with earlier DFT calculations [12]. For polygermyne, a bandgap of 1.7 eV is found with the minimum of the conduction band at the A point in the Brillouin zone, which is folded back to the zone center in the experimentally found tr6 modification leading to the prediction of a direct bandgap of 1.7 eV in good agreement with the experimental results.

While the optical properties of sheet polymers near the band gap are studied rather well, this is not the case for higher energies, also for Si sheet polymers. As a first attempt, the diffuse reflection of polygermyne and siloxene are compared in figure 6 to the theoretical joint density of states obtained for polygermyne and polysilyne from the band structure shown in figure 5 using a broadening of 0.3 eV. A comparison of polysilyne and siloxene is allowed in this case since the varying concentration of OH-groups predominantly influences the band gap while the

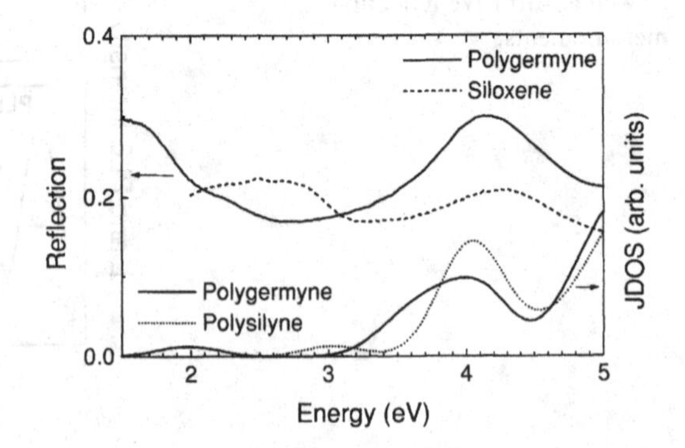

Figure 6. Diffuse reflection of polygermyne and siloxene compared to the joint density of states (JDOS) obtained from the bandstructure shown in figure 5.

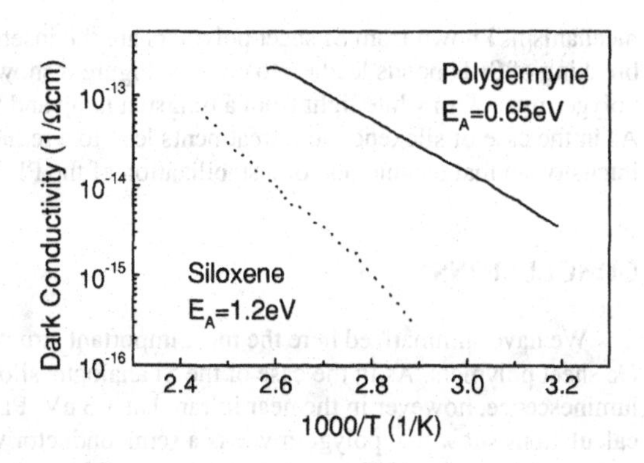

Figure 7. Dark conductivity of epitaxial polygermyne and siloxene films, respectively. An Ag top contact on the epitaxial films was used as the front contact, the Si or Ge substrates were used as back contacts.

optical properties at higher energies are governed by the backbone. The strong absorption around 4 eV expected from the parallel lowest conduction and highest valence bands between the Γ and M and the A and L points in the Brillouin zone is experimentally observed as a peak in the reflection. The maximum in the reflection of siloxene is shifted by 0.2 eV to higher energies with respect to that in polygermyne. While exact matrix elements have not been used for the prediction of the optical properties, the similar shift of 0.2 eV between the maxima in the JDOS of polygermyne and polysilyne adds further confidence in the validity of the band structure calculations.

ELECTRICAL PROPERTIES AND STABILITY

In figure 7, the thermal activation of the dark conductivity of Si- and Ge-based sheet polymers is compared. The activation energies found are 0.65 ± 0.05 eV for polygermyne and 1.2 ± 0.1 eV for siloxene. These values are about one half of the band gaps of these materials as obtained from optical characterization. This indicates that the Fermi-level is near the middle of the band gap so that the sheet polymers prepared via topotactic transformation of epitaxial Zintl-phases are undoped semiconductors.

While the layered structure of the sheet polymers leads to changes in the electronic bandstructure which are desired for materials used to construct light emitting devices, this structure also makes the polymers more vulnerable to degradation. Possible degradation

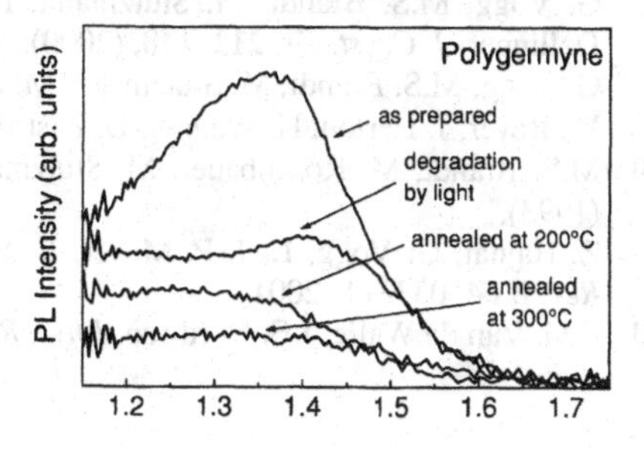

Figure 8. Potoluminescence intensity of epitaxial polygermyne after prolonged illumination and thermal anneal.

mechanisms known from Si sheet polymers are the insertion of oxygen into the backbone and the breaking of Si-H bonds leading to defects. Figure 8 shows the effect of prolonged illumination of polygermyne with white light from a tungsten lamp and thermal anneal in ambient atmosphere. As in the case of siloxene, both treatments lead to a reduction of the photoluminescence intensity, so that techniques for a stabilization of the PL appear necessary.

CONCLUSIONS

We have summarized here the most important structural, electronic and optical properties of Ge sheet polymers. As in the case of the Si analogue siloxene, polygermyne has a very efficient luminescence, however in the near infrared at 1.3 eV. PLE measurements and band structure calculations show that polygermyne is a semiconductor with a direct band gap. Since crystalline and amorphous Si and Ge are completely miscible, it should be possible to produce mixed SiGe sheet polymers which should also have a direct band structure and should allow to tune the emission continuously. Recent experiments have shown that this is indeed the case, allowing to obtain group-IV sheet polymers with luminescence maxima between 2.4 and 1.3 eV.

ACKNOWLEDGEMENTS

The authors acknowledge the financial support of the Deutsche Forschungsgemeinschaft through Schwerpunktprogramm "Silicium-Chemie" (Br 1585/2-3 and Fr 889/9-3).

REFERENCES

[1] F. Wöhler, *Lieb. Ann.* **127**, 257 (1863).
[2] I. Hirabayashi, K. Morigaki, S. Yamanaka, *J. Non.-Cryst. Solids* **59&60**, 645 (1983).
[3] M.S. Brandt, H.D. Fuchs, A. Höpner, M. Rosenbauer, M. Stutzmann, J. Weber, M. Cardona, H.J. Queisser, *Mat. Res. Soc. Symp. Proc.* **262**, 849 (1992).
[4] U. Dettlaff-Weglikowska, W. Hönle, A. Molassioti-Dohms, S. Finkenbeiner, J. Weber, *Phys. Rev. B* **56**, 13132 (1997).
[5] G. Schott, *Z. Chemie (Leipzig)* **6/7**, 194 (1962).
[6] G. Vogg, M.S. Brandt, M. Stutzmann, *Adv. Mat.* **12**, 1278 (2000).
[7] G. Vogg, M.S. Brandt, M. Stutzmann, I. Genchev, A. Bergmaier, L. Görgens, G. Dollinger, *J. Cryst. Gr.* **212**, 148, (2000).
[8] G. Vogg, M.S. Brandt, M. Stutzmann, M. Albrecht, *J. Cryst. Gr.* **203**, 570 (1999).
[9] W. Beyer, J. Herion, H. Wagner, U. Zastrow, *Philos. Mag. B* **63**, 269 (1991).
[10] M.S. Brandt, M. Rosenbauer, M. Stutzmann, *Mat. Res. Soc. Symp. Proc.* **298**, 301 (1993).
[11] Z. Hajnal, G. Vogg, L. J. P. Meyer, B. Szücs, M. S. Brandt, T. Frauenheim, *Phys. Rev. B* **64**, 033311 (2001).
[12] C.G. Van de Walle, J.E. Northrup, *Phys. Rev. Lett.* **70**, 1116 (1993).

Nanocrystalline Materials

Mat. Res. Soc. Symp. Proc. Vol. 667 © 2001 Materials Research Society

Luminescence of $Lu_2O_3:Tm^{3+}$ nanoparticles

Celso de Mello Donegá[1,3], Eugeniusz Zych[2], and Andries Meijerink[1]
[1]Debye Institute, Dept. of Physics and Chemistry of Condensed Matter, Utrecht University, Princetonplein 1, 3508 TA Utrecht, The Netherlands.
[2]Faculty of Chemistry, Wroclaw University, 14 Joliot-Curie Street, 50-383 Wroclaw, Poland.
[3]On leave of absence from Dept. of Fundamental Chemistry, UFPE, Recife-PE, Brazil

ABSTRACT

This paper presents for the first time a comparison between the luminescence properties (*viz.* emission and excitation spectra, lifetimes, and concentration quenching) of nanocrystalline and microcrystalline $Lu_2O_3:xTm^{3+}$ (x= 0.1– 5 mol%). The results show that the most important difference between the two size regimes is the higher defect concentration in the nanoparticles. These defects give rise to a broadband emission (λ_{max} = 430 nm), and to partial quenching of the Tm^{3+} emission, in addition to the expected concentration quenching by cross-relaxation between Tm^{3+} ions. The defect concentration seems to be similar in all nanocrystalline samples, so that those with the lowest Tm^{3+} concentrations experience the most pronounced quenching. The nature of these defects is as yet unknown. The local structure around the Tm^{3+} ions is not noticeably different in the two size regimes. No evidence of phonon confinement or quantum size effects was observed in the Tm^{3+} luminescence.

INTRODUCTION

Nanostructured materials have attracted great interest in recent years because their properties are markedly different from those of bulk materials [1,2]. These differences arise from several effects (*viz.* larger surface to volume ratio, quantum confinement of charge carriers and geometrical confinement of phonons), which may have various impacts depending on the material and property under consideration [1,2]. The vast majority of the investigations have been focused on metals and semiconductors, rather than on insulators, even though many important commercial applications involve materials belonging to this latter class (*e.g.* phosphors for lighting and displays, lasers, scintillators, etc.). However, in the last few years there has been a growing interest in the properties of nanocrystalline lanthanide-doped insulating phosphors [3]. The most studied nanoparticulate material is $Y_2O_3:Eu^{3+}$ [3-7], but other well-known phosphors, such as $Y_2SiO_5:Eu^{3+}$ [8], $YVO_4:Eu^{3+}$ [9-10], $Y_2O_3:Tb^{3+}$ [11], and $LaPO_4:Eu^{3+}$ or Ce^{3+},Tb^{3+} [9], prepared in a nanocrystalline form, have also been investigated. In this work we carried out for the first time a comparison between the luminescence of $Lu_2O_3:Tm^{3+}$ in the bulk and nanocrystalline size regimes.

EXPERIMENTAL DETAILS

Nanocrystalline $Lu_2O_3:xTm^{3+}$ (x= 0.1– 5 mol%) samples were prepared by a combustion technique using urea as a fuel and $(Lu,Tm)(NO_3)_3$ as an oxidizer. Typically, 4 mmol of $(Lu,Tm)(NO_3)_3$ and 6 mmol of urea were mixed with 1 mL water and the resultant solution was placed into a preheated furnace at 833 K for about 5 minutes in an ambient air atmosphere. The product was a foamy porous body, which could be easily ground to a fine powder. The microcrystalline samples were obtained by subjecting the nanocrystalline powders to a thermal treatment at 1473 K for 4 h under air, followed by grinding and a second treatment at 1273 K during 10 h. The samples were characterized by diffuse reflectance spectroscopy (Perkin-Elmer Lambda 16) and X-ray powder diffraction (Philips PW1729 X-ray diffractometer, CuKα). Some nanocrystalline samples were also analyzed by transmission electron microscopy (TEM) and selected area electron diffraction (SAED). Luminescence spectra were recorded on a SPEX Fluorolog Spectrophotometer model F2002, equipped with two double grating 0.22 m SPEX monochromators and a 450 W Xe lamp as the excitation source. Luminescence lifetime measurements under 355 nm excitation were performed down to 4.2 K by using a set-up described in detail elsewhere [12].

DISCUSSION

The average crystallite sizes estimated from the TEM images for the combustion-made raw samples ranged from 11-13 nm. The SAED patterns could be univocally indexed to cubic Lu_2O_3. The X-ray powder diffraction patterns recorded for all the investigated samples confirmed the SAED results showing the expected lines for cubic Lu_2O_3, with broader peaks for the nanocrystalline samples. The mean particle diameters were estimated from the full width at half maximum of the peaks using the Scherrer's formula [12], and were found to be ~20 nm for the combustion-made raw samples, in reasonable agreement with the TEM results.

As a representative example, figure 1 compares the emission and excitation spectra of the nanocrystalline and microcrystalline Lu_2O_3:1%Tm^{3+} samples. The emission spectra of all samples under UV excitation consist of several narrow lines, assigned to the $^1D_2 \rightarrow {}^3F_4$ transition of Tm^{3+}. The $^1G_4 \rightarrow {}^3H_6$ emission lines, expected at about 500 nm, are hardly observable. A broad emission band with maximum at 430 nm, which is ascribed to radiative recombination at defects, is also observed.

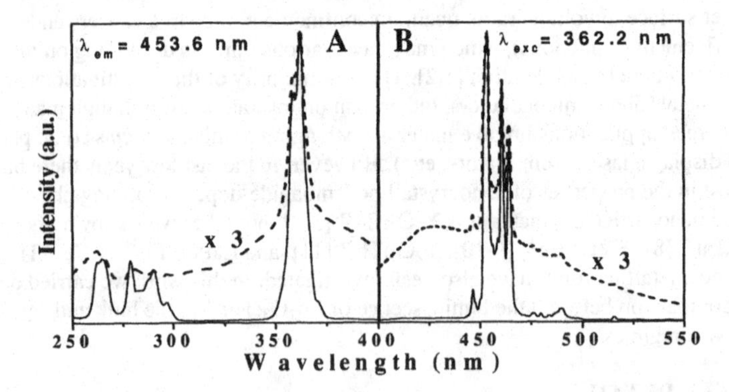

Figure 1. (**A**) The excitation spectra (for λ_{em}= 453.6 nm) and (**B**) the emission spectra (for λ_{exc}= 362.2 nm) of the nanocrystalline (dashed line) and microcrystalline (solid line) Lu_2O_3:1%Tm^{3+} samples at room temperature. The intensities have been normalized to the strongest peak intensity. The spectra are corrected for the instrumental response and were obtained under the same instrumental conditions.

The intensity of the broad band is independent of the Tm^{3+} concentration ($[Tm^{3+}]$) and decreases by one order of magnitude upon particle growth (by thermal treatment) to the microcrystalline size regime, while at the same time the total Tm^{3+} emission intensities increase by a factor ranging from 4 to 13, depending on the $[Tm^{3+}]$ (lower $[Tm^{3+}]$ shows larger enhancements after the thermal treatment). Similarly, the broad band observed in the excitation spectra at about 380 nm, ascribed to defect absorption, is no longer observable after the treatment. These observations can be ascribed to a decrease of the defect concentration after the thermal treatment. We note that the excitation band of the defect luminescence is resonant with the $^3H_6 \rightarrow {}^1D_2$ excitation transition of Tm^{3+}. There are no observable differences between the Tm^{3+} emission (*viz.* spectral position, line widths and branching ratios) in the nanocrystalline and microcrystalline samples, showing that the local structure is the same in both size regimes. Nevertheless, considerable differences in the branching ratios of observed lines can be noted between the excitation spectra of nanoparticles and microparticles, especially for wavelengths shorter than 300 nm ($^3H_6 \rightarrow {}^1I_6, {}^3P_{2,1,0}$ transitions of Tm^{3+}). These differences can be ascribed to competitive absorption by defects. In agreement with most of the results reported for Eu^{3+} in nanocrystalline hosts (e.g. [3-6,8-10]), size effects on the f-f emission spectra are not observed, as expected considering the very localized and shielded nature of the $4f^n$ shell.

Despite the lower emission intensities for the nanocrystalline samples, the concentration dependence of the total Tm^{3+} emission intensity shows a similar behavior for both size regimes: the intensity increases with $[Tm^{3+}]$ up to 1 mol% and then decreases as $[Tm^{3+}]$ increases. This concentration quenching is well-known for Tm^{3+} ions, and results from energy transfer via cross-relaxation in pairs of Tm^{3+} ions [13], in which one ion relax from the $^1D_2$ level to a lower excited level while the other one is excited from the ground $^3H_6$ level to an intermediate excited level. However, if the concentration-normalized intensities (integrated luminescence intensity divided by Tm concentration) are compared, a very important difference becomes evident. Figure 2 shows that in the microcrystalline samples $\Delta I/[Tm^{3+}]$ decreases continuously as the concentration increases, showing that cross-relaxation processes are already active at $[Tm^{3+}]$ above 0.1% mol%. In striking contrast, $\Delta I/[Tm^{3+}]$ in the nanocrystalline samples increases up to 1 mol% Tm^{3+}, and only then decrease with increasing concentration.

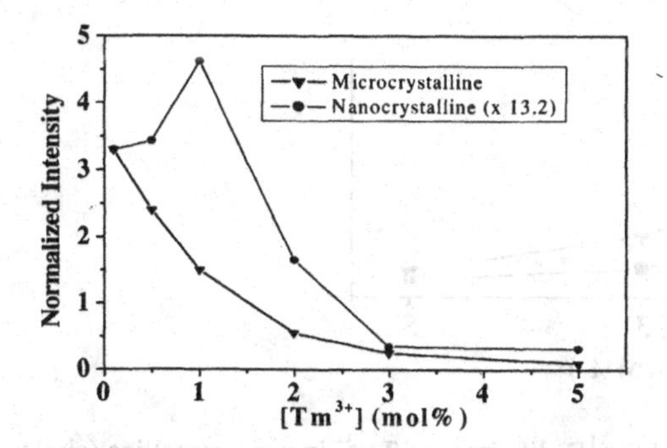

Figure 2. The concentration dependence of the concentration-normalized Tm^{3+} luminescence intensity ($\Delta I/[Tm^{3+}]$) in nanocrystalline (circles) and microcrystalline (triangles) Lu_2O_3:xTm^{3+} samples at room temperature. The solid lines were drawn for clarity. All the spectra were measured under identical instrumental conditions and powder layer thickness.

Although the concentration quenching has been reported to become less effective for Eu^{3+} in certain nanocrystalline hosts (e.g. $Y_2SiO_5:Eu^{3+}$ [8]), this cannot be expected to happen for Tm^{3+}. For Eu^{3+} doped materials the concentration quenching arises as a consequence of energy migration over the $^5D_0$ level of the Eu^{3+} ions, which eventually leads to non-radiative losses at quenching sites (defects or impurities). This energy diffusion process can be greatly reduced in nanoparticles due to the hindering effects of the particle boundaries to energy transfer processes [8]. However, the concentration quenching for Tm^{3+} is a consequence of a single-step energy transfer between neighboring ions over short distances (< 1.0 nm), and thus cannot be strongly affected by the nanoparticle boundaries.

In order to better understand the differences in the concentration dependence of the luminescence intensities of Tm^{3+} in nanoparticles and microparticles, the lifetimes of the $^1D_2$ level were determined at room temperature for all samples. Figure 3 summarizes the results. Except for the 0.1 mol% Tm^{3+} microcrystalline sample, the $^1D_2$ decay transients have a non-exponential beginning. In the microcrystalline $Lu_2O_3:xTm^{3+}$ the $^1D_2$ decay transients become progressively faster and to a higher degree non-exponential with increasing x, already above 0.1 mol%, in line with the observed decrease in the concentration normalized luminescence intensity. This behavior can be ascribed to the increase in the non-radiative cross-relaxation rates as the Tm^{3+} concentration increases, since this increases the concentration of Tm^{3+} ion pairs. Although there is no literature data on bulk $Lu_2O_3:Tm^{3+}$, the spectroscopy of Tm^{3+} in the isostructural Y_2O_3 has been investigated in detail [14]. The reported $^1D_2$ lifetimes for Tm^{3+} in Y_2O_3 [14] are in good agreement with those observed here. The $^1D_2$ lifetimes of Tm^{3+} in Lu_2O_3 are shorter in the nanoparticles than in the microparticles. Figure 3 shows that the concentration dependences in both size regimes are similar for $x \geq 1$ %. For $x < 1$ %, however, the behavior is drastically different. In the microparticles the $^1D_2$ lifetimes are longer for lower $[Tm^{3+}]$ (less effective cross-relaxation), whereas in the nanoparticles the opposite is observed.

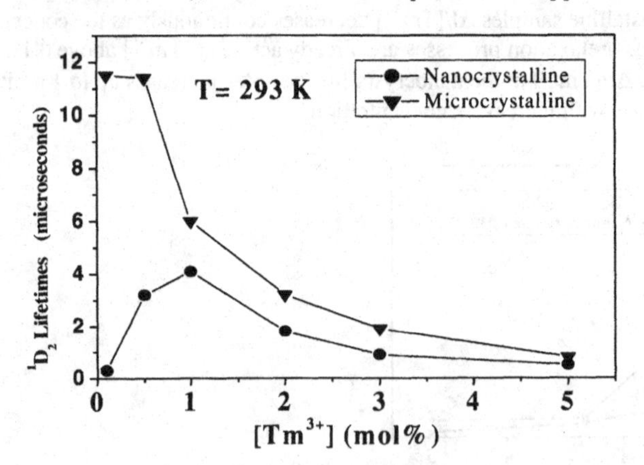

Figure 3. The concentration dependence of the $^1D_2$ lifetimes of Tm^{3+} in nanocrystalline (circles) and microcrystalline (triangles) $Lu_2O_3:xTm^{3+}$ samples. The emission was monitored at 453.1 nm under 355 nm excitation. The solid lines are drawn for clarity.

These results show that, in addition to the expected concentration quenching by cross-relaxation between Tm^{3+} ions, the Tm^{3+} luminescence is strongly quenched in the nanoparticles by an additional process, which can be ascribed to energy transfer to defect states. In this connection, we note that the excitation spectra of the nanocrystalline Lu_2O_3:xTm^{3+} samples (figure 1) showed that the defects responsible for the broad emission band at 430 nm have excited states which are resonant with the $^1D_2$ level of the Tm^{3+} ion, making an energy transfer process between them very likely. Considering that the lifetime of the broad band emission is much shorter ($\leq$200 ns) than the radiative lifetime of the $^1D_2$ level of Tm^{3+} in Lu_2O_3 (*viz.* 12 μs, [14]), the energy transfer process $^1D_2 \rightarrow$defect states should be the dominant one. This additional quenching mechanism is more evident for the lower [Tm^{3+}], what is consistent with an almost constant defect concentration in all samples, as suggested by the small variation of the broad band emission intensity between different samples. The temperature dependence of the $^1D_2$ lifetime was also determined for the nanocrystalline and microcrystalline samples containing 1% Tm^{3+} (figure 4). The $^1D_2$ lifetime is essentially temperature independent in microcrystalline Lu_2O_3:1%Tm^{3+}, whereas in nanocrystalline Lu_2O_3:1%Tm^{3+} it increases as the temperature decreases. This shows that lowering the temperature reduces the rates for the energy transfer process $^1D_2 \rightarrow$defect states.

Although the lower quantum efficiency often observed for nanocrystalline lanthanide doped phosphors [4,9,10] is usually ascribed to quenching by defects, this is, to the best of our knowledge, the first direct observation of such a process. Unfortunately, the data reported here do not allow a conclusive statement concerning the nature of these defects. Considering that the preparation of the nanoparticles involves a fast combustion reaction, it could be expected that reaction residues such as CO_2 and NO_x would be present, as indeed observed for samples prepared using glycine as fuel [15]. However, IR vibrational spectra of the samples investigated here show no evidence of such reaction residues [16]. Moreover, such residues would mostly be at the surface and thus could not strongly affect bulk Tm^{3+} ions. Most importantly, they could not give rise to a temperature dependent quenching of the Tm^{3+} emission, since the most likely quenching process induced by such species, *viz.* multiphonon relaxation by high frequency phonons, is known to be essentially temperature independent in the range investigated here [17].

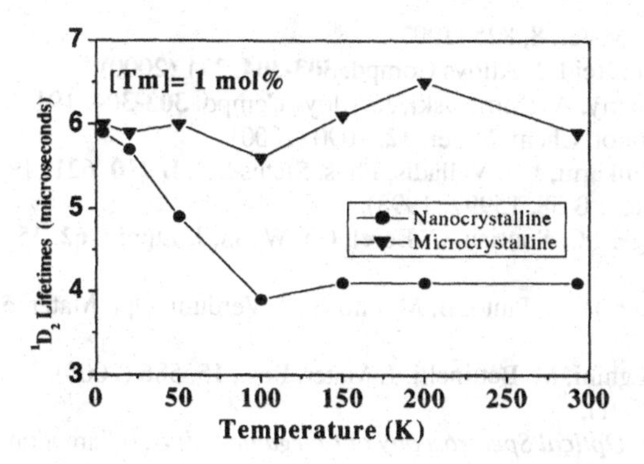

Figure 4. The temperature dependence of the $^1D_2$ lifetime of Tm^{3+} in nanocrystalline (circles) and microcrystalline (triangles) Lu_2O_3:1 mol%Tm^{3+} samples. The emission was monitored at 453.1 nm under 355 nm excitation. The solid lines are drawn for clarity.

CONCLUSIONS

The differences observed between the luminescence properties of nanocrystalline and microcrystalline Lu_2O_3:xTm^{3+} can be ascribed to the higher defect concentration in the nanoparticles. These defects give rise to a broad band emission with maximum at ~430 nm, and to partial quenching of the $^1D_2$ emission of the Tm^{3+} ion, in addition to the expected concentration quenching by cross-relaxation between Tm^{3+} ions. This additional quenching process occurs both by competitive absorption and by energy transfer from the $^1D_2$ level of Tm^{3+} to defect states. The nature of these defects is as yet unknown. The local structure around the Tm^{3+} ions is not noticeably different in the two size regimes. No evidence of quantum size or phonon confinement effects in the Tm^{3+} luminescence was observed.

ACKNOWLEDGMENTS

We gratefully acknowledge the financial support of the Netherlands Organization for Scientific Research (NWO) and of the Council for Chemical Sciences (CW). Also, support of NATO under Grant No. PST.CLG.976212 and of Committee for Scientific Research, KBN, under Grant No. 3 T09B 031 16, in various aspects of this work is appreciated.

REFERENCES

1. S.V. Gaponenko; *Optical Properties of Semiconductor Nanocrystals*, (Cambridge University Press, Cambridge, 1998).
2. A. P. Alivisatos, J. Phys. Chem **100,** 13226 (1996).
3. B.M. Tissue, Chem. Mater. **10**, 2837 (1998)
4. R. Schmechel, M. Kennedy, H. von Seggern, H. Winkler, M. Kolbe, R.A. Fischer, L. Xaomao, A. Benker, M. Winterer, H. Hahn, J. Appl. Phys. **89,** 1679 (2001).
5. R.S. Meltzer, S.P. Feofilov, B.M. Tissue, H.B. Yuan, Phys. Rev. B **60,** R14012 (1999).
6. H-S. Yang, K.S. Hong, S.P. Feofilov, B.M. Tissue, R.S. Meltzer, W.M. Dennis, J. Lumin. **83&84**, 139 (1999).
7. Q. Li, L. Gao, D. Yan, Nanostruc. Mater. **8**, 825 (1997).
8. C.-K. Duan, M. Yin, K. Yan, M. F. Reid, J. Alloys Compd. **303-304**, 371 (2000).
9. M. Haase, K. Riwotzki, H. Meyssamy, A. Kornowski, J. Alloys Compd. **303-304,** 191 (2000).
10. A. Huignard, T. Gacoin, J.-P. Boilot, Chem. Mater. **12**, 1090 (2000).
11. R.N. Bhargava, V. Chabra, B. Kulkarni, J.V. Veliadis, Phys. Status Sol. B **210**, 621 (1998).
12. A. A. Bol, A. Meijerink, Phys. Rev. B **58,** 15997 (1998).
13. C. Li, A. Lagriffoul, R. Moncorge, J.C. Souriau, C. Borel, Ch. Wyon, J. Lumin. **62**, 157 (1994).
14. Y. Guyot, R. Moncorge, L.D. Merkle, A. Pinto, B. McIntosh, H. Verdum, Opt. Mater. **5**, 127 (1996).
15. S. Polizzi, G. Fagherazzi, A. Speghini, M. Bettinelli, J. Mater. Res., **15**, 586 (2000).
16. E. Zych, Opt. Mater., **16**, 445 (2001).
17. B. Henderson and G.F. Imbusch, *Optical Spectroscopy of Inorganic Solids*, (Clarendon Press, Oxford, 1989).

Mat. Res. Soc. Symp. Proc. Vol. 667 © 2001 Materials Research Society

Synthesis and luminescence properties of colloidal lanthanide doped YVO_4

Arnaud Huignard, Thierry Gacoin, Frédéric Chaput and Jean-Pierre Boilot
Laboratoire de Physique de la Matière Condensée, UMR CNRS 7643, Ecole Polytechnique, 91128 Palaiseau, France
Patrick Aschehoug and Bruno Viana
Laboratoire de Chimie Appliquée de l'Etat Solide, UMR CNRS 7574, ENSCP, 75005 Paris, France

ABSTRACT

Aqueous colloidal solutions of well dispersed YVO_4:Ln (Ln = Eu, Nd) nanoparticles are synthesized through precipitation reactions at room temperature. In the case of YVO_4:Eu, a luminescence quantum yield of 15% is found, which is not as high as in the bulk due to the existence of residual crystalline defects and nonradiative relaxations from the hydroxylated surface. Appropriate hydrothermal annealing and deuteration of the surface allow to rise the yield up to 38%. Incorporation of the nanocrystals into a transparent silica matrix is achieved through preliminary coating of the particles with a functionnalized silicon alkoxide and further dispersion into a sol-gel precursor solution. Such sol-gel materials doped with YVO_4:Nd nanocrystals are transparent and exhibit the typical emission at 1.06 μm of the Nd^{3+} ion.

INTRODUCTION

Intensive research on the luminescent properties of chalcogenide nanomaterials has been carried out during this last decade. It has led to the synthesis of core-shell nanostructures (CdSe / ZnS [1]) or doped nanoparticles (CdS:Mn [2]) with narrow size distributions and high luminescence quantum yields. Oxide materials, which are extensively used in most of the devices devoted to the production of light like lamps, lasers, TV phosphors or X ray detectors [3], have recently been studied at the nanometer scale [4], [5]. Promising applications are expected in the fields of biological labels [6] and integrated optical systems.

The basic idea of this work is to explore the potentialities of lanthanide (Ln) doped yttrium orthovanadate (YVO_4) nanoparticles synthesized through colloidal chemistry. The choice of this material is explained by the excellent phosphor (Ln = Eu) and laser (Ln = Nd) properties exhibited by bulk YVO_4 as an host matrix. In this paper, we first expatiate upon the optical properties of colloidal Ln (Ln = Eu, Nd) doped YVO_4 nanoparticles. In a second part, we report on the synthesis of composite materials in which the nanocrystals are dispersed in a transparent sol-gel matrix. This kind of elaboration has proved its efficiency in the case of CdS nanoparticles [7] and could be an interesting alternative route to the difficult growth of high quality YVO_4:Nd monocrystals for laser applications.

EXPERIMENTAL SECTION

Synthesis of the aqueous colloidal solutions

The basic principle of the colloidal synthesis of the particles relies on an aqueous precipitation reaction at room temperature from $Y(NO_3)_3$, $Nd(NO_3)_3$ or $Eu(NO_3)_3$, and $NaVO_3$ at pH=12.5. The obtained precipitate is subsequently dialyzed, sonicated and further stabilized with sodium hexametaphosphate. More details concerning this synthesis may be found in a previous paper [5].

Silica coating of the nanoparticles

The nanoparticles are coated with a functionnalized silicon alkoxide by adapting a procedure previously developed by Philipse in the case of bohemite nanoparticles [8]. 50 ml of a 10^{-2} mol.l^{-1} $Y_{1-x}Ln_xVO_4$ colloidal solution is first added to 50 ml of a 3% w/w sodium silicate solution. After a 24 hour aging time, this aqueous mixture is dialyzed, concentrated down to 50 ml and subsequently added dropwise to 150 ml of ethanol containing 5 equivalents of 3-(methacryloxylpropyl) trimethoxysilane (TPM). This mixture is then refluxed for 24 hours and transferred into propan-1-ol by azeotropic distillation. The grafted nanoparticles are centrifuged to eliminate unreacted TPM. The sediment is finally dispersed into ethanol with a concentration up to 0.5 mol.l^{-1} solutions.

Synthesis of the sol-gel matrices doped with the nanoparticles

The ethanol dispersion of TPM grafted $Y_{1-x}Ln_xVO_4$ (Ln = Eu, Nd) nanoparticles can be mixed with an hydrolysed sol of vinyl triethoxysilane (VTEOS) following a process previously described [9]. The sol is then either concentrated to make thin films by spin coating or slowly dried to elaborate transparent monoliths.

Characterization of the nanoparticles

High Resolution Transmission Electron Microscopy (HRTEM) observations were carried out on an AKASHI TOPCON 002B microscope operating at 160 kV (point resolution 1.8Å, Cs = 0.4 nm). Absorption spectra were recorded on a Shimadzu 1600 A spectrophotometer. Visible luminescence spectra as well as lifetime measurements were recorded on a Hitachi F-4500 spectrofluorometer. The quantum yields were determined by comparing the integrated emission of the colloidal solutions with the emission from a Rhodamine 6G solution in ethanol having the same optical density and excited at the same wavelength (280 nm). IR luminescence spectra is recorded at room temperature using a titanium-sapphire laser pumped by an argon laser as the excitation source, and a PbS cell for the detection.

DISCUSSION

Structural characterizations of the bare nanoparticles

The crystalline phase of the nanoparticles as well as the coherence length were determined by X-ray diffraction (XRD) measurements. The recorded diagram for $Y_{0.95}Eu_{0.05}VO_4$ nanoparticles was found to be in agreement with the Zircon type structure of YVO_4 or $EuVO_4$ [10]. Moreover, it reveals that the size of the crystalline domains, as determined by the Scherrer formula, varies from 10 to 20 nm. XRD was also used to check the homogeneous incorporation of the lanthanide dopant into the YVO_4 matrix. The method consists of plotting the cell volume, calculated from the refined cell parameters, versus the fraction of the dopant initially present in the reaction mixture. In the case of europium, the incorporation is homogeneous whatever x varies from 0 to 1. On the contrary, the incorporation of neodymium is only homogeneous up to x = 0.2. This can be explained by the large difference of ionic radii between Nd^{3+} (r = 1.109 Å) and Y^{3+} (r = 1.019 Å) in the coordination 8 [11].

The size and the morphology of the nanoparticles were determined by HRTEM. Figure 1 shows that the nanoparticles are mainly anisotropic, with two characteristic dimensions of around 15 and 30 nm. However, the contrast observed within the particles suggests that they are polycrystalline with numerous defects. These latter may be attributed to the low temperature synthesis and aggregation of primary particles.

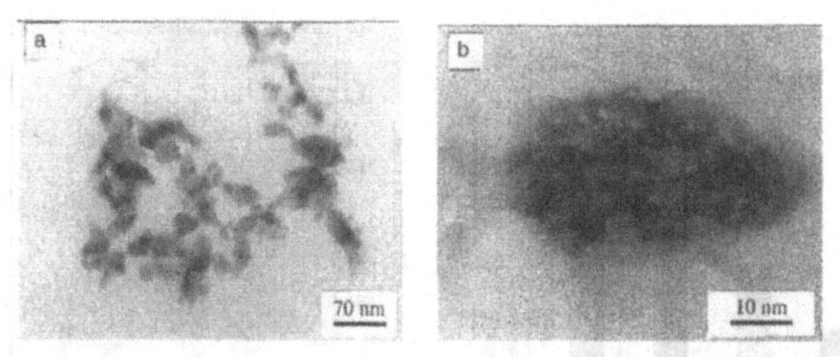

Figure 1 : (a) HRTEM picture of an aqueous $Y_{0.95}Eu_{0.05}VO_4$ colloid, (b) HRTEM of an isolated nanoparticle.

Optical properties of the aqueous $Y_{1-x}Eu_xVO_4$ colloids

The optical properties of a $Y_{0.95}Eu_{0.05}VO_4$ colloid are reported on Figure 2. Upon excitation in the VO_4^{3-} absorption band in the UV region, an energy transfer between the vanadate groups and the europium ions occurs, resulting in the Eu^{3+} luminescence. The most intense emission band at 617 nm is attributed to the $^5D_0$-$^7F_2$ transition. In fact, the observed luminescence mechanism as well as the branching ratios of the different emission bands are very similar to what is observed in the bulk material.

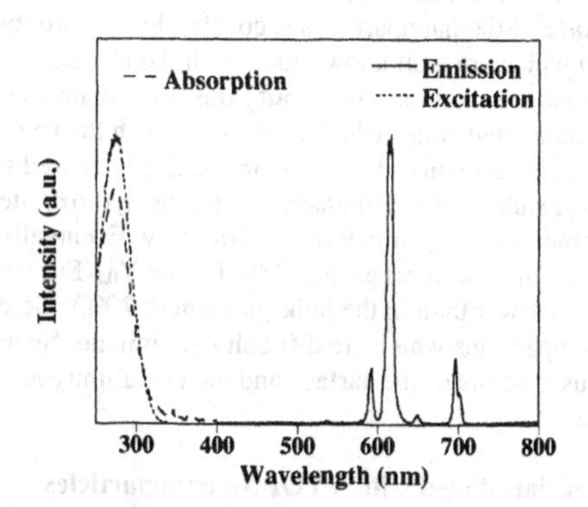

Figure 2 : Absorption, excitation (λ_{em} = 617 nm) and luminescence spectra (λ_{ex} = 280 nm) for the $Y_{0.95}Eu_{0.05}VO_4$ colloid.

Concerning the efficiency of the luminescence of the $Y_{1-x}Eu_xVO_4$ colloids, the quantum yield was found to reach a maximum of 15% for x = 0.3. This is in contrast with the bulk material for which the maximum quantum yield is 70% for x= 0.05 [12]. In order to understand these differences, which may obviously be due to the microstructure of the nanoparticles, several treatments on the colloids were carried out (Figure 3).

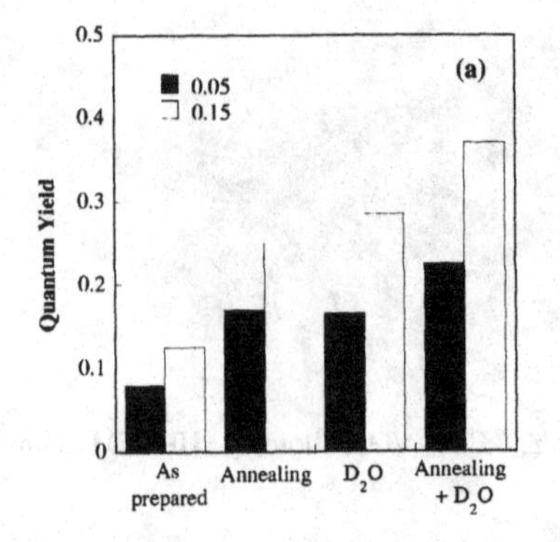

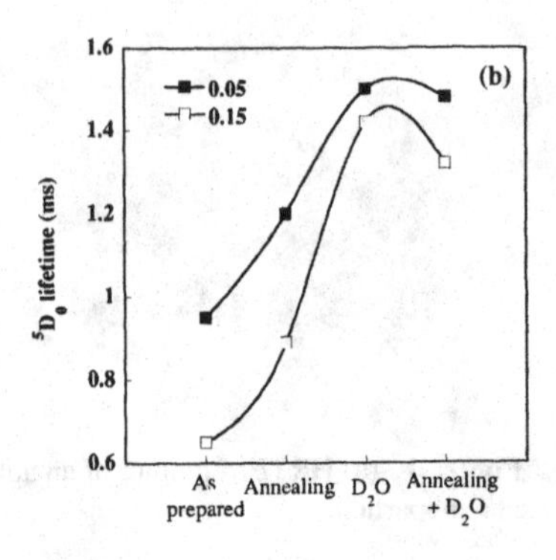

Figure 3 : Effect of hydrothermal annealing (130°C) and D_2O treatment on $Y_{1-x}Eu_xVO_4$ colloids with x = 0.05 and x = 0.15. (a) quantum yields, (b) lifetime of the Eu $^5D_0$ level

First of all, the colloids were hydrothermally annealed at 130°C for 14 hours. This was achieved to reduce the number of crystalline defects inside the nanoparticles which may act as nonradiative centers. The temperature of 130°C was chosen because it does not change neither the size of the nanoparticles nor their coherence lengths. This treatment leads to a significant increase of the quantum yield as well as the europium $^5D_0$ lifetime, which means that some crystalline defects acting as nonradiative centers are eliminated. Despite this annealing, the luminescence behavior of the colloids is still different from the one observed in the bulk. It must be kept in mind however that the nanoparticles are synthesized and dispersed in water. As a consequence, the surface of the nanoparticles is covered by OH groups (either chemically bonded or just adsorbed) which are well known to quench the luminescence of rare-earth ions [13] through multiphonon relaxation. To quantify this nonradiative effect, the nanoparticles were transferred into deuterated water which does not quench the lanthanide luminescence. Whatever the europium composition, the quantum yield is increased twofold and the $^5D_0$ lifetime lengthened. This result clearly demonstrates that the hydroxylated surface of the nanoparticles is a critical parameter in the luminescence efficiency. Eventually, both treatments were cumulated. The quantum yield reaches then 38% for the $Y_{0.85}Eu_{0.15}VO_4$ composition. However, as this value is lower than in the bulk (maximum 70%), there are still distortions and defects inside the nanoparticles which are difficult to eliminate. Nevertheless, these experiments prove that cautious controls of the surface and the cristallinity lead to rather efficient luminescent nanomaterials.

Synthesis of transparent materials doped with YVO_4:Nd nanoparticles

In order to incorporate the nanoparticles inside sol-gel based silica materials (thin films or monoliths), it is necessary to preliminary coat them with functionnalized silica. TPM was used ensuring the compatibility of the nanoparticles with the sol-gel process. The efficiency of the grafting was checked by TEM (Figure 4). Core-shell $Y_{0.9}Nd_{0.1}VO_4$ / silica nanostructures are clearly visible with a shell thickness of around 10 nm. Such particles are highly stable in ethanol and can be easily incorporated into hydrolyzed solutions of vinyltriethoxysilane precursors of the host matrix. Subsequent sol-gel processing leads either to transparent monoliths of a few cm^3 or thin films of a few microns deposited on glass substrates by spin-coating.

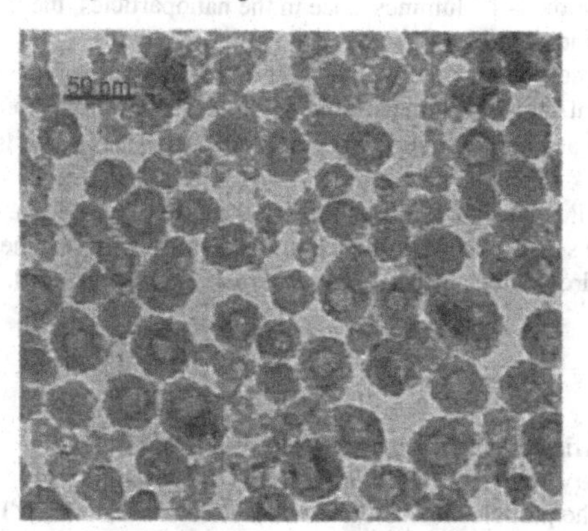

Figure 4 : TEM picture of a core/shell $Y_{0.9}Nd_{0.1}VO_4$ / silica colloid

The optical properties of a VTEOS monolith doped with $Y_{0.9}Nd_{0.1}VO_4$ nanoparticles are shown on Figure 5.

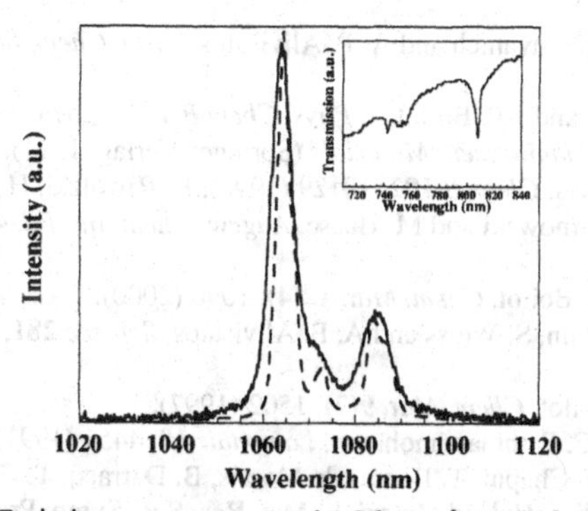

Figure 5 : Emission spectrum around 1.06 µm (λ_{ex} = 808 nm) of a $Y_{0.9}Nd_{0.1}VO_4$ VTEOS monolith and comparison with bulk material (dashed line). The inset shows the transmission spectrum.

The two bands on the transmission spectrum (inset) are respectively attributed to the $^4I_{9/2} - {}^4F_{3/2}, {}^4F_{7/2}$ (750 nm) and $^4I_{9/2} - {}^4F_{5/2}, {}^2H_{11/2}$ (808 nm) transitions of Nd^{3+} ions. The position of these absorption bands is in total agreement with results reported in the bulk YVO_4:Nd material [14]. The emission spectrum was obtained by excitation of the $^4I_{9/2} - {}^4F_{5/2}, {}^2H_{11/2}$ band (808 nm). The two emission bands at 1.064 µm and 1.085 µm are due to the $^4F_{3/2} - {}^4I_{11/2}$ neodymium transition splitted by the crystal field. A slight broadening is observed compared to the bulk material, which must be related to the lower crystallinity of the nanoparticles.

In order to assess the efficiency of the Nd^{3+} luminescence in the nanoparticles, the lifetime of the $^4F_{3/2}$ emitting level was measured for different neodymium contents ($0.01<x<0.1$). The decays appear to be non-exponential at the short times, indicating Nd^{3+} - Nd^{3+} interactions. Besides, measurements at longer times for the various Nd^{3+} compositions enabled us to estimate the $^4F_{3/2}$ lifetime at infinite dilution to be around 10 μs. This value is lower than the 90 μs reported in the bulk [15]. The nature of the non-radiative pathways which explains such a difference is similar to what was described in the case of europium. However, as neodymium is much more sensitive to OH quenching than europium [16], the nonradiative relaxation through the hydroxylated surface of the nanoparticles is enhanced.

CONCLUSION

We have reported a simple and efficient way to elabrate YVO_4:Ln (Ln = Eu, Nd) nanoparticles through colloidal chemistry. We have also demonstrated the possibility of grafting functionnalized silica around the particles to elaborate core/shell nanostructures. This allows the dispersion of the particles into sol-gel silica matrices to obtain either bulk materials or thin films. The study of the optical properties of the nanocrystallites as well as their ability to be incorporated into silica transparent materials are encouraging results in the outlook of conceiving integrated optical devices. Moreover, the YVO_4:Ln grafted nanoparticles could also be useful as ultrasensitive and selective biological labels.

REFERENCES

1. X. Peng, M. C. Schlamp, A. V. Kadavanich and A. P. Alivisatos *J. Am. Chem. Soc.* **119**, 7019 (1997).
2. G. Counio, S. Esnouf, T. Gacoin and J-P. Boilot, *J. Phys. Chem B* **102**, 5257 (1998).
3. G. Blasse and B. C. Grabmeier, *Luminescent Materials* (Springer Verlag, 1994).
4. K. Riwotzki and M. Haase, *J. Phys. Chem B* **102**, 10129 (1998), K. Riwotzki, H. Meissamy, H. Schnablegger, A. Kornowski and M. Haase, *Angew. Chem. Int. Ed.* **40(3)**, 573 (2001).
5. A. Huignard, T. Gacoin and J. P. Boilot, *Chem. Mat.* **12(4)**, 1090 (2000).
6. M. Bruchez Jr, M. Moronne, P. Gin, S. Weiss and A. P. Alivisatos, *Science* **281**, 2013 (1998).
7. T. Gacoin, L. Malier and J-P. Boilot, *Chem. Mat.* **9(7)**, 1502 (1997).
8. A. P. Philipse, A. Nechifor and C. Pathmamanoharan, *Langmuir* **10**, 4451 (1991).
9 J-P. Boilot, J. Biteau, A. Brun, F. Chaput, T. Dantas de Morais, B. Darracq, T. Gacoin, K. Lahlil, J-M. Lehn, Y. Levy, L. Malier, *Hybrid Materials*, Mat. Res. Soc. Symp. Proc. **519**, 227 (1998)
10. Schwarz, Z. *Anorg. Allg. Chem.* **322**, 143 (1963).
11. Shannon and Prewitt, *Acta Cryst.* **A32**, 785 (1976).
12. R. C. Ropp, J. Electrochem. Soc. Solid State Science. 115(9), 940 (1968).
13. G. Blasse, *Prog. Solid State Chem.* **18**, 119 (1988).
14. O. Guillot – Noël, B. Viana, G. Aka, D. Gourier, A. Kahn – Harari and D. Vivien, *J. Lum.* **72 – 74**, 195 (1997).
15. O. Guillot-Noel, B. Viana, B. Bellamy, D. Gourier, G.B.Zogo-MBoulou, S. Jandl, *Optical Materials* **13(4),** 427 (2000).
16. G. Stein and E. Würzberg, *J. Chem. Phys.* **62(1)**, 208 (1975).

Mat. Res. Soc. Symp. Proc. Vol. 667 © 2001 Materials Research Society

Luminescence of nanocrystalline $ZnS:Pb^{2+}$

Ageeth A. Bol, Andries Meijerink
Debye Institute, Physics and Chemistry of Condensed Matter, Utrecht University
P.O. Box 80 000, 3508 TA Utrecht, The Netherlands
Fax: +31 30 253 2403, E-mail: a.a.bol@phys.uu.nl

ABSTRACT

Nanocrystalline $ZnS:Pb^{2+}$ is synthesized via a precipitation method. The luminescence is studied and the influence of the size of the nanocrystals on the luminescence properties is investigated. Nanocrystalline $ZnS:Pb^{2+}$ shows a white emission under UV excitation. At least two luminescence centers are involved. One center is identified as a Pb^{2+} ion located on a regular Zn^{2+} site and gives a red emission under 480 nm excitation. The luminescence properties of this emission are characteristic for transitions on Pb^{2+} ions. The other centers are not as well defined and give a broad green emission band under 380 nm excitation and also show luminescence properties typically observed for Pb^{2+}. The green emission probably originates from a charge-transfer like D-band emission of Pb^{2+} in ZnS close to a defect (*e.g.* an S^{2-} vacancy or an O^{2-} ion on an S^{2-} site). A relation between the temperature quenching of the emissions and the band gap is observed and indicates that photoionization occurs.

INTRODUCTION

Efficiently luminescing undoped and doped semiconductor nanocrystals have received considerable attention during the last decades [1-4]. The changes in the electronic structure as a function of particle size (quantum size effects) are intriguing and have stimulated fundamental research on various types of nanocrystalline semiconductors. Up till now, research of doped nanocrystals has been focussed on the luminescence of nanocrystalline $ZnS:Mn^{2+}$. Various studies have reported an influence of quantum size effects on the luminescence of luminescing dopants incorporated in the nanocrystals [*e.g.*, 5]. Recent studies have provided convincing evidence against the presence of an influence of quantum size effects on the lifetime and efficiency of the dopant emission in $ZnS:Mn^{2+}$ [6, 7]. In this article the luminescence of a different quantum sized material, nanocrystalline $ZnS:Pb^{2+}$, is described.

The Pb^{2+} ion (outer electron configuration: $6s^2$) belongs to the group of s^2 ions. The s^2 configuration gives a non-degenerate ground state, $^1S_0$. The first excited state with s-p configuration is split by spin-orbit coupling into three triplet states ($^3P_0$, $^3P_1$ and $^3P_2$) and at higher energy there is one singlet state ($^1P_1$). In the literature the transitions between the ground state ($^1S_0$) and these excited states are labeled with A ($^3P_1$), B ($^3P_2$) and C ($^1P_1$). A transition to a fourth excited state, the so-called D-band, has also been observed for Pb^{2+} in several compounds, and is considered to be a charge transfer type transition [8]. Charge transfer transitions involve the transfer of an electron from the ligand to the metal ion or *vice versa*. These transitions are characterized by intense broad absorption and emission bands and large Stokes shifts [9]. The position of this band is strongly dependent on the host lattice.

In this contribution the influence of the size of the nanocrystals on the luminescence of nanocrystalline $ZnS:Pb^{2+}$ is investigated and the origin of the observed luminescence is discussed.

EXPERIMENTAL DETAILS

The ZnS:Pb^{2+} nanoparticles are made using a well-known precipitation method [10]. 10.2 g $Na(PO_3)_n$ (Aldrich) was dissolved in 80 ml distilled water. 10 ml of 1 M $Zn(ClO_4)_2.6H_2O$ and 0.5 ml of 0.01 M $Pb(ClO_4)_2.3H_2O$ were added to this solution. After stirring for 10 minutes, 10 ml 1 M $Na_2S.9H_2O$ was injected. Immediately a white suspension is formed. The suspension was centrifuged and washed with distilled water and ethanol. The nanoparticles were dried in vacuum (sample A). To increase the particle size a small amount of sample A was heated in a nitrogen atmosphere for 10 minutes at 450 °C (sample B). Sample C was prepared using 5 ml of 1 M $Na_2S.9H_2O$ (S^{2-} deficient) and sample D using 15 ml of 1 M $Na_2S.9H_2O$ (S^{2-} excess).

The set-ups used for X-ray powder diffraction and luminescence spectroscopy are described in reference [11]. X-ray powder diffraction patterns of the ZnS:Pb^{2+} nanoparticles show broad lines at positions that are in agreement with the zincblende modification of ZnS.

RESULTS AND DISCUSSION

In table I the average particle diameter of sample A, B, C and D is shown. The average particle diameter was calculated from the XRD pattern of the sample using Scherrer's formula [12]. When sample A is subjected to a heat treatment (10 minutes at 450 °C in a nitrogen atmosphere) the average particle size increases from 4.3 nm (sample A) to 5.5 nm (sample B). As can be deduced from table I the sulfide concentration used in the synthesis have a large influence on the particle size. The sample made with a deficiency of S^{2-} (sample C) has the smallest average particle size (3.0 nm), while the sample made with an excess of S^{2-} (sample D) has an average particle size of 5.5 nm. A relation between the particle size and the Zn/S precursor ratio has been reported before by Suyver *et al.* [13].

Figure 1 shows excitation and emission spectra of sample A measured at liquid helium temperatures (~ 4 K). A very broad emission extending from 400 nm to 800 nm (Full width half maximum (FWHM) ~ 6300 cm^{-1}) was obtained for excitation at 380 nm (figure 1a). The excitation spectrum for this broad emission shows a broad band centered around 380 nm. Around 320 nm a small shoulder is present. The position of this shoulder corresponds to the position of the excitation maximum of the nanocrystalline ZnS host. From the weakness of the host lattice excitation band in the excitation spectrum depicted in figure 1a it can be deduced that the energy transfer process from the ZnS host to the Pb^{2+} impurity is not efficient.

Table I. Average particle size derived from XRD and quenching temperature of the green emission (λ_{exc} = 380 nm) of sample A, B, C and D

Sample	Average diameter XRD (nm)	T_q^{green} (K)
A	4.3	230
B	5.5	195
C	3.0	250
D	5.5	200

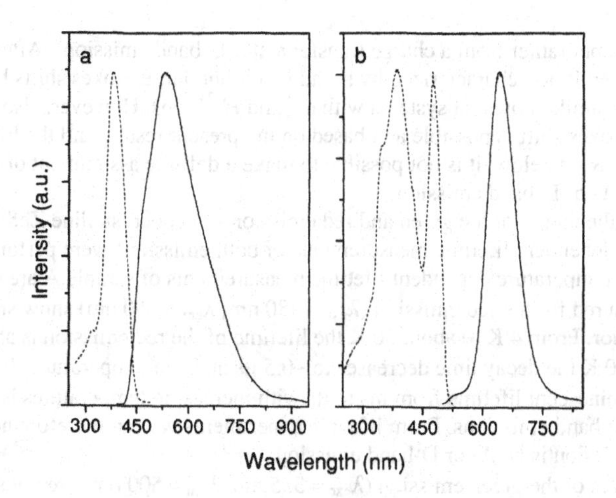

Figure 1. Excitation (dashed lines) and emission spectra (solid lines) measured at 4 K of nanocrystalline ZnS:Pb^{2+} (sample A). (a). Excitation spectra measured for λ_{em}=520 nm, emission spectra measured for λ_{exc} = 380 nm. (b). Excitation spectra measured for λ_{em} = 650 nm, emission spectra measured for λ_{exc} = 480 nm.

The Pb^{2+} ions can be most efficiently excited with radiation of 380 nm, which corresponds to a localized transition involving the Pb^{2+} ion. The broad green emission resembles the emission of Pb^{2+} ions in CaO, if Pb^{2+} has a d^{10} ion, like Cd^{2+} and Zn^{2+} as nearest neighbor [14]. The observed green emission was assigned to a transition from a charge-transfer state involving Cd^{2+} or Zn^{2+} and Pb^{2+} [14]. Based on the large FWHM of this emission band, the large Stokes shift and the temperature dependence of the lifetime (see below), it is difficult to make a clear identification of the centers responsible for the green luminescence in ZnS:Pb^{2+}. In analogy with the luminescence of Pb^{2+} in CaO:Zn^{2+}, Pb^{2+} it may be assigned to a D-band emission of Pb^{2+} in ZnS close to a defect (*e.g.* an S^{2-} vacancy or O^{2-} on an S^{2-} site). Probably, various types of perturbed Pb^{2+} centers contribute.

A red luminescence band (FWHM ~ 2000 cm^{-1}) centered around 650 nm can be selectively excited with 480 nm radiation (figure 1b). Just as for the broad band emission discussed above, the red emission can be excited most efficiently by sub-band gap excitation of the Pb^{2+} impurity. The 650 nm emission was also reported for bulk ZnS:Pb^{2+}. Uehara [15] and later Scharmann [16] assigned this emission to the $^3P_{0,1} \rightarrow {}^1S_0$ transition (A-band) originating from a Pb^{2+} center on a regular Zn^{2+} site (surrounded by four S^{2-} ligands). Evidence for the assignment of the emission to $^3P_{0,1}$ emission was obtained by the analysis of the excitation spectrum. A splitting of the $^1S_0 \rightarrow {}^3P_{0,1}$ excitation band into two bands is typically observed for s^2 ions and is due to a Jahn-Teller deformation in the excited state. In the presently studied nanocrystalline ZnS:Pb^{2+} samples Jahn-Teller splitting is not observed in the $^3P_{0,1}$ excitation spectrum, probably due to inhomogeneous broadening.

The Stokes shift of the red emission is large (~6000 cm^{-1}). Also there is a clear offset between the long wavelength edge of the relatively narrow excitation band (edge at 520 nm) and the onset of the broad emission band (onset at 570 nm). This indicates that the emission may not originate from the $^3P_{0,1}$ state, but rather from a charge transfer state (D-band emission). After $^1S_0 \rightarrow {}^3P_{0,1}$ excitation, D-band emission characterized by broad bands and large Stokes shifts has been reported before for similar covalent systems with d^{10} and Pb^{2+} ions. However, also for A-band emission a large Stokes shift is possible and based on the present results and the lifetime measurements discussed below, it is not possible to make a definite assignment of the red emission to either A- or D-band emission.

To investigate the nature of the green and red emission of nanocrystalline ZnS:Pb^{2+} in more detail temperature dependent lifetime measurements of both emissions were performed. In figure 2 the results of the temperature dependent lifetime measurements of sample A are depicted. The decay curves measured for the red emission (λ_{exc} = 480 nm, λ_{em} = 650 nm) show single exponential behavior. From 4 K to about 50 K the lifetime of the red emission is about 3 ms (see figure 2). Above 50 K the decay time decreases to ~0.5 μs at room temperature. A strong decrease of the luminescent lifetime from ms to μs with increasing temperatures is typically found for A- and D-band emissions. From lifetime experiments we can therefore not deduce whether the red emission is an A- or D-band emission.

The decay curves of the green emission (λ_{exc} = 355 nm, λ_{em} = 500 nm) are not single-exponential. Fitting with a double exponential decay function gave a good agreement results. The observation of a non-single exponential decay is in agreement with the assumption that the green emission is due to various types of Pb^{2+} centers. The lifetime of the green emission also shows a behavior typical of Pb^{2+}. At low temperature (up till 40 K) the decay times of the white emission are constant with components in the ms range (figure 2a). Above 40 K the decay time decreases to ~0.5 μs at room temperature.

The temperature dependence of the broad emission band, observed at an excitation wavelength of 380 nm, of sample A is depicted in figure 3a. From figure 3a it can be concluded that the shape of this emission is dependent on the temperature. The intensity of the green spectral range (around 550 nm) is stable going from 5 to about 50 K. Above 50 K the green emission starts to quench. Similar behavior was found for sample B, C and D.

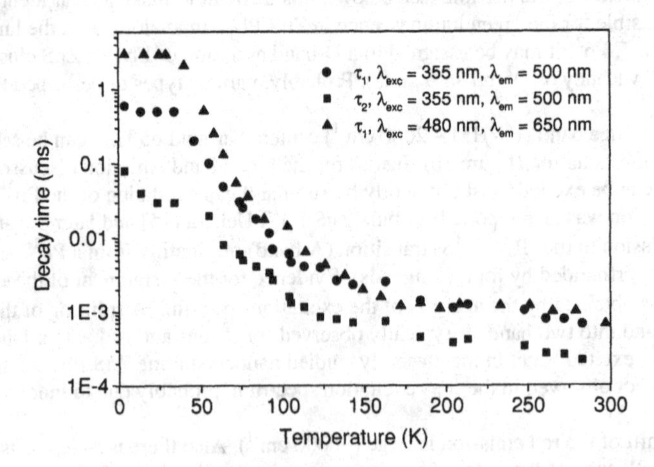

Figure 2. Luminescence decay times as a function of temperature of nanocrystalline ZnS:Pb^{2+} (sample A).

A measure for the temperature quenching is T_q, which is defined as the temperature at which the intensity of the emission has decreased to half of its maximum intensity. Table I shows the quenching temperatures of the green emission measured at 550 nm of sample A, B, C and D. As can be seen in table I T_q varies from sample to sample. Sample C (average particle diameter 3.0 nm) shows the highest quenching temperature (T_q = 250 K), followed by sample A (d_{avg} = 4.3 nm, T_q = 230 K). Sample B has a quenching temperature of 200 K and an average particle diameter of 5.5 nm. This suggests that the quenching temperature is dependent on the particle size: with increasing particle size the quenching temperature decreases. If quenching occurs by thermally stimulated excitation of a charge carrier from an excited state to the conduction band (photoionization) the quenching temperature depends on the energy difference between the excited state of the impurity and the band edge of the conduction band. The size dependence of T_q suggests that the quenching mechanism is thermally activated photoionization. Due to quantum size effects the conduction band shifts to higher energy and as a result the energy difference between the localized states of Pb^{2+} (not influenced by quantum size effects) and the conduction band will increase, resulting in a higher T_q for Pb^{2+} in smaller ZnS particles.

The temperature dependence of the red emission is depicted in figure 3b. In figure 3b the temperature dependence of the emission spectra of sample A is shown. For sample B and D similar results were obtained. Since sample C shows hardly any red luminescence the temperature dependence of the red emission could not be measured. For the red emission similar behavior as for the green emission is observed: the quenching temperature for the Pb^{2+} emission in the sample with the larger particles (sample B) is lower than for the smaller particles (sample A), implying that the quenching of the luminescence at higher temperatures is caused by photoionization.

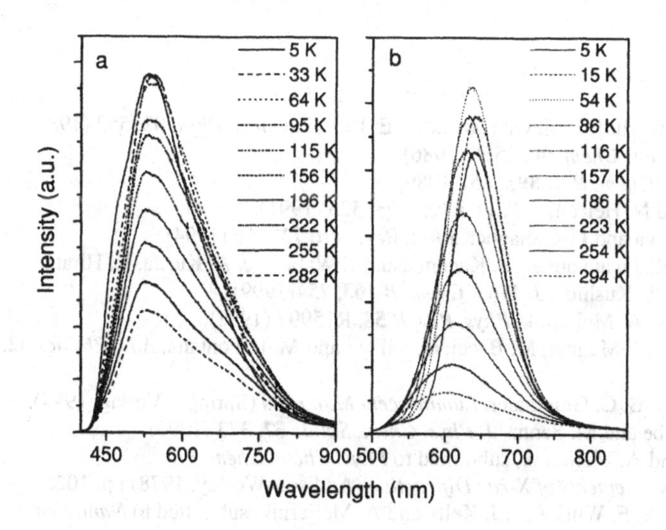

Figure 3. Temperature dependence of (a) the green emission (λ_{exc} = 380 nm) and (b) the red emission (λ_{exc} = 480 nm) of nanocrystalline ZnS:Pb^{2+} (Sample A).

CONCLUSION

Nanocrystalline ZnS:Pb^{2+} gives a very broad emission under UV excitation. At least two types of luminescent centers are observed. One center gives a red emission upon 480 nm excitation. Based on our results and previous publications the red emission is assigned to a luminescence from a Pb^{2+} center on a regular Zn^{2+} site (surrounded by four S^{2-} ligands). The red emission shows luminescence properties typically observed for A-band D-band transitions of Pb^{2+} ions. It is not possible to make a definite assignment of the red emission to either A- or D-band emission.

The other center(s) give a very broad green emission upon 380 nm excitation, which is not as well defined as the red emission. In analogy with the luminescence of Pb^{2+} in CaO:Zn^{2+}, Pb^{2+} the green emission can be assigned to a D-band emission of Pb^{2+} in ZnS close to a defect (*e.g.* an S^{2-} vacancy or O^{2-} on an S^{2-} site). Probably, various types of perturbed Pb^{2+} centers contribute.

With increasing temperature both emissions quench. The quenching temperature is dependent on the band gap of the host material and is therefore sensitive to the particle size of the nanocrystalline semiconductor. A photoionization model explains the relation between band gap and quenching temperature.

ACKNOWLEDGEMENTS

We would like to thank Dr. Alok Srivastava (GE Corporate Research, Niskayuna NY, USA) for his suggestion to incorporate Pb^{2+} in ZnS nanocrystals. Financial support of Philips Lighting NV is gratefully acknowledged.

REFERENCES

[1] R. Rossetti, R. Hull, J. M. Gibson and L. E. Brus, *J. Chem. Phys.* **82**, 552 (1985).
[2] L. Brus, *J. Phys. Chem.* **90**, 2555 (1986).
[3] A. Henglein, *Chem. Rev.* **89**, 1861(1989).
[4] Y. Wang and N. Herron, *J. Phys. Chem.* **95**, 525 (1991).
[5] R. N. Bhargava and D. Gallagher, *Phys. Rev. Lett.* **72**, 416 (1994).
[6] N. Murase, R. Jagannathan, Y. Kanematsu, M. Watanabe, A. Kurita, K. Hirata, T. Yazawa and T. Kushida, *J. Phys. Chem. B* **103**, 754(1999).
[7] A. A. Bol and A. Meijerink, *Phys. Rev. B* **58**, R15997 (1998).
[8] A. Ranfagni, M. Mugnai, M. Bacci, G. Viliani and M. P. Fontana, *Adv. Physics* **32**, 823 (1982).
[9] G. Blasse and B. C. Grabmaier, *Luminescent Materials* (Springer Verlag, 1994).
[10] I. Yu, T. Isobe and M. Senna, *J. Phys. Chem. Solids* **57**, 373 (1996).
[11] A. A. Bol and A. Meijerink, submitted to *Phys. Chem. Chem. Phys.*
[12] B. D. Cullity, *Elements of X-ray Diffraction* (Addison-Wesley, 1978) pp. 102.
[13] J. F. Suyver, S. F. Wuister, J. J. Kelly and A. Meijerink, submitted to *Nano Lett.*
[14] H. F. van den Brand-Folkerts, *Luminescence of the Lead Ion in Solids*, (Thesis Utrecht University, 1996) Ch. 8.
[15] Y. Uehara, *J. Chem. Phys.* **51**, 4385 (1969).
[16] A. Scharmann, D. Schwabe and D. Weyland, *J. Lumin.* **12/13**, 479 (1976).

Mat. Res. Soc. Symp. Proc. Vol. 667 © 2001 Materials Research Society

Luminescence of doped nanocrystalline ZnSe

J. F. Suyver[1], S. F. Wuister, T. van der Beek, J. J. Kelly and A. Meijerink
Debye Institute, Physics and Chemistry of Condensed Matter, Utrecht University, P.O. Box 80.000, 3508 TA Utrecht, The Netherlands.

ABSTRACT

Luminescence of nanocrystalline ZnSe:Mn^{2+} and ZnSe:Cu^{2+} prepared via an organic chemical synthesis method are described. The spectra show distinct ZnSe, Mn^{2+} and Cu^{2+} related emissions, all of which are excited via the host lattice. The Mn^{2+} emission wavelength depends on the concentration of Mn^{2+} incorporated into the ZnSe lattice, which is attributed to the presence of Mn^{2+} pair-states at higher concentrations. The ZnSe:Cu^{2+} luminescence was studied as a function of the crystal-size. Temperature-dependent photoluminescence spectra and photoluminescence lifetime measurements are also presented and the results are compared to those of Mn^{2+} and Cu^{2+} in bulk ZnSe.

INTRODUCTION

Nanocrystalline semiconductors such as CdS, CdSe and ZnS have been studied extensively [1–5] and the luminescence of nanocrystalline semiconductors with optically active ions has received considerable attention in relation to possible applications [1, 2]. A simple wet chemical synthesis is attractive for device applications since large amounts of well-defined nanocrystals (NC) can be obtained in this way [5]. Efficiently luminescing ZnS:Mn^{2+} nanocrystals (quantum efficiency (QE) of more than 10 %) can easily be made and are considered for application in low voltage electroluminescent devices [6, 7]. Due to the unfavourable position of the ZnS valence bandedge better results are expected for doped ZnSe nanocrystals (valence band at higher energy with respect to ZnS) [8]. Recently, two reports on the synthesis and luminescence properties of nanocrystalline ZnSe doped with a luminescent ion (Mn^{2+}) have appeared [9, 10].

In this paper the preparation of nanocrystalline ZnSe particles doped with Mn^{2+} or Cu^{2+} is discussed. The ZnSe:Cu^{2+} product is studied as a function of particle size. Photoluminescence (PL) emission and excitation spectra are presented and discussed. Temperature-dependent luminescence and luminescence lifetime measurements are also reported and compared to results obtained for the luminescence in bulk ZnSe.

EXPERIMENTAL

The synthesis route that was used to prepare the doped ZnSe nanocrystals is a variation of the TOP/TOPO synthesis [9, 11, 12] and was performed in the protective dry-nitrogen atmosphere of a glovebox. A 45 ml sample of hexadecylamine (HDA) was heated to 310°C. Either a variable amount of

[1]Corresponding author. Tel.: +31 – 30 – 253 2214; Fax: +31 – 30 – 253 2403; E-mail: j.f.suyver@phys.uu.nl

manganese cyclohexanebuterate powder or 4.8 mg of $Cu(CH_3COO)_2$ was dissolved in 12 ml of tri-n-octylphosphine (TOP). To this mixture 3 ml 1 M TOPSe (prepared by dissolving Se-powder in TOP) was added. After stirring, 0.32 g di-ethylzinc was added. This solution was shaken and then added to the hot HDA.

In the case of ZnSe:Mn^{2+}, the nanocrystals were grown at 275°C for 4 hours. The mixture was then cooled to 70°C and the nanocrystals were precipitated by addition of 25 ml of anhydrous 1-butanol followed by 30 ml of anhydrous methanol. The residue was centrifuged, decanted and washed with anhydrous methanol. Finally, these samples were dried in a vacuum desiccator, and a fine white powder was obtained. This powder consists of the ZnSe:Mn^{2+} nanocrystals with a HDA and TOP capping.

In the case of the ZnSe:Cu^{2+} nanocrystals, a 2 ml (liquid) sample was taken every 30 minutes over a period of 4.5 hours after the injection. Each of these samples was precipitated in 5 ml of anhydrous 1-butanol. The final mixture was redissolved in 1-octanol and all measurements discussed for the ZnSe:Cu^{2+} nanocrystals were performed on these clear suspensions.

The chemical composition of the samples was determined using a Perkin-Elmer Optima-3000 inductively coupled plasma (ICP) spectrometer. To determine the particle diameter, X-ray powder diffraction (XRD) spectra were measured with a Philips PW 1729 X-ray diffractometer using Cu K_α radiation (1.542 Å). From the broad peaks in the XRD spectrum the diameter of the nanocrystals was calculated.

Photoluminescence (PL) emission and excitation spectra were recorded with a SPEX Fluorolog spectrofluorometer, model F2002, equipped with two monochromators (double-grating, 0.22 m) and a 450 W xenon lamp as the excitation source. All PL emission spectra were corrected for the spectral response of the emission monochromator and the PM tube. PL lifetimes were measured using the third harmonic (355 nm) of a Quanta-ray DCR Nd:YAG laser as an excitation source. The emission light was transported through a fibre optics cable to a monochromator (Acton SP-300i, 0.3 m, 150 lines/mm grating, blazed at 500 nm). The signals were recorded using a thermoelectrically cooled photomultiplier tube in combination with a Tektronix 2430 oscilloscope. Temperature-dependent PL emission and lifetime measurements were recorded using a liquid helium flow-cryostat equipped with a sample heater to stabilize the temperature at different temperatures between 4 K and room temperature.

RESULTS FOR ZnSe:Mn^{2+} NANOCRYSTAL POWDERS

The amounts of Mn, Zn and Se present in the ZnSe:Mn^{2+} samples were measured using ICP analysis. X-ray powder diffraction analysis indicated that the particles had a radius of 3 to 4 nm, and no relation between size and the Mn^{2+} concentration was observed.

Figure 1(a) shows the PL spectra of several nanocrystalline ZnSe:Mn^{2+} samples with Mn^{2+} concentrations ranging from 0 % to 0.9 %. The measurement conditions were identical and relative intensities can be compared. The PL peak

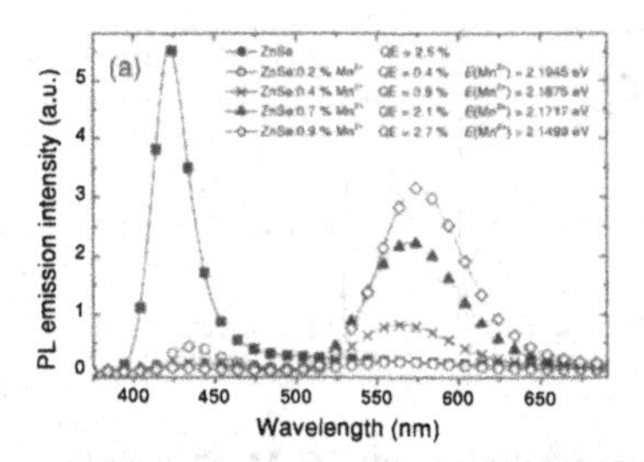

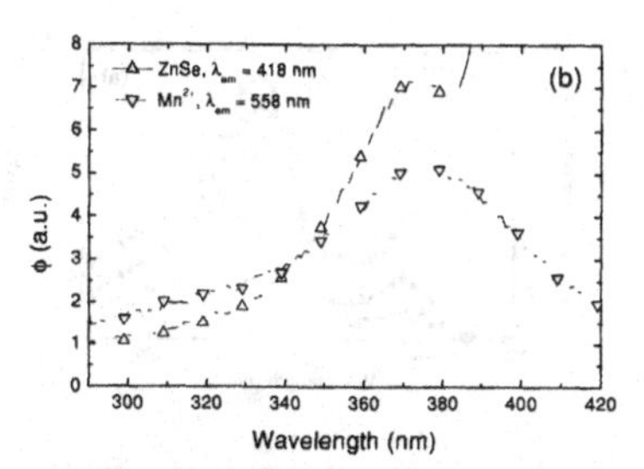

Figure 1: (a) PL spectra of ZnSe:Mn^{2+} nanocrystals for different Mn^{2+} concentrations (indicated in the figure). Measured at room temperature with an excitation wavelength of 330 nm. (b) Excitation spectra for the ZnSe: 0.2 % Mn^{2+} sample. Measured at room temperature.

around ~420 nm is attributed to a sub-bandgap luminescence of the ZnSe host lattice [11]. The peak at ~570 nm is attributed to the $^4T_1 \rightarrow {}^6A_1$ transition in the Mn^{2+} ion and is similar to Mn^{2+} in bulk ZnSe [13]. For samples containing 0.9 % Mn^{2+} the total luminescence quantum efficiency (dominated by the Mn^{2+} PL) has become comparable to that of the undoped sample (dominated by the ZnSe PL). The observed Mn^{2+} PL emission energies are shown in figure 1(a). A shift of the emission maximum to lower energy is observed for increasing Mn^{2+} concentrations. This shift is explained by the formation of pairs of Mn^{2+} ions at higher concentrations [14]. Lifetime measurements, indicating a decrease in Mn^{2+} lifetime for increasing Mn^{2+} concentrations, provide additional evidence for the formation of these Mn^{2+} pairs in the ZnSe nanocrystals [9].

PL excitation spectra of both the ZnSe and the Mn^{2+} related emission show a clear maximum for excitation at ~370 nm, as is shown in figure 1(b). The steep increase beyond 390 nm in the ZnSe excitation spectrum is due to the detection of scattered excitation light, not rejected by the monochromator. The position of the excitation maximum varies from sample to sample, and is explained by variation in particle size. The fact that both ZnSe and Mn^{2+} have the same excitation maximum proves that the excitation of the divalent manganese takes place through energy transfer from the ZnSe host lattice. This agrees with the observation in figure 1(a): incorporation of Mn^{2+} results in a decrease of the ZnSe related PL and a concomitant increase in the Mn^{2+} PL. Energy transfer to Mn^{2+} incorporated in ZnSe has been studied in bulk ZnSe:Mn^{2+} and is known to be efficient [15].

Figure 2(a) shows temperature-dependent PL measurements. The quenching of the ZnSe and Mn^{2+} PL has also been reported for bulk ZnSe:Mn^{2+} [16]. A shift of the ZnSe related PL emission energy to longer wavelengths is observed for increasing temperature. These shifts are fitted using the standard Varshni-equation for the temperature dependence of the bandgap [18], as is shown in figure 2(b). From these fits values were found that compare well with those found for bulk ZnSe.

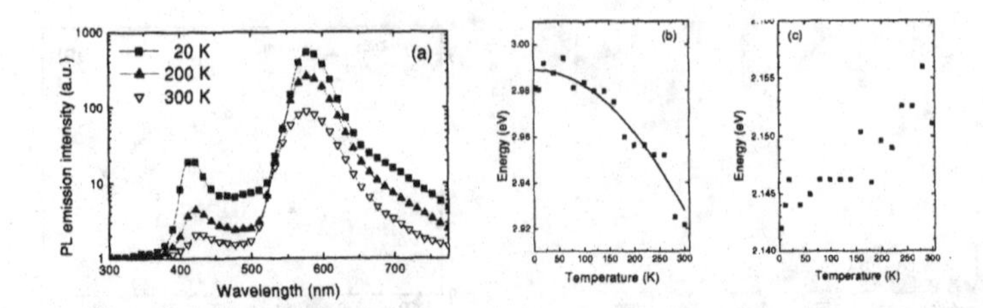

Figure 2: (a) Temperature-dependent PL spectra for the ZnSe:0.7 % Mn^{2+} sample. Note the logarithmic intensity axis. Excitation was at 330 nm. (b) Temperature dependence of the ZnSe emission. The line through the data is a fit using the Varshni equation. (c) Maximum of the Mn^{2+} related PL for different temperatures.

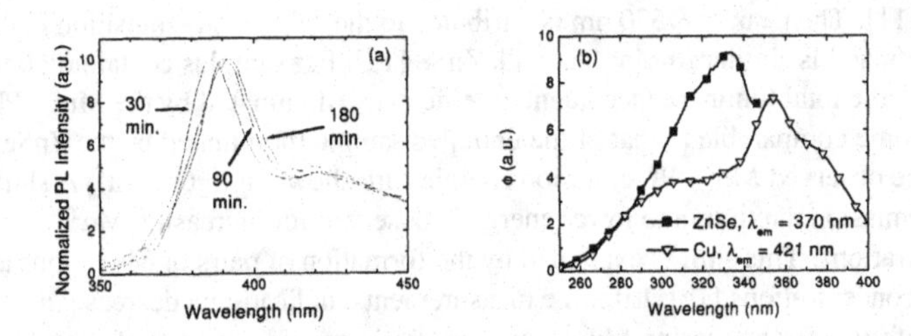

Figure 3: (a) ZnSe:Cu^{2+} suspensions at differnent times. All spectra were measured at room temperature and excited at 330 nm. (b) Excitation spectra for the ZnSe:Cu^{2+} sample (reaction time = 30 minutes). The spectra were measured at 4 K and the emission wavelengths are indicated in the figure.

Figure 2(c) shows the Mn^{2+} PL energy as a function of temperature. The shift of the Mn^{2+} emission energy to shorter wavelength has also been observed in bulk ZnSe:Mn^{2+} [17], but the blue-shift for nanocrystalline ZnSe:Mn^{2+} (12 meV) is less than in the bulk system (23 meV). This may be due to a slightly different lattice expansion for a ZnSe nanocrystal compared to bulk ZnSe.

RESULTS FOR ZnSe:Cu^{2+} NANOCRYSTAL SUSPENSIONS

The PL spectra of several ZnSe:Cu^{2+} suspensions are shown in figure 3(a). The time indicated in the figures represents the time after the injection in the HDA (i.e. the actual synthesis time). The spectra show two distinct emission bands: the ZnSe defect luminescence (~390 nm), and a transition of a conduction band electron to a localised copper defect at ~425 nm [19]. For different synthesis-times, a shift to lower energy of both the ZnSe and Cu^{2+} PL is observed. These shifts are consistent with the growth of the nanocrystal. The

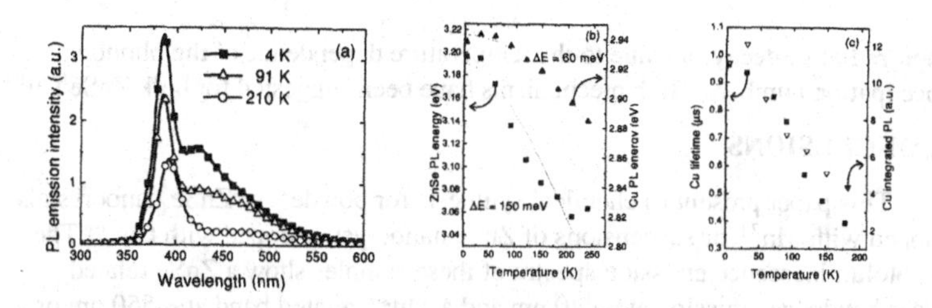

Figure 4: (a) Temperature-dependent PL spectra of a ZnSe:Cu^{2+} sample (reaction time = 60 minutes). Excitation was at 330 nm. (b) Temperature dependence of the ZnSe and Cu^{2+} emissions. The lines through the data are fits using the Varshni equation. (c) Temperature dependence of the Cu^{2+} lifetime. Excitation was at 330 nm.

radius of these NC as obtained from XRD measurements was between 1 and 3 nm, depending on the synthesis time. This agrees with the fact that the ZnSe PL is at higher energy than was observed for the ZnSe:Mn^{2+} crystals (radius of 4 nm). A typical excitation spectrum of both the ZnSe and Cu^{2+} PL peaks, shown in figure 3(b), reveals two excitation bands for the Cu^{2+} luminescence. The higher energy excitation band is due to excitation of the Cu^{2+} via the ZnSe host-lattice (band at 320 nm). Furthermore, the Cu^{2+} ions can also be excited by direct excitation of the Cu^{2+}-center at 350 nm. The energy difference between the two excitation maxima, 0.3 eV, is similar to the energy difference between the two emission peaks shown in figure 3(a) and agrees with the position of a known Cu^{2+} level in bulk ZnSe located 0.3 eV above the top of the valence band [19].

Figure 4(a) shows the PL spectra of a suspension (reaction time = 60 minutes) at different temperatures. The blue-shift of the ZnSe and Cu^{2+} peaks as a function of temperature is shown in figure 4(b). The data can be fitted very well using the Varshni-equation, just like the measurements presented in figures 2(b) and (c). The fact that the Cu^{2+} emission can be fitted well is a direct result of the nature of this emission: as the temperature changes, the conduction band edge energy will shift. Since the Cu^{2+} emission is a transition between the conduction band and a localized state in the bandgap (not temperature dependent), this results in a shift of the Cu^{2+} emission equal to the shift in the energy of the conduction band edge. From the change in energy of the ZnSe bandgap (150 meV) and the Cu^{2+} or conduction band (60 meV) the change in valence-band energy was calculated to be 90 meV.

Figure 4(c) shows the temperature dependence of the Cu^{2+} PL lifetime and the integrated photoluminescence intensity (i.e. the quantum efficiency). From this figure it is clear that the quenching of the lifetime and the luminescence are comparable (the same quenching temperature). This temperature quenching can be explained by backtransfer from the Cu^{2+} to the ZnSe, or by an increased non-radiative decay rate due to the temperature dependence of the phonon occupation numbers. Both mechanisms have been suggested for bulk ZnSe:Cu^{2+}.

CONCLUSIONS

This paper presents a chemical synthesis for powders of ZnSe nanocrystals doped with Mn^{2+} or suspensions of ZnSe nanocrystals doped with Cu^{2+}. The photoluminescence emission spectra of these samples show a ZnSe related near-bandedge emission at ~440 nm and a Mn^{2+} related band at ~550 nm or a Cu^{2+} related band at ~470 nm. The wavelength of the Mn^{2+} related emission (excited through the ZnSe host lattice) shifts to lower energy for increasing amounts of Mn^{2+} incorporated into the ZnSe lattice. The ZnSe related luminescence has a temperature-induced shift similar to that of bulk ZnSe. The temperature dependence of the Cu^{2+} PL maximum is related to the temperature-induced shift of the ZnSe conduction band.

ACKNOWLEDGEMENTS

This work is part of the Research Program of the Priority Program for new Materials (PPM) and was made possible by financial support from the Dutch Association for Scientific Research (NWO).

References

[1] N. C. Greenham, X. Peng and A. P. Alivisatos, *Phys. Rev. B*, 1996, **54**, 17628.
[2] B. O. Dabbousi, M. G. Bawendi, O. Onitsuka and M. F. Rubner, *Appl. Phys. Lett.*, 1995, **66**, 1316.
[3] L. Brus, *J. Phys. Chem.*, 1986, **90**, 2555.
[4] R. N. Bhargava, D. Gallagher, X. Hong and A. Nurmikko, *Phys. Rev. Lett.*, 1994, **72**, 416.
[5] I. Yu, T. Isobe and M. Senna, *J. Phys. Chem. Solids*, 1996, **57**, 373.
[6] J. F. Suyver, R. Bakker, A. Meijerink and J. J. Kelly, *Phys. Stat. Sol. B* **224**, 307 (2001).
[7] J. Leeb, V. Gebhardt, G. Müller, D. Su, M. Giersig, G. McMahon and L. Spanhel, *J. Phys. Chem. B*, 1999, **103**, 7839.
[8] G. H. Schoenmakers, E. P. A. M. Bakkers and J. J. Kelly, *J. Electrochem. Soc.*, 1997, **144**, 2329.
[9] J. F. Suyver, S. F. Wuister, J. J. Kelly and A. Meijerink, *Phys. Chem. Chem. Phys.*, 2000, **2**, 5445.
[10] D. J. Norris, N. Yao, F. T. Charnock and T. A. Kennedy, *Nano Lett.*, 2001, **1**, 3.
[11] M. A. Hines and P. Guyot-Sionnest, *J. Phys. Chem. B*, 1998, **108**, 3655.
[12] C. B. Murray, D. J. Norris and M. G. Bawendi, *J. Am. Chem. Soc.*, 1993 **115**, 8706.
[13] H. Waldmann, C. Benecke, W. Busse, H.-E. Gumlich and A. Krost, *Semicond. Sci. Technol.*, 1989, **4**, 71.
[14] C. R. Ronda and T. Amrein, *J. Lumin.*, 1996, **69**, 245.
[15] U. Stutenbäumer, H.-E. Gumlich and H. Zuber, *Phys. Stat. Sol. B*, 1989, **156**, 561.
[16] J. F. MacKay, W. M. Becker, J. Spałec and U. Debska, *Phys. Rev. B*, 1990, **42**, 1743.
[17] J. Xue, Y. Ye, F. Medina, et al., *J. Lumin.*, 1998, **78**, 173.
[18] Y. P. Varshni, *Physica*, 1967, **34**, 149.
[19] G. Jones and J. Woods, *J. Lumin.*, 1974, **9**, 389.

Mat. Res. Soc. Symp. Proc. Vol. 667 © 2001 Materials Research Society

Pressure Tuning Spectroscopy of Mn^{2+} in Bulk and Nanocrystalline Sulfide Semiconductors

Randy J. Smith, Yongrong Shen, and Kevin L. Bray
Department of Chemistry, Washington State University, Pullman, WA 99164

ABSTRACT

We report the results of high pressure luminescence studies of the emission of Mn^{2+} in bulk $Zn_{0.55}Ga_{0.30}S$ up to ~214 kbar, bulk ZnS up to ~184 kbar, and nanocrystalline ZnS (~8 nm particle sizes) up to ~300 kbar. We observed a strong redshift with pressure (-30(3) cm^{-1}/kbar) for the $^4T_1 \rightarrow {}^6A_1$ emission transition of Mn^{2+} in all three systems. We also observed emission quenching at high pressure in all three systems and attribute the quenching to a pressure induced phase change of the ZnS host lattice to a rocksalt (NaCl) phase.

INTRODUCTION

Mn^{2+}:ZnS is widely used as a phosphor in cathode-ray tubes and electroluminescent displays. The high efficiency of Mn^{2+}:ZnS has motivated efforts to shift or broaden the yellow $^4T_1 \rightarrow {}^6A_1$ emission of Mn^{2+} to achieve a wider range of colors for phosphor or filtered white light displays. Most attempts to tune the emission properties of Mn^{2+} in sulfide semiconductors have focused on chemically modifying the ZnS host lattice or preparing the nanocrystalline form. Recent work has shown that alloying ZnS with Ga leads to a red shift and broadening of the Mn^{2+} emission [1] and that higher excitation efficiencies are achievable in nanocrystalline Mn^{2+}:ZnS [2].

In this paper, we discuss high pressure luminescence studies of the emission of Mn^{2+} in bulk ZnS, $Zn_{1-3x/2}Ga_xS$ (x = 0.30) and nanocrystalline ZnS. Our objective is to use pressure to systematically vary the coordination geometry, crystal field strength and electronic states of Mn^{2+} as well as the band structure of the host lattice in an attempt to understand the chemical and physical factors responsible for determining the emission properties of Mn^{2+} in sulfide semiconductors. In nanocrystalline material, pressure also provides a way to vary particle size and to explore size-dependent properties.

MATERIAL SYNTHESIS

Nanocrystalline Mn^{2+}:ZnS powders were synthesized by a solution phase method [4,5]. We first prepared an aqueous solution containing 380 mL of 0.13 M $Zn(CH_3COO)_2$, 50 g sodium polyphosphate ($Na(PO_3)_n$) as a particle stabilizer, and 20 mL of 0.05 M $Mn(CH_3COO)_2$. After stirring at room temperature for ~15 min., a stoichiometric amount of 1 M Na_2S was slowly added. The solution became opaque and white as nanocrystalline Mn^{2+}:ZnS formed. After centrifuging and washing several times, the resulting nanocrystalline Mn^{2+}:ZnS powder was dried in a rotovap at ~45 degrees. X-ray diffraction analysis indicated that the synthesized nanoparticle powders had a cubic zincblende structure. Bulk 1% Mn^{2+}:$Zn_{0.55}Ga_{0.3}S$ and 1% Mn^{2+}:ZnS powders were prepared by a high temperature solid state reaction technique [3] and were kindly provided by Prof. D. A. Keszler. X-ray powder diffraction measurements on the as-received samples indicated that both powders possessed a hexagonal wurtzite structure.

EXPERIMENTAL DETAILS

Luminescence excitation measurements at ambient conditions were completed with a fluorimeter using solution samples prepared by dispersing the powders in either distilled water or in methanol. High pressure luminescence measurements were performed with a continuous-wave laser system that included a 1 m monochromator, photon counting electronics, argon laser and photomultiplier tube detectors. High pressure was generated by a diamond-anvil cell. Samples used in the high pressure experiments were ~30 µm thick with lateral dimensions of 50–100 µm cut from pellets prepared by pressing the powders. A poly(dimethylsiloxane) fluid was used as the pressure transmitting medium and ruby was used as a pressure calibrant.

AMBIENT PRESSURE RESULTS

Figure 1 shows emission spectra of bulk Mn^{2+}:$Zn_{0.55}Ga_{0.3}S$, bulk Mn^{2+}:ZnS, and two different particle sizes of nanocrystalline Mn^{2+}:ZnS at ambient conditions. The broad emission of bulk Mn^{2+}:ZnS is due to the $^4T_1 \rightarrow {}^6A_1$ transition of Mn^{2+} in sites with tetragonal symmetry. The emission peak is located at ~583 nm at ambient conditions. The two nanocrystalline Mn^{2+}:ZnS samples showed a slight redshift in the emission relative to the bulk sample. The emission band of bulk Mn^{2+}:$Zn_{0.55}Ga_{0.3}S$ exhibited a significant red shift (~50 nm) and was inhomogeneously broadened relative to bulk Mn^{2+}:ZnS. The broadening is a consequence of alloying with Ga and the accompanying statistical variation of Ga and Zn in the second nearest neighbor coordination shell of Mn^{2+}. Li and Keszler [3] observed a systematic decrease in the unit cell dimensions of $Zn_{1-3x/2}Ga_xS$ with increasing Ga content. The lattice constants (a, c) and unit cell volume (V) decrease from a = 3.827 Å, c = 6.266 Å, and V = 79.479 Å^3 for x = 0 to a = 3.789 Å, c = 6.189 Å, and V = 77.102 Å^3 for x = 0.4. The structural variation indicates that the Mn^{2+}-S^{2-} bond length decreases in Mn^{2+}:$Zn_{1-3x/2}Ga_xS$ with increasing Ga content. As a result, the crystal field experienced by Mn^{2+} strengthens with increasing Ga content and the $^4T_1 \rightarrow {}^6A_1$ emission of Mn^{2+} shifts red as a consequence of the increased crystal field splitting.

Excitation spectra of the $^4T_1 \rightarrow {}^6A_1$ emission of Mn^{2+} at ambient conditions are shown in Fig. 2. The excitation spectra were obtained by monitoring the peak emission intensity for each

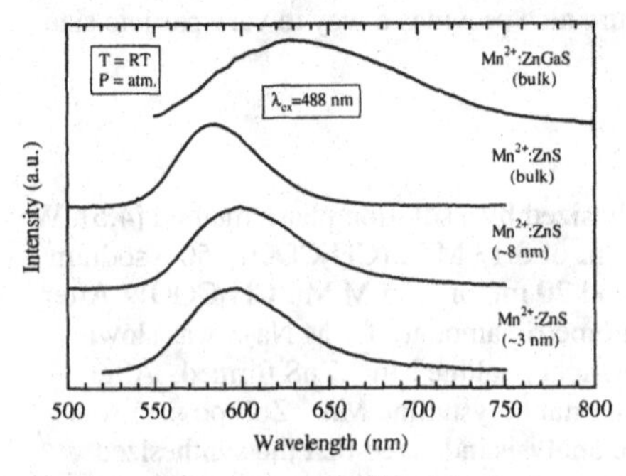

Figure 1. Corrected, normalized emission spectra of bulk Mn:ZnGaS, bulk Mn:ZnS, ~8 nm Mn:ZnS, and ~3 nm Mn:ZnS at ambient conditions.

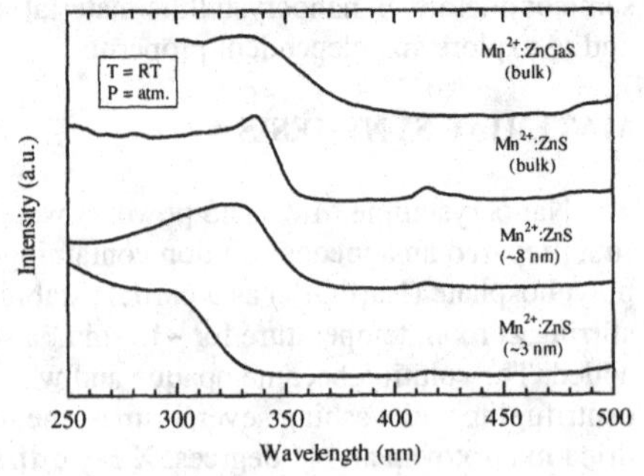

Figure 2. Excitation spectra of bulk Mn^{2+}:ZnS, bulkMn^{2+}:$Zn_{0.55}Ga_{0.3}S$, ~8 nm Mn^{2+}:ZnS, ~3 nm Mn^{2+}:ZnS at ambient conditions.

sample as the excitation wavelength was scanned. The d-d transitions of Mn^{2+} are observed between 400 nm and 500 nm. Below 400 nm, strong excitation bands associated with the host lattice bandedge were observed. The onset of the bandedge excitation occurred at ~355 nm for bulk Mn^{2+}:ZnS, ~360 nm for 8 nm Mn^{2+}:ZnS, ~330 nm for 3 nm Mn^{2+}:ZnS, and ~390 nm for Mn^{2+}:$Zn_{0.55}Ga_{0.3}S$. The quantum confinement effect [6] is readily evident in nanocrystalline Mn^{2+}:ZnS.

HIGH PRESSURE RESULTS

Bulk Mn^{2+}:ZnS

High pressure luminescence spectra of bulk Mn^{2+}:ZnS were measured up to 184 kbar. Figure 3 shows representative emission spectra of bulk Mn^{2+}:ZnS at room temperature. We observed a strong pressure-induced redshift (-31(2) cm^{-1}/kbar) of the ${}^4T_1 \rightarrow {}^6A_1$ emission band of Mn^{2+} in bulk ZnS. The observed shift rate is similar to the one reported by House and Drickamer [7]. The emission intensity increased by a factor of ~4 between ambient pressure and ~102 kbar. House and Drickamer [7] observed a factor of ~2 increase in emission intensity over a 100 kbar pressure range and attributed it to an enhanced radiative decay rate because of the accompanying decrease in the emission lifetime of Mn^{2+}.

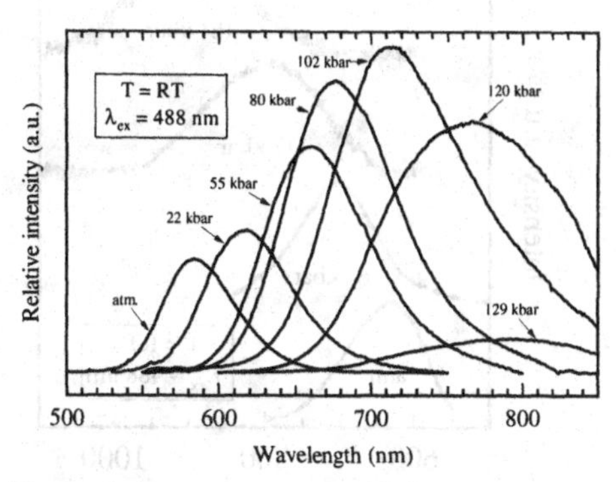

Figure 3. Corrected luminescence spectra of bulk Mn^{2+}:ZnS.

Above ~102 kbar, the emission intensity of Mn^{2+} unexpectedly decreased and was completely quenched at pressures slightly above ~129 kbar. Attempts to induce Mn^{2+} emission above ~129 kbar by varying excitation wavelength (457.9 nm and 514.5 nm) or lowering the temperature to 20 K failed. The ${}^4T_1 \rightarrow {}^6A_1$ emission of Mn^{2+} reappeared upon release of pressure.

Bulk Mn^{2+}:$Zn_{0.55}Ga_{0.3}S$

Luminescence studies of bulk Mn^{2+}:$Zn_{0.55}Ga_{0.3}S$ were completed up to 214 kbar. Figure 4 shows representative luminescence spectra of bulk Mn^{2+}:$Zn_{0.55}Ga_{0.3}S$ and the variation of the peak emission energy as a function of pressure. Between ambient pressure and ~170 kbar, the ${}^4T_1 \rightarrow {}^6A_1$ emission band exhibited a redshift (-27(1) cm^{-1}/kbar) and a weak decrease in intensity. A discontinuous change in shift rate and pronounced decrease in emission intensity were observed above ~170 kbar. The emission was completely quenched above ~214 kbar and we were unable to induce emission by varying the excitation wavelength (328 nm, 457.9 nm, 514.5 nm, and 628.5 nm) or lowering the temperature to 20 K. The pressure-induced emission quenching in bulk Mn^{2+}:$Zn_{0.55}Ga_{0.3}S$ is similar to that observed in bulk Mn^{2+}:ZnS.

Nanocrystalline Mn^{2+}:ZnS

Figure 5 shows the emission spectra observed from nanocrystalline Mn^{2+}:ZnS (~8 nm particle size) up to 172 kbar. A linear redshift (-30(1) cm^{-1}/kbar) was observed for the emission band of Mn^{2+} over this pressure range. Between ambient pressure and ~100 kbar, the emission band showed no significant change in intensity. A decrease in emission intensity, however, was observed above ~100 kbar with complete quenching occurring above ~200 kbar.

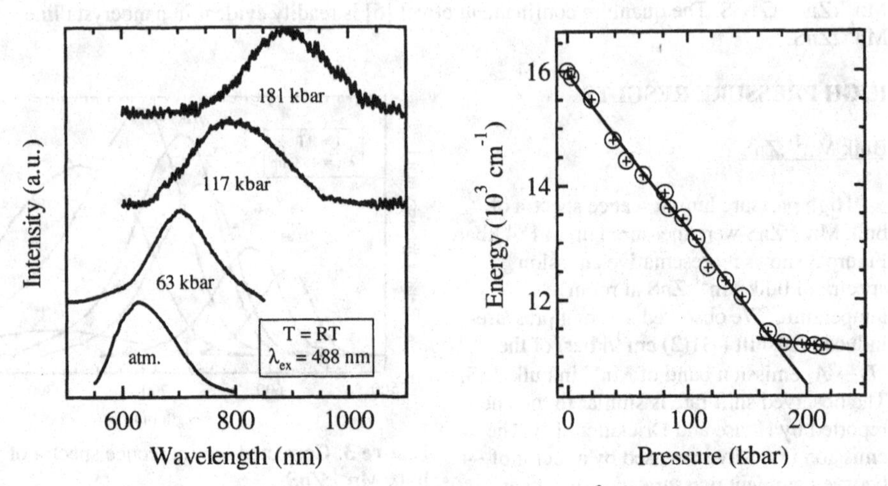

Figure 4. Corrected, normalized luminescence spectra of bulk Mn^{2+}:$Zn_{0.55}Ga_{0.3}S$ (left) and variation of peak emission energy with pressure (right).

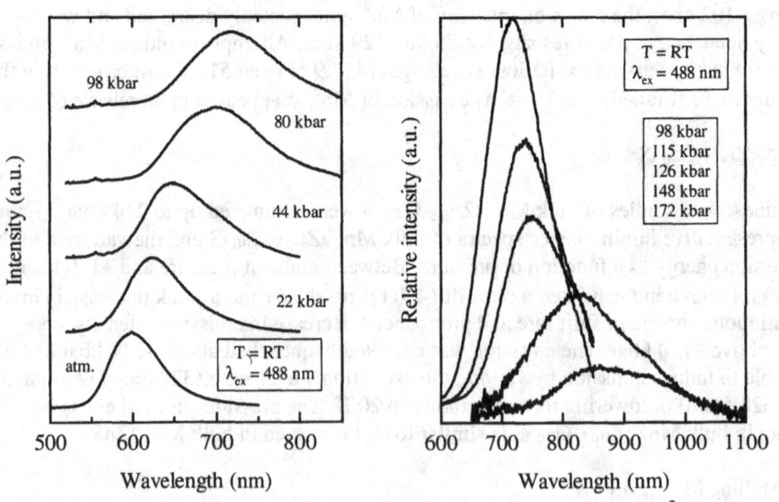

Figure 5. Corrected, normalized luminescence spectra of nanocrystalline Mn^{2+}:ZnS (~8 nm particle size) up to 98 kbar (left). Corrected, unnormalized luminescence spectra above 98 kbar (right).

DISCUSSION

Bulk ZnS is known to undergo a structural transformation from a cubic (*B*3, zincblende) or hexagonal (*B*4, wurtzite) structure to a NaCl structure (*B*1, rocksalt) at a reported transition pressure, P_t, between 120 and 150 kbar [8-11]. More recently, Jiang *et al.* [12] and Qadri *et al.* [13] found that the transition pressure of nanocrystalline ZnS increases with decreasing particle size. In a synchrotron x-ray diffraction study, Qadri *et al.* [13] obtained P_t = 190 kbar for 2.8 nm cubic ZnS and P_t = 150 kbar for 25.3 nm hexagonal ZnS. Jiang *et al.* [12] reported P_t = ~190 kbar for 36 nm size cubic ZnS and P_t = ~205 kbar for 11 nm size cubic ZnS in an electrical resistance study.

Ves *et al.* [10] employed high pressure absorption to study the band structure of bulk cubic ZnS and found that the band gap increased from ~3.7 eV at ambient pressure to ~4.3 eV immediately before the zincblende → rock salt phase transition that they observed at ~150 kbar. They further reported a change in band structure from direct-to-indirect during the phase change and that the indirect gap of ZnS in the rock salt phase exhibits a broad absorption edge with an onset near 2 eV (~620 nm) that varies weakly with pressure.

Based on the reported phase behavior of bulk and nanocrystalline ZnS, we preliminarily attribute the observed quenching of the ${}^4T_1 \rightarrow {}^6A_1$ emission of Mn^{2+} in bulk ZnS, bulk $Zn_{0.55}Ga_{0.30}S$, and ~8 nm nanocrystalline ZnS at high pressure to a pressure-induced phase transition from the cubic or hexagonal structure to the NaCl structure. The quenching pressures, P_q, observed for the emission of Mn^{2+} (P_q = ~129 kbar for bulk Mn^{2+}:ZnS, P_q = ~214 kbar for bulk Mn^{2+}:$Zn_{0.55}Ga_{0.30}S$, and P_q = ~200 kbar for ~8 nm nanocrystalline Mn^{2+}:ZnS) are consistent with reported phase transition pressures. Visual observation during the high pressure experiments showed a rapid change in the appearance of the samples from transparent below the quenching pressure to opaque above the quenching pressure.

Two possible effects could explain the quenching of the ${}^4T_1 \rightarrow {}^6A_1$ emission of Mn^{2+} in the high pressure rocksalt ZnS phase. First, in the rocksalt ZnS phase, Mn^{2+} occupies a six-fold coordinate site with O_h symmetry. The presence of inversion symmetry makes all $d \leftrightarrow d$ transitions formally forbidden. Since inversion symmetry is absent in the initial zincblende or wurtzite phases, this effect is consistent with the strong reduction observed for the emission intensity at high pressure. Second, the reduction in ZnS bandgap that accompanies the transition to the rocksalt phase may force the ${}^4T_1$ emitting state of Mn^{2+} to become resonant with the conduction band. Promotion of an electron into a resonant excited state normally leads to delocalization and transfer of the electron to the host conduction band and a non-radiative return of the electron to the ground electronic state of the emitting center.

Finally, we note that the quenching of Mn^{2+} emission occurred over a much wider pressure range in nanocrystalline Mn^{2+}:ZnS than in bulk Mn^{2+}:ZnS or Mn^{2+}:$Zn_{0.55}Ga_{0.30}S$ and that the onset of the reduction in intensity in the nanocrystalline sample occurred at pressures well below those associated with the transparent-to-opaque visual transformation. These observations suggest that a quenching mechanism independent of the phase change is operable in the nanocrystalline sample. One possibility is a pressure-induced interaction between Mn^{2+} and surface states of the nanocrystalline ZnS host. Luminescence experiments using band-to-band and above bandedge excitation are currently underway to further understand the quenching behavior of nanocrystalline Mn^{2+}:ZnS. We are also initiating high pressure studies of other particle sizes.

SUMMARY

High pressure luminescence studies of Mn^{2+} in bulk $Zn_{0.55}Ga_{0.30}S$, bulk ZnS, and nanocrystalline ZnS (~8 nm particle size) revealed a strong redshift (-30(3) cm^{-1}/kbar) for the $^4T_1 \rightarrow ^6A_1$ emission band of Mn^{2+} and a pronounced decrease in emission intensity. The pressure range over which the intensity decreased varied with the system: ~100 - 129 kbar for bulk Mn^{2+}:ZnS, ~170 - 214 kbar for bulk Mn^{2+}: $Zn_{0.55}Ga_{0.30}S$, and ~100 - ~210 kbar for nanocrystalline Mn^{2+}:ZnS (~8 nm particle size). We preliminarily attribute the quenching to a pressure-induced phase transition of the ZnS host lattice from a zincblende or wurtzite structure to a rocksalt (NaCl) structure. In bulk Mn^{2+}:ZnS and bulk Mn^{2+}: $Zn_{0.55}Ga_{0.30}S$, the sample began to turn opaque at the onset of emission quenching and was fully opaque when complete quenching occurred. In nanocrystalline Mn^{2+}:ZnS (~8 nm particle size), the sample appearance also changed from transparent to opaque; but the onset of emission quenching preceded the first indications of opaqueness by ~100 kbar. This observation suggests that pressure enhances non-radiative processes unique to the nanocrystalline phase prior to the phase change.

ACKNOWLEDGMENTS

Acknowledgment is made to the donors of the Petroleum Research Fund, administered by the ACS, for partial support of this research. Additional financial support was provided by the National Science Foundation. We thank Prof. D. A. Keszler for generously supplying the bulk Mn^{2+}-doped samples used in this study and also G. Cunningham for completing x-ray powder diffraction experiments. RJS further acknowledges mini-grant and travel grant support from the College of Sciences and the Graduate School of Washington State University.

REFERENCES

1. V. Dimitrova, A. Draeseke, J. Tate, T. Yokoyama, B. L. Clark, and D. A. Keszler, *Appl. Phys. Lett.* **75**, 2353 (1999).
2. R. N. Bhargava, D. Gallagher, X. Hong, and A. Nurmikko, *Phys. Rev. Lett.* **72**, 416 (1994).
3. D. Li, Ph.D. thesis, Oregon State University, (1999).
4. A. A. Bol and A. Meijerink, *Phys. Rev. B* **58**, R15997 (1998).
5. I. Yu, T. Isobe, and M. Senna, *J. Phys. Chem. Solids* **57**, 373 (1996).
6. P. E. Lippens and M. Lannoo, *Phys. Rev. B* **39**, 10935 (1989).
7. G. L. House and H. G. Drickamer, *J. Chem.. Phys.* **67**, 3230 (1977).
8. G. A. Samara and H. G. Drickamer, *J. Phys. Chem. Solids* **23**, 457 (1962).
9. G. J. Piermarini, S. Block, J. D. Barnett, and R. A. Forman, *J. Appl. Phys.* **46**, 2774 (1975).
10. S. Ves, U. Schwarz, N. E. Christensen, K. Syassen, and M. Cardona, *Phys. Rev. B* **42**, 9113 (1990).
11. S. Desgreniers, L. Beaulieu, and I. Lepage, *Phys. Rev. B* **61**, 8726 (2000).
12. J. Z. Jiang, L. Gerward, D. Frost, R. Secco, J. Peyronneau, and J. S. Olsen, *J. Appl. Phys.* **86**, 6608 (1999).
13. S. B. Qadri, E. F. Skelton, A. D. Dinsmore, J. Z. Hu, W. J. Kim, C. Nelson, and B. R. Ratna, *J. Appl. Phys.* **89**, 115 (2001).

Synthesis and Processing

Mat. Res. Soc. Symp. Proc. Vol. 667 © 2001 Materials Research Society

Cerium Doped Garnet Phosphors for Application in White GaN-based LEDs

Jennifer L. Wu[1], Steven P. Denbaars[2], Vojislav Srdanov[3], Henry Weinberg[4]
[1]Department of Chemical Engineering, [2]Department of Materials, [3]Center for Polymers and Organic Solids, University of California, Santa Barbara
Santa Barbara, CA, 93106
[4]Symyx Technologies, 3100 Central Expressway
Santa Clara, CA, 95051

ABSTRACT

Recently, renewed interest has emerged for the development of visible light, down-converting phosphors for application in white light emitting diodes (LEDs). In such devices, a blue GaN LED can act as a primary light source, serving as an efficient pump to excite photoluminescence in a phosphor with subsequent emission occurring at lower energies. The combination of blue light from the LED chip and emission from the phosphor(s) produces white light. It was recently reported that a combinatorial approach to synthesize and screen potential inorganic phosphors for use in white LEDs could aid in identifying improved phosphors for blue to yellow down conversion. Solid state thin-film arrays (libraries) based on the garnet structure $(A_{1-x},B_x)_{3-z}(C_{1-y},D_y)_5O_{12}:Ce^{3+}{}_z$, where A, B = Y, Gd, Lu, La; C, D = Al, Ga; x and y = 0 to 1.0; and z = 0.03, were synthesized, and x-ray diffraction was used to select library samples of the crystalline garnet phase. Libraries of these various garnets were then characterized spectroscopically and their properties compared to traditionally prepared bulk powder phosphors of similar composition. Emission and excitation trends show that as larger cations are substituted for Y (A = Y), emission and excitation are red-shifted and as larger cations are substituted for Al (C = Al), emission and excitation are blue-shifted. If smaller cations are substituted for Y and Al an opposite trend is observed. Temperature dependence photoluminescence measurements and emission/excitation trends are also examined.

INTRODUCTION

Several methods have been used to develop white LEDs, which include use of conjugated polymers [1], organic dye molecules [2], and inorganic phosphors [3] as luminescent converters of blue LED light. In such a device, an InGaN blue LED can serve as the primary light source, acting as a pump to generate photoluminescence (PL) in organic or inorganic luminescent materials in which subsequent photon emission occurs at lower energies [1-5], cf., Fig. 1. $Y_3Al_5O_{12}:Ce^{3+}$ (YAG:Ce^{3+}) emission is well suited for such applications, since when properly mixed its yellow emission under blue-light excitation yields white light [2,3]. The excitation and emission of YAG:Ce^{3+} have been well studied [6,7]. It is known that substitution of Gd^{3+} and Ga^{3+} for Y^{3+} and Al^{3+}, respectively, in the garnet host shifts the emission of YAG:Ce^{3+} so that different shades of white light can be realized [2-5]. The addition of the larger ion Gd^{3+} for Y^{3+} red-shifts YAG:Ce^{3+} emission and substitution of Ga^{3+} for Al^{3+} tends to blue shift the characteristic yellow emission. A complex compositional space, such as the substituted YAG host or ternary systems, lends itself well to the combinatorial chemistry approach.

The purpose of this work is to use the combinatorial chemistry technique to understand and examine spectral trends in phosphors for application in blue to yellow conversion LEDs. Both YAG:Ce^{3+} and various garnet hosts ($A_{3-x}B_xC_{5-y}D_yO_{12}:Ce^{3+}$) were synthesized and characterized. Library and bulk phosphor excitation and emission trends, as well as color point are compared to demonstrate that the combinatorial approach can be used to rapidly screen potential phosphors for use in luminescence down conversion LEDs.

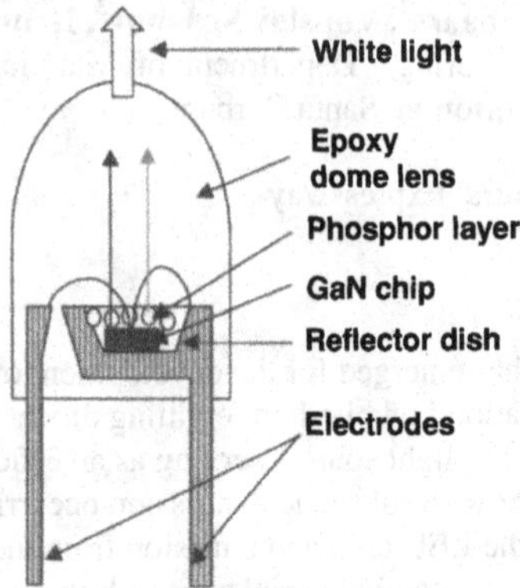

Figure 1. Structure of a GaN-based white light emitting diode.

EXPERIMENTAL

The library synthesis and characterization methodology has been described previously [10]. Bulk garnet powders of the type $(A_{1-x},B_x)_{3-z}(C_{1-y},D_y)_5O_{12}:Ce^{3+}{}_z$, where A, B = Y, Gd, Lu, La; C, D = Al, Ga; x and y = 0 to 1.0; and z = 0.03, were synthesized using a mixed oxide solid-state approach [9]. The starting materials were Y_2O_3, Gd_2O_3, Lu_2O_3, Al_2O_3, Ga_2O_3, and CeO_2. These were milled in the appropriate stoichiometric amounts and fired at 1450°C for 6 hours in air or a hydrogen/argon mixture. The product was then reground. X-ray diffraction was performed on the bulk samples to verify crystallization of the garnet structure.

Emission and excitation spectra were collected on a scanning fluorimetry setup, employing two SPEX model 1680 double spectrometers and a Xe lamp excitation source. The collected signal was detected with a photon counting Hamamatsu R928HA photomultiplier tube, along with a Pacific Instruments AD-126 amplifier/discriminator and Model 126 photometer. Temperature dependent photoluminescence measurements were collected using the 458-nm line from a continuous wave Ar^+ laser as an excitation source. Photoluminescence was spectrally resolved by a 0.5 m monochromator and CCD detector. An Oxford Instruments Optistat^{CF} dynamic continuous flow cryostat was used for the temperature dependent measurements.

RESULTS AND DISCUSSION

A preliminary blue to yellow phosphor screening using CCD imaging was utilized to examine the intensities of the various garnet phosphors [10]. These high throughput screens of the $A_{3-x}B_xC_{5-y}D_yO_{12}$:Ce libraries revealed variations in luminescence intensity due to compositional changes. The most intense elements of a library are dependent on the excitation wavelength. This is observed in the $Gd_{3-x}Lu_xAl_{5-y}Ga_yO_{12}$:Ce library shown in Figure 2. At 465-nm excitation, $Gd_{2.4}Lu_{0.6}Al_5O_{12}$, element (1,3) using a (row, column) notation, and $Gd_{2.7}Lu_{0.3}Al_{4.5}Ga_{0.5}O_{12}$, element (2,2), are the brightest compositions. As the excitation shifts to lower wavelengths, Ga-substituted elements increase in intensity and element (4,1), $Gd_3Al_{3.5}Ga_{1.5}O_{12}$:Ce, becomes the most intense at 430 nm. Similarly, other garnet libraries were characterized and the most intense garnet compositions from those libraries were identified. The most intense library elements were then synthesized in bulk to compare bulk and thin-film trends and to more closely examine their spectroscopic properties. Those results will be reported in a future publication [11].

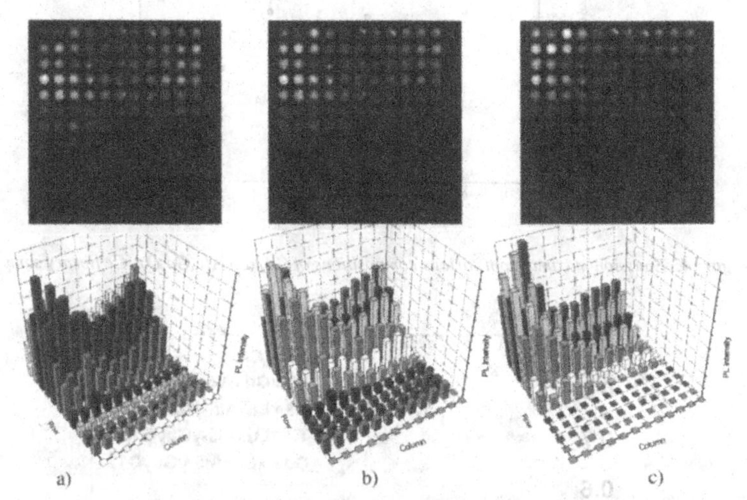

Figure 2. High throughput images of $Gd_{3-x}Lu_xAl_{5-y}Ga_yO_{12}$:Ce library along with histograms of corresponding intensities at a) 430-nm, b) 450-nm and c) 465-nm excitation.

Data show that emission shifts are not purely a function of ionic radii; the emission maximum clearly depends on the crystalline site and degree of host lattice substitution. Along with shifting peak emission, $\lambda_{ex,max}$ decreases as larger cations are substituted in the Al site of YAG, while substitution in the Y site slightly increases $\lambda_{ex,max}$, cf., Fig. 3 and 4. The opposite is true when smaller cations are substituted in those respective garnet sites. Figure 5 shows a CIE plot of the various garnet libraries synthesized, illustrating the range of colors we can achieve through appropriate substitution in the garnet lattice.

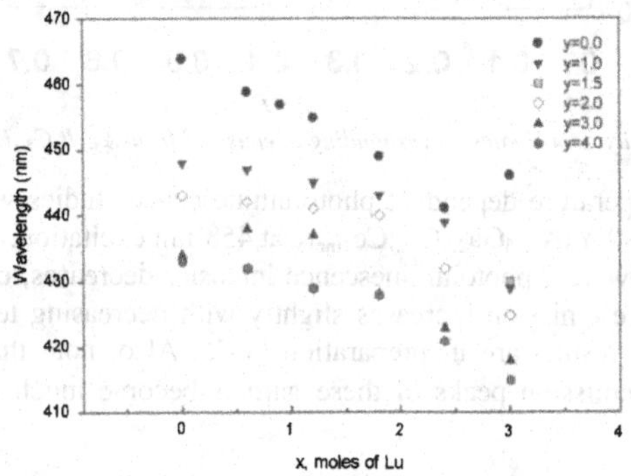

Figure 3. a) Excitation maxima from various compositions of $Gd_{3-x}Lu_xAl_{5-y}Ga_yO_{12}$:Ce library samples.

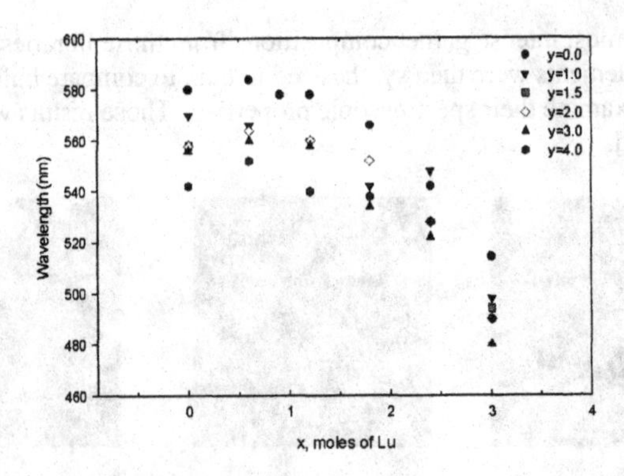

Figure 4. *Emission maxima from various compositions of $Gd_{3-x}Lu_xAl_{5-y}Ga_yO_{12}$:Ce library samples.*

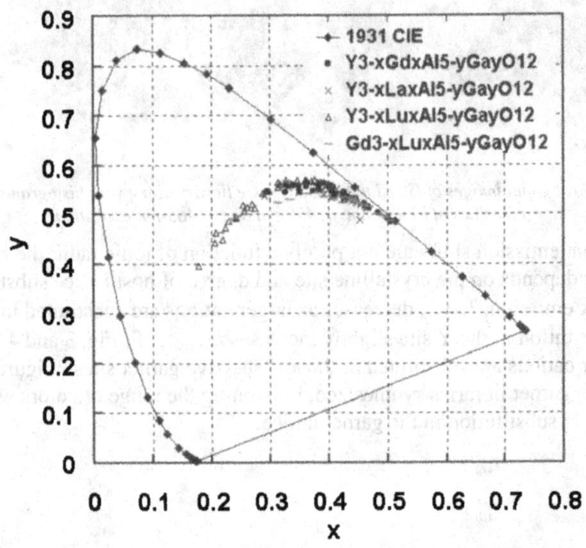

Figure 5. *CIE chromaticity coordinates corresponding to emission from $A_{3-x}B_xC_{5-y}D_yO_{12}$:Ce garnet libraries.*

In addition, temperature dependent photoluminescence studies were performed for bulk powder YAG:$Ce_{1mol\%}$ and $Y_3Al_{2.5}Ga_{2.5}O_{12}$:$Ce_{1mol\%}$ at 458-nm excitation. In the case of YAG:Ce, as the temperature is lowered, photoluminescence intensity decreases, cf. Fig. 6. On the other hand, $Y_3Al_{2.5}Ga_{2.5}O_{12}$:Ce emission increases slightly with decreasing temperatures, cf., Fig. 7. Future details on these results are in preparation [11]. Also, note that in both systems, the $^2D \rightarrow F_{5/2}$ and $^2D \rightarrow F_{7/2}$ emission peaks of these garnets become much more distinguishable at lower temperatures.

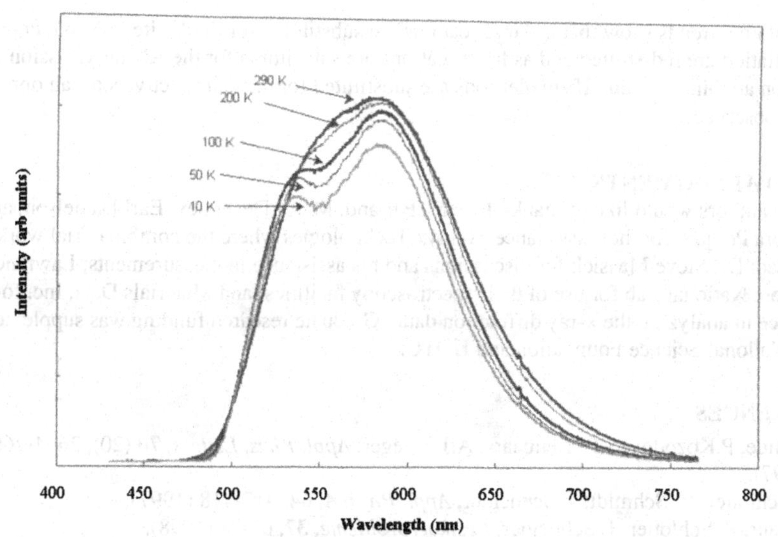

Figure 6. *Emission spectra of YAG:Ce under 458-nm excitation at various temperatures.*

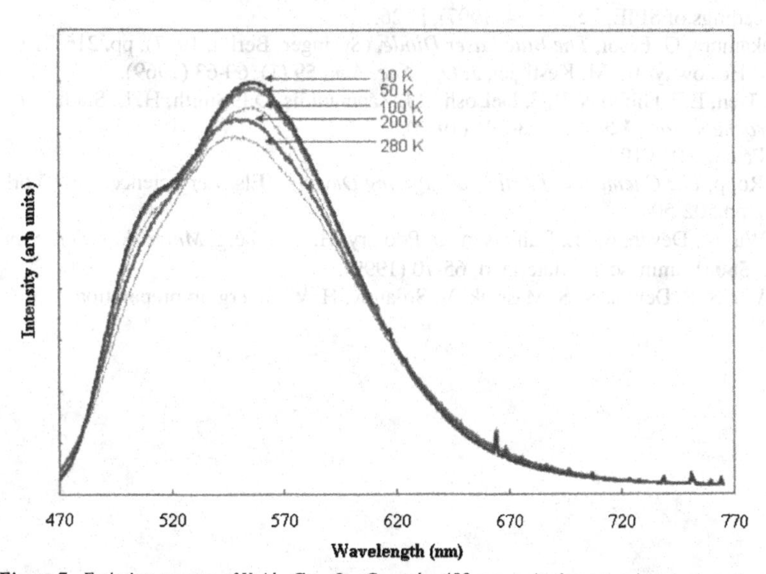

Figure 7. *Emission spectra of $Y_3Al_{2.5}Ga_{2.5}O_{12}$:Ce under 458-nm excitation at various temperatures.*

CONCLUSIONS

The thin film combinatorial approach was demonstrated to be effective in qualitatively evaluating the luminescence properties of blue to yellow down-conversion phosphors. Emission and excitation trends show that as larger cations are substituted for the Y site in YAG, emission and excitation are red-shifted and as larger cations are substituted for the Al site, emission and excitation are blue-shifted. If smaller ions are substituted for those respective sites an opposite trend is observed.

ACKNOWLEDGMENTS

The authors would like to thank: Eric McFarland, Marty Devenney, Earl Danielson, and Damadora Poojary for their assistance; Symyx Technologies where the combinatorial work was conducted; Dr. Steve Massick for discussions and his assistance in measurements; Lawrence Livermore National Lab for use of their spectroscopy facilities; and Materials Data, Inc. for their assistance in analyzing the x-ray diffraction data. Graduate research funding was supplemented by the National Science Foundation and HOTC.

REFERENCES

1. F. Hide, P.Kozodoy, S.P. Denbaars, A.J. Heeger, *Appl. Phys. Letters*, **70** (20), 2664-2666 (1997).
2. P. Schlotter, R. Schmidt, J. Schneider, *Appl. Phys. A*, **64**, 417-418 (1997).
3. J. Baur, P. Schlotter, J. Schneider, *Festkoerprobleme*, **37**, 67-78 (1998).
4. S. Nakamura, in *Light-Emitting Diodes: Research, Manufacturing, and Applications*, (Proceedings of SPIE, Feb 13-14, 1997), p. 26.
5. S. Nakamura, G. Fasol, *The Blue Laser Diode*, (Springer, Berlin, 1997), pp. 216-219.
6. W. W. Holloway, Jr., M. Kestigan, *J. Opt. Soc. Am.*, **59** (1), 60-63 (1969).
7. T. Y. Tien, E.F. Gibbons, R.G. DeLosh, P.J. Zacmanidis, D.E.Smith, H. L. Stadler, *J. Electrochem. Soc.*, **120** (2), 278-281 (1973).
8. U.S. Patent 6,013,199.
9. R.C. Ropp, *The Chemistry of Artificial Lighting Devices*, (Elsevier Science, New York, 1993), pp.502-504.
10. J. L. Wu, M. Devenney, E. Danielson, D. Poojary, H. Weinberg, *Mater. Res. Soc. Symp. Proc.*, **560** (Luminescent Materials), 65-70 (1999).
11. J. L. Wu, S. P. Denbaars, S. Massick, V. Srdanov, H. Weinberg, in preparation.

Mat. Res. Soc. Symp. Proc. Vol. 667 © 2001 Materials Research Society

Structural and Optical Properties of ZnS:Mn Films Grown by Pulsed Laser Deposition

K. M. Yeung, S. G. Lu, C. L. Mak* and K. H. Wong
Department of Applied Physics and Materials Research Centre, The Hong Kong Polytechnic University, Hung Hom, Hong Kong, China.
* corresponding author: apaclmak@polyu.edu.hk

ABSTRACT

High-quality manganese-doped zinc sulfide (ZnS:Mn) thin films have been deposited on various substrates using pulsed laser deposition (PLD). Effects of back-filled Ar pressure and substrate temperature on the structural as well as optical properties of ZnS:Mn films were studied. Structural properties of these films were characterized by X-ray diffraction (XRD) and scanning electron microscopy (SEM). Photoluminescence (PL) and optical transmittance were used to characterize the optical properties of these films. Our results reveal that ZnS:Mn films were polycrystalline with a mixed phase structure consisting of both wurtzite and zinc-blende structure. The ratio of these two structures was strongly depended on the change of substrate temperature. Low substrate temperature facilitated the formation of zinc-blende structure while the wurtzite phase became dominant at high substrate temperature. ZnS:Mn films with preferred wurtzite structure were obtained at a substrate temperature as low as 450°C. An orange-yellow emission band was observed at ~590 nm. As the substrate temperature increased, the peak of this PL band shifted to a shorter wavelength. Furthermore, shifts in the absorption edge and the energy gap due to the change in substrate temperature were also observed. The variation in these optical properties will be correlated to their structural change.

INTRODUCTION

Zinc sulfide (ZnS) thin films have received much attention in view of their wide applications to opto-electronic devices, in particular, electroluminescent (EL) devices. Different techniques have been employed to prepare ZnS films, such as sputtering [1], metallo-organic chemical vapour deposition [2], molecular beam epitaxy [3], atomic layer epitaxy[4], sol-gel[5], spray pyrolysis[6] and pulsed laser deposition (PLD)[7, 8]. Among various techniques, PLD is a promising method for thin film deposition because of its low substrate temperature, good stoichiometry, fast deposition rate and relative simple setup.

In this study, high quality ZnS:Mn films were fabricated by PLD method. Effects of different processing parameters (back-filled Ar pressure and substrate temperature) on the structural properties of these ZnS:Mn films were investigated. Changes in optical transmittance and photoluminescence properties associated with the structural change were also studied.

EXPERIMENTAL DETAILS

Laser ablation of ZnS:Mn was carried out with an ArF excimer laser (Lambda Physik, 193 nm, 20 ns) focused on the rotating target at an incidence angle of 45°. The laser fluence was ~ 2 J/cm^2. The ZnS target used in this work were doped with Mn of 1% by weight and obtained from Target Materials, Inc. Deposition experiments were performed in a vacuum chamber which was evacuated to a base pressure of 1.5 x 10^{-4} Torr and then back filled with various Ar pressures ranging from 300 mTorr to 1.5 Torr. Films were grown on either (001) Si or ITO-coated glass substrates which were mounted on a substrate holder equipped with a heater capable to heat the substrate up to 500°C. The distance between the substrate holder and the target surface was fixed at 4.5 cm in our experiment. The influence of several deposition parameters (substrate temperature and back-filled Ar pressure) on the structural and the optical properties of the films has been investigated.

The crystallography of the as-grown ZnS:Mn films was studied by XRD (Philips X'pert) using Cu K_α radiation. The surface morphology and the grain size were studied using SEM (Leica, stereoscan 440). In our PL measurement, ZnS:Mn films were excited by a 488 nm laser light from an argon ion gas laser (Coherent Innova 70) with power of 100 mW. PL spectra were recorded by a double grating monochromator (Spex 1403) equipped with the standard photon counting equipment. Optical transmittance measurement was performed by an UV-VIS scanning spectrophotometer (Shimadzu UV – 2101 PC).

RESULTS

Structural characterization

Figure 1a shows the XRD patterns of ZnS:Mn films grown on (001) Si with different substrate temperatures. At low substrate temperature (150°C), only diffraction peaks arisen from zinc-blende (cubic) structure with a preferred (111) orientation of ZnS were observed. As substrate temperature increases, peaks corresponded to wurtzite (hexagonal) structure were appeared. As a result, the films contained a mixed phase of both zinc-blende and wurtzite structure, and became polycrystalline. As the temperature further increased above 450°C, the wurtzite phase without preferred orientation became dominant. A similar substrate temperature dependence was also observed in ZnS:Mn films grown on ITO-coated glass (Fig. 1b). This demonstrated that low substrate temperature facilitated the formation of zinc-blende structure with a preferred (111) orientation while wurtzite phase became dominant at high substrate temperature. ZnS:Mn films grown on (001) Si prepared under different back-filled Ar pressure were also examined by XRD. The substrate temperature was fixed at 450°C. The XRD patterns, in general, showed a mixture of both zinc-blende and wurtzite structure indicating that polycrystalline films of mixed phase structure were obtained. No significant change in the structure was observed by varying the back-filled Ar pressure ranging from 300 mTorr to1.5 Torr.

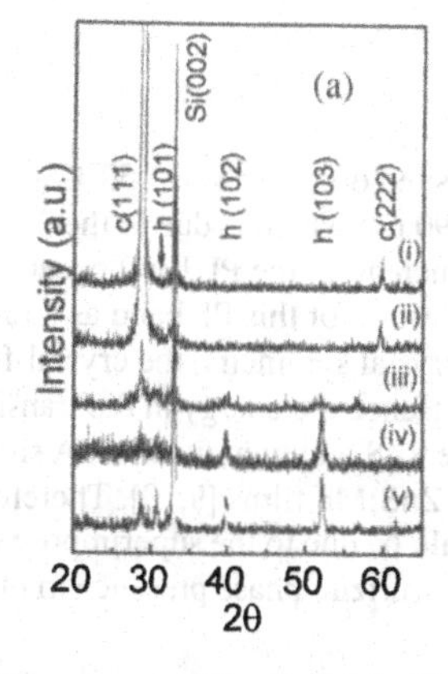

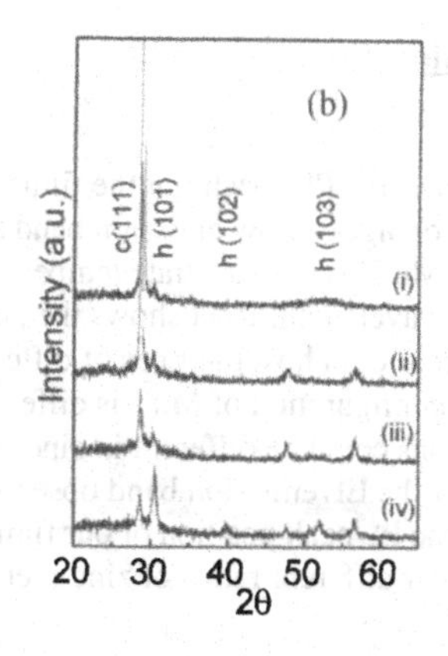

Figure 1. XRD patterns of the ZnS:Mn films grown on (a) (001) Si with substrate temperatures of (i) 150°C, (ii) 250°C, (iii) 350°C, (iv) 450°C, and (v) 500°C; on (b) ITO –coated glass with temperatures of (i) 150°C, (ii) 250°C, (iii) 350°C, and (iv) 450°C.

In addition to the crystallinity orientation, we have studied the grain size and surface morphology of our ZnS:Mn films. The SEM images in Fig. 2 reveal that the deposited films are, in general, smooth and crack-free with grain size of ~100 nm. However, there is a change in the shape of the grain as substrate temperature increased. For films grown at lower substrate temperature (150°C) as shown in Fig. 2a, the grains were circular in shape with a narrow size distribution. As the deposition temperature increased, the grains were elongated and showing an irregular grain growth with some degree of grain agglomeration, resulting in a much wider size distribution. Similar temperature dependence of grain size was also observed in ZnS:Mn films grown on ITO-coated glass. On the other hand, the films grown on Si substrates with various Ar pressures did not show significant change in the films' surface morphology. Our results demonstrated that the structure of ZnS:Mn films are strongly dependent on substrate temperature. However, the back-filled Ar pressure seemed to have no significant effect on the structural properties of ZnS:Mn films.

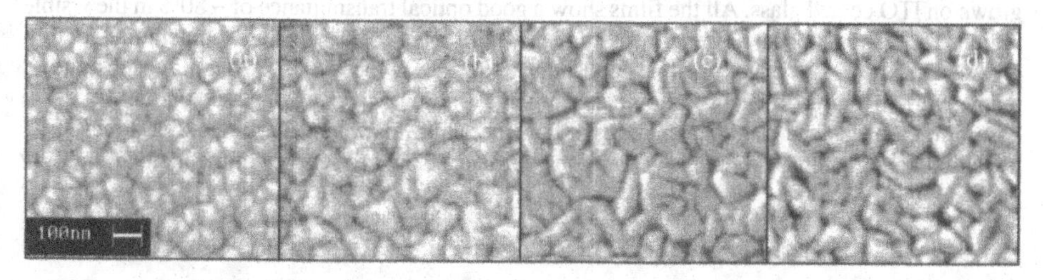

Figure 2. SEM images of ZnS:Mn films grown on (001) Si with 500 mTorr back-filled Ar pressure and at substrate temperatures of (a) 150°C, (b) 250°C, (c) 350°C, and (d) 450°C.

Optical properties

Figure 3 shows the PL spectra of the films deposited on (001) Si at 150°C and 450°C. The PL spectra show an orange-yellow emission band at ~590 nm which is due to the $^4T_1(G) - {}^6A_1(S)$ transition of the Mn^{2+}[8]. As substrate temperature increased, the PL band became narrower and shifted to lower wavelength. Inset shows the peak position of this PL band as a function of substrate temperature. In host matrices of different crystal symmetry, the crystal-field splitting of the $3d^5$ electron configuration of Mn^{2+} is difference. Hence the energy in the transition of $^4T_1(G) - {}^6A_1(S)$ of Mn^{2+} ions could be different in zinc-blende ZnS and wurtzite ZnS. A similar shift have been reported for the EL emission band observed in ZnS:Mn films [9,10]. Therefore, we suggest that the shift in the PL peak position of our films could be due to the superimpose of the emissions of Mn^{2+} ions under different ratios of zinc-blende to wurtzite phase presence in our ZnS films.

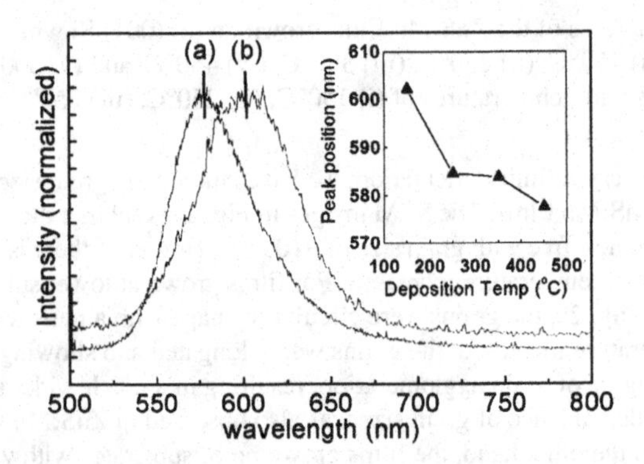

Figure 3. PL spectra of ZnS:Mn films on (001) Si deposited at (a) 150 C and (b) 450 C. Inset is the peak position of the PL band as a function of substrate temperature.

Figure 4 shows the substrate temperature dependence of the transmittance of ZnS:Mn films grown on ITO coated glass. All the films show a good optical transmittance of ~ 80% in the visible region. The transmittance, however, was slightly deteriorated as the substrate temperature got to ~ 450°C. The absorption edge for films deposited at 150°C was located at ~370 nm. As substrate temperature increased, this absorption edge shifted to shorter wavelength. Based on these absorption profiles, the band gap energies of our ZnS:Mn films were determined. For direct transition, the relation between the absorption coefficient α and the energy of the incident light $h\nu$ is given by:[6,11,12]

$$(\alpha h\nu)^2 = B(h\nu - E_g) \qquad (1)$$

where B is a constant and E_g is the band gap energy. Figure 5 shows the plot of $(\alpha h\nu)^2$ vs $h\nu$. By extrapolating the linear portion of the curve to zero, films with band gap energies of 3.50, 3.47, 3.44 and 3.28 eV were obtained at deposition temperatures of 450, 350, 250, 150°C, respectively. For ZnS, zinc-blend structure has a smaller energy gap than that of the wurtzite structure.

Therefore at higher deposition temperature, the films possess a higher wurtzite to zinc-blende ratio resulting in a larger band gap. Therefore, as suggested by our XRD data, the shifts of both absorption edge and band gap energy are related to the change of wurtzite to zinc-blende ratio in ZnS:Mn films with different deposition temperatures.

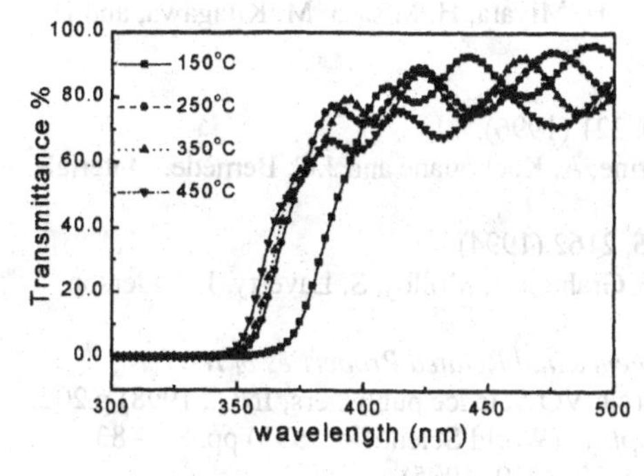

Figure 4. Transmittance of ZnS:Mn films on ITO-coated glass deposited with various substrate temperatures.

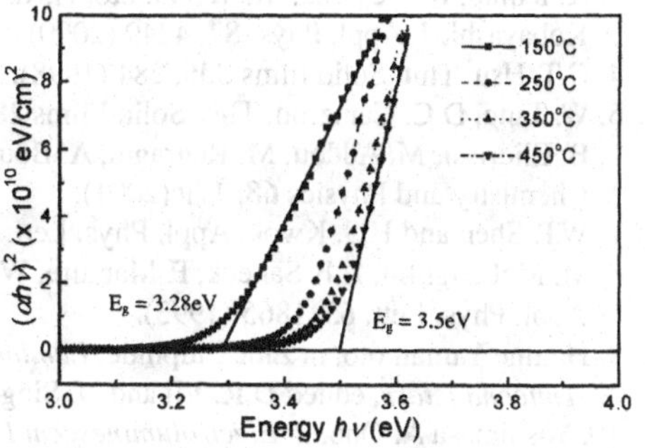

Figure 5. The dependencies of $(\alpha h\nu)^2$ on $h\nu$ of ZnS:Mn films.

CONCLUSIONS

ZnS:Mn films were successfully growth on (001) Si and ITO-coated glass by PLD method. The deposited ZnS:Mn films were smooth and crack-free. As the substrate temperature increased, the crystallite changed from a circular to an elongated shape with a wider size distribution. Meanwhile, the structure changed from a zinc-blende dominated structure to a mixture of zinc-blende and wurtzite structure. ZnS:Mn films with preferred wurtzite structure were obtained at substrate temperatures ≥ 450°C. However, the back-filled Ar pressure had no significant effect on the structural properties of ZnS:Mn films. Apart from the structural changes, the peak of the orange-yellow PL band shifted to a shorter wavelength with increasing substrate temperature. Besides, shifts in the absorption edge as well as the energy gap due to the structural changes were also observed.

ACKNOWLEDGMENTS

We would like to thank W.L. Tsui for his helpful discussion. The work described in this paper was supported by Centre for Smart Materials of the Hong Kong Polytechnic University (under grant no. A316). K.M. Yeung was supported by a M.Phil studentship of the Hong Kong Polytechnic University.

REFERENCES

1. S.K. Mandal, S. Chaudhuri, A.K. Pal, Thin Solid Films **350,** 209 (1999).
2. J. Fang, P. H. Holloway, J.E. Yu, K.S. Jones, Appl. Surf. Sci. **70/71,** 701 (1993)
3. K. Ichino, K. Ueyama, M. Yamamoto, H.Kariya, H. Miyata, H. Misasa, M. Kitagawa, and H. Kobayashi, J. Appl. Phys. **87**, 4249 (2000).
4. C.T. Hsu, Thin Solid films **335**, 284 (1998).
5. W. Tang, D.C. Cameron, Thin Solid Films **280**, 221 (1996).
6. B. Elidrissi, M. Addou, M. Regragui, A. Bougrine, A. Kachouane and J.C. Bernède, Materials Chemistry and Physics **68**, 179 (2001).
7. W.P. Shen and H.S. Kwok, Appl. Phys. Lett. **65**, 2162 (1994).
8. M. McLaughlin, H.F. Sakeek, P. Maguire, W.G. Graham, J. Molloy, S. Laverty, J. Anderson, Appl. Phys. Lett. **63**, 1865 (1993).
9. Hajime Yamamoto, in Zinc Sulphide, *Luminescence and Related Properties of II-VI semiconductors*, edited D.R. Vij and N. Singh (NOVO Science publishers, INC., 1998) p.202.
10. Yoshimasa A. Ono, in *Electroluminescent Displays* (World Scientific, 1995) pp. 82 – 83.
11. T.He, P. Ehrhart and P. Meuffels, J. Appl. Phys. **79**, 3219 (1995).
12. C.H. Kam, S.D. Cheng, Y. Zhou, K. Pita, X.Q. Han, W.X. Que, H.X. Zhang, Y.L. Lam, Y.C. Chan, Z. Sun, and X.Shi,. SPIE **3896**, 466 (1999).

Mat. Res. Soc. Symp. Proc. Vol. 667 © 2001 Materials Research Society

Red Emitting Electroluminescent Devices Using Ga_2O_3 Phosphor Thin Films Prepared by Sol-Gel Process

Tadatsugu Minami, Tetsuya Shirai and Toshihiro Miyata
Optoelectronic Device System R&D Center, Kanazawa Institute of Technology,
7-1 Ohgigaoka, Nonoichi, Ishikawa, 921-8501, JAPAN

ABSTRACT

High-luminance red-emitting thin-film electroluminescent (TFEL) devices have been developed using Ga_2O_3 phosphor thin films prepared by a sol-gel deposition method. Single-insulating-layer-type TFEL devices were fabricated by depositing a Cr- or a Eu-activated Ga_2O_3 phosphor thin film onto a thick $BaTiO_3$ ceramic sheet insulator. The Ga_2O_3:Cr or Ga_2O_3:Eu thin-film emitting layer was prepared by a sol-gel process using gallium acethylacetonate ($Ga(C_5H_7O_2)_3$) as the Ga source with $Cr(C_5H_7O_2)_3$ or $EuCl_3$ as the dopant source, respectively. A high red luminance of 622 cd/m^2 was obtained for a Ga_2O_3:Cr TFEL device driven by a sinusoidal wave voltage at 1 kHz.

INTRODUCTION

Oxide phosphors feature a higher chemical stability than the sulfide phosphors which are widely used as the screen of cathode ray tubes (CRT) and the emitting layer of conventional thin-film electroluminescent (TFEL) devices [1]. As a result, oxide phosphors have become widely used in various emissive flat panel displays such as plasma display panels (PDP) and field emission displays (FED) [2-4]. Recently, we have reported that high-luminance TFEL devices are realizable using various oxide phosphors [5,6]. In addition, we reported that a stable long term operation in the atmosphere could be achieved in TFEL devices using oxide phosphors such as $ZnGa_2O_4$:Mn and Ga_2O_3:Mn [7,8]. In particular, TFEL devices with an oxide phosphor thin-film emitting layer can be fabricated in the presence of water. Recently, we demonstrated that high luminance oxide phosphor TFEL devices could be fabricated by chemical deposition using either a solution coating technique [9-11] or a sol-gel process [12-15], both eliminating the need for vacuum processes. TFEL devices using Ga_2O_3, SnO_2 and SnO_2-Ga_2O_3 phosphor thin films prepared by a sol-gel process produced higher luminances than devices prepared by r.f. magnetron sputtering using these thin films [16-20]. Since the sol-gel process eliminates the need for a heat treatment with a rapid temperature rise as well as the need for vacuum processes, it can be suited for large area thin-film deposition and inexpensive device fabrication.

In this paper, we describe newly developed high-luminance red-emitting TFEL devices using either Cr- or Eu-activated Ga_2O_3 phosphor thin films prepared by the sol-gel deposition method. Single-insulating-layer-type TFEL devices were fabricated by depositing the Ga_2O_3 phosphor thin-film emitting layer onto a thick $BaTiO_3$ ceramic sheet insulating layer.

EXPERIMENTAL DETAILS

The TFEL devices were fabricated by depositing Ga_2O_3 phosphor thin films onto thick sintered $BaTiO_3$ insulating ceramic sheets (thickness, about 0.2 mm), as shown in Fig.1 [21]. The Ga_2O_3:Cr or Ga_2O_3:Eu thin-film emitting layer was prepared by a sol-gel process using gallium acethylacetonate ($Ga(C_5H_7O_2)_3$) as the Ga source with $Cr(C_5H_7O_2)_3$ or $EuCl_3$ as the dopant source; the Cr or Eu dopant content (Cr/(Cr+Ga) or Eu/(Eu+Ga) atomic ratio) was varied from 0.1 to 4 atomic%. $Ga(C_5H_7O_2)_3$ and $Cr(C_5H_7O_2)_3$ or $EuCl_3$ were dissolved in methanol (CH_3OH) by stirring for 30 minutes at room temperature (RT) in a N_2 gas atmosphere. Subsequently, H_2O

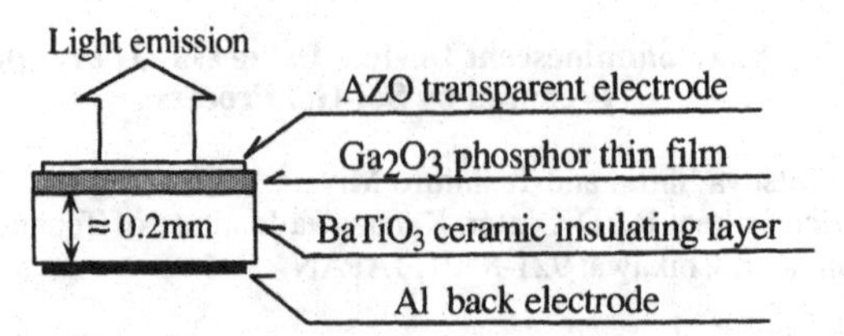

Fig.1 Schematic cross-section of a thick ceramic-insulating-layer-type TFEL device.

and HCl were added into the solution and stirred for an additional 5 hours at 50°C in the atmosphere; the resulting solution exhibited a pH of about 3.4. The $BaTiO_3$ ceramic sheets were first immersed in the solution and then dried about 5 min in air at RT before being heated in a furnace for 10 min in air at 600-1000°C. This heat-treatment temperature is referred to as the depositing temperature in this report. This procedure was repeated 2-25 times in order to obtain a film thickness of approximately 1.5-2 μ m. The film thickness is represented by the dipping number (number of times dipped into the solution) in this report. In order to improve crystallinity as well as luminescent properties, most of the Ga_2O_3 phosphor thin films were postannealed in Ar for 1 h at temperatures in the range from 870-1070°C. The heat-treatment temperature used in this annealing process is referred to as the annealing temperature. In the final procedure of TFEL device fabrication, an Al-doped ZnO transparent conducting film and an Al film were deposited as the transparent electrode and the back electrode, respectively. The electroluminescent (EL) characteristics of the TFEL devices driven by a sinusoidal wave voltage at 1 kHz (or 60 Hz) were measured using a Sawyer-Tower circuit, an ac power meter and a conventional luminance meter.

RESULTS AND DISCUSSION

Ga_2O_3:Cr TFEL Devices

The obtained EL characteristics in the TFEL devices using a Ga_2O_3:Cr thin film as the emitting layer were strongly dependent on the thin-film preparation conditions such as the depositing and annealing temperatures and the dipping number, *i.e.*, representing thin-film emitting layer thickness, as well as the Cr content. As an example, Fig. 2 shows luminance (L) and transferred charge density (Q) as functions of applied voltage (V) for TFEL devices using Ga_2O_3:Cr thin films prepared with various depositing temperatures and driven at 1 kHz: Thin-film emitting layer thickness, about 2 μ m; dipping number, 25; Cr content, 0.3 at.%; and annealed at 1020°C. As can be seen in Fig.2, the obtained maximum luminance was relatively independent of the depositing temperature in the range from 600 to 1000°C. However, the L-V and the Q-V characteristics were considerably affected by the depositing temperature. As can be seen in Fig.2, the depositing temperature dependence of the L-V characteristics are correlated to that of the Q-V characteristics.

Figure 3 shows X-ray diffraction patterns of Ga_2O_3:Cr thin-film emitting layers prepared with and without heat treatment at an annealing temperature of 1020°C. These films with a thickness of about 2 μ m were prepared at a depositing temperature of 900°C with a dipping number of 25 and a Cr content of 0.3 at.%. Annealing at 1020°C considerably improved the crystallinity of Ga_2O_3:Cr thin films as evidenced by the observed peaks which are identified as diffraction from the β -Ga_2O_3 lattice. In general, the EL characteristics of TFEL devices using Ga_2O_3:Cr thin films were significantly improved by annealing at temperatures above about 900°C. In particular, devices fabricated at an annealing temperature in the range from 900 to 1070°C exhibited a high luminance emission.

In contrast, EL emissions could be observed in TFEL devices using a Ga_2O_3:Cr thin film

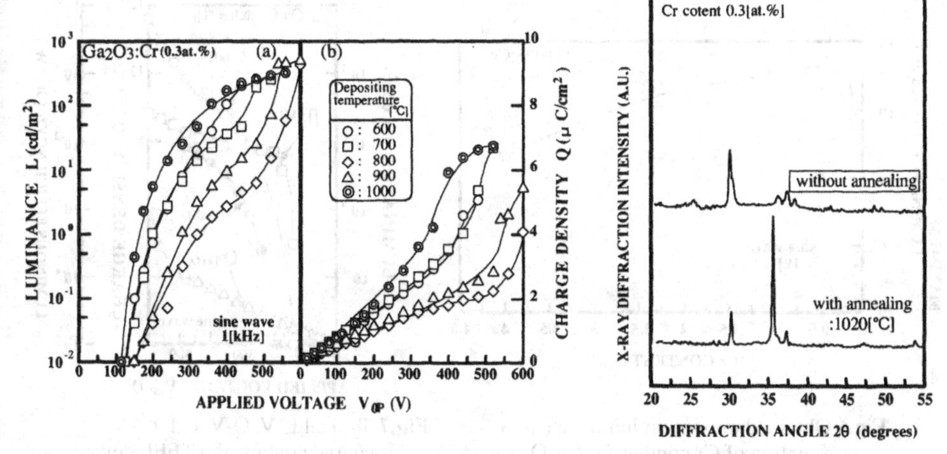

Fig.2 L-V and Q-V characteristics as functions of depositing temperature for Ga_2O_3:Cr TFEL devices.

Fig.3 X-ray diffraction patterns of Ga_2O_3:Cr thin-film emitting layers prepared with and without annealing.

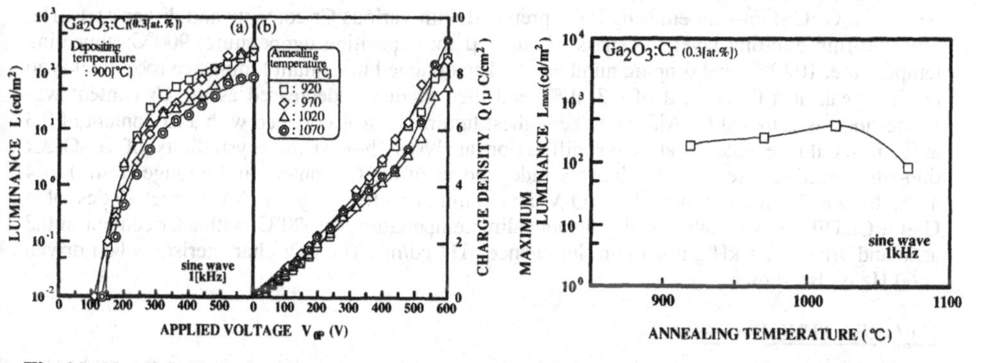

Fig.4 L-V and Q-V characteristics as functions of annealing temperature for Ga_2O_3:Cr TFEL devices.

Fig.5 Obtained maximum luminance as a function of annealing temperature for Ga_2O_3:Cr TFEL devices.

prepared without annealing; however, the obtained luminance was lower than that found in devices fabricated with annealing at a temperature above about 900°C. L-V and Q-V characteristics and obtained maximum luminance as functions of the annealing temperature are shown in Figs.4 and 5, respectively, for TFEL devices using Ga_2O_3:Cr thin films prepared at a depositing temperature of 900°C and annealed at temperatures ranging between 900 and 1070°C and driven at 1 kHz. These thin films with a thickness of about 2 μ m were prepared with a dipping number of 25 and a Cr content of 0.3 at.%. The highest luminance was obtained in a device with a Ga_2O_3:Cr thin-film emitting layer annealed at 1020°C. It was found that the annealing temperature dependence of the L-V characteristics are correlated to that of the Q-V characteristics. However, X-ray diffraction analyses showed that the crystallinity of Ga_2O_3:Cr thin-film emitting layers was relatively independent of annealing temperature above about 900°C.

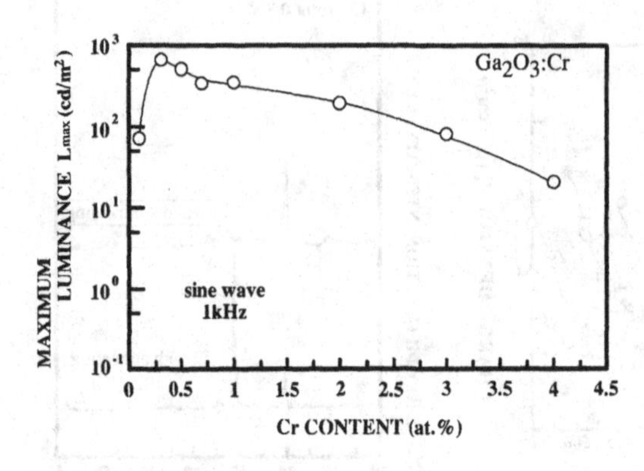

Fig.6 Obtained maximum luminance as a function of Cr content for Ga_2O_3:Cr TFEL devices.

Fig.7 Typical L-V, Q-V and η -V characteristics of a TFEL device using a Ga_2O_3:Cr thin film prepared under optimal conditions.

Figure 6 shows obtained maximum luminance as a function of Cr content for TFEL devices with a Ga_2O_3:Cr thin-film emitting layer prepared with various Cr contents and driven at 1 kHz: The thin-film emitting layer thickness, about 2 μ m; depositing temperature, 900°C; annealing temperature, 1020°C; and dipping number, 25. The obtained maximum luminance rose quickly to reach a peak at a Cr content of 0.3 at.%, and then gradually decreased as the Cr content was further increased to 4 at.%. Although the highest luminance was obtained with a Cr content of 0.3 at.%, it should be noted that X-ray diffraction analyses showed the crystallinity of Ga_2O_3:Cr thin-film emitting layers to be relatively independent of the Cr content in the range from 0 to 4 at.%. Figure 7 shows typical L-V, Q-V and luminous efficiency (η -V) characteristics of a Ga_2O_3:Cr TFEL device fabricated at an annealing temperature of 1020°C with a Cr content of 0.3 at.% and driven at 1 kHz: maximum luminance, 622 cd/m^2. The L-V characteristic when driven at 60 Hz is also shown.

Ga_2O_3:Eu TFEL Devices

Ga_2O_3:Eu TFEL devices fabricated in the same manner as the Ga_2O_3:Cr TFEL devices described above also produced a high luminance. As an example, L-V and Q-V characteristics and obtained maximum luminance as functions of the dipping number are shown in Figs.8 and 9, respectively, for TFEL devices with a Ga_2O_3:Eu thin-film emitting layer prepared with various Eu contents and driven at 1 kHz: depositing temperature, 900°C; annealing temperature, 1020°C; and Eu content, 0.3 at.%. As can be seen in Fig.8, the L-V characteristics, in general, improved as the dipping number was increased up to about 5. In addition, the dipping number dependence of the L-V characteristics is correlated to that of the Q-V characteristics. As can be seen in Fig.9, the obtained maximum luminance reached a peak at a dipping number of 5, and then gradually decreased as the dipping number was further increased from 7 to 25: From X-ray diffraction analyses, it was found that the crystallinity of the Ga_2O_3:Eu thin-film emitting layers improved as the dipping number was increased from 2 to 25. However, these X-ray diffraction patterns also showed that the positions and intensities of the observed diffraction peaks markedly changed

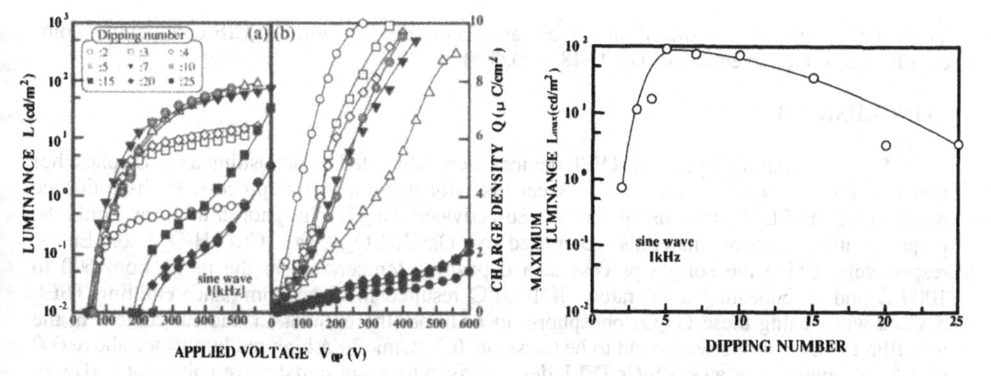

Fig.8 L-V and Q-V characteristics as functions of dipping number for Ga_2O_3:Eu TFEL devices.

Fig.9 Obtained maximum luminance as a function of dipping number for Ga_2O_3:Eu TFEL devices.

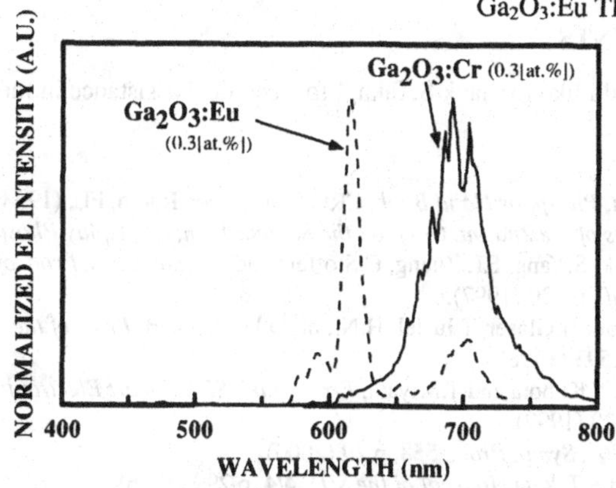

Fig.10 EL emission spectra from Ga_2O_3:Cr and Ga_2O_3:Eu TFEL devices.

when the thin-film emitting layer was prepared with a dipping number of 5 or greater. These data show that the L-V characteristics of TFEL devices using Ga_2O_3:Eu thin films prepared by the sol-gel process were complexly dependent on both the thickness and the crystallinity of the thin-film emitting layer used.

TFEL devices were fabricated using Ga_2O_3:Eu thin films prepared with various Eu contents up to 4 at.%. The highest luminance was obtained in a device using a Ga_2O_3:Eu thin film prepared with a Eu content of 0.3 at.%.

Color and spectra of EL emissions

TFEL devices fabricated using a Ga_2O_3:Cr or a Ga_2O_3:Eu thin film exhibited a red EL emission color. Figure 10 shows typical EL emission spectra from Ga_2O_3:Cr and Ga_2O_3:Eu TFEL devices. These emission spectra were independent of preparation conditions and Cr or Eu content up to 4 at.%. The emission spectra from the Ga_2O_3:Cr TFEL devices, consisting of multiple peaks, may be a result of the ^2E-$^4A_2$ transition in the Cr^{3+} ion [1]: CIE chromaticity color coordinates of the red emission, (x=0.639, y=0.336). The main peak in the emission spectra from the Ga_2O_3:Eu

TFEL devices may be a result of the ^5D-^7F transition in the Eu^{3+} ion [1]: CIE chromaticity color coordinates of the red emission, (x=0.548, y=0.405).

CONCLUSIONS

Single-insulating-layer-type TFEL devices were fabricated by depositing a Ga_2O_3 phosphor thin film onto a thick $BaTiO_3$ ceramic sheet insulator using a sol-gel process. High-luminance red-emitting TFEL devices using Cr- or Eu-activated Ga_2O_3 phosphor thin films could be prepared using source materials composed of $Ga(C_5H_7O_2)_3$ and $Cr(C_5H_7O_2)_3$ or $EuCl_3$, respectively. Using the sol-gel process at a depositing temperature in the range from 600 to 1000°C and an annealing temperature of 1020°C resulted in higher luminance emitting TFEL devices when using these Ga_2O_3 phosphors. In addition, the optimal Cr and Eu content in the thin-film emitting layers was found to be the same, 0.3 atomic%. A high red luminance above 600 cd/m^2 was obtained for a Ga_2O_3:Cr TFEL device driven by a sinusoidal wave voltage at 1 kHz. In addition, Ga_2O_3:Eu TFEL devices fabricated in the same manner also produced a high red luminance.

ACKNOWLEDGMENTS

The authors would like to thank S.Takagi for technical assistance in the experiments.

REFERENCES

1. A.Shionoya, W.M.Yen, *Phosphor Hand Book*, CRC Press, Boca Raton, FL, (1998).
2. A.Vecht, *Ext. Abstracts of the 2nd Int. Conf. on the Sci.and Tech. of Display Phosphors*, p.247 (1996).
3. B.K.Wagner, J.Penczek, S.Yang, F.L.Zhang, C.Stoffers and C.J.Summers, *Proc. of the 17th Int. Display Research Conf.*, p.330 (1997).
4. H.Bechtel, W.Czarnojan, H.Glaser, T.Justel, H.Nikol, D.U.Wiechert, *Proc. of the 5th Int. Display Workshops*, p.593 (1998).
5. T.Minami, H.Yamada, Y.Kubota and T.Miyata, *Proc. of the SPIE Smart Electronics and MEMS*, **3242**, p.229 (1997).
6. T.Minami, *Mat. Rrs. Soc. Symp. Proc.*, **558**, p.29 (2000).
7. T.Minami, Y.Kuroi and S.Takata, *Journal of the SID*, **4/4**, p.299 (1996).
8. T.Miyata, T.Nakatani and T.Minami, *Proc. of the 6th Int. Display Workshops*, p.865 (1999).
9. T.Minami, Y.Kuroi, Y.Sakagami and T.Miyata, *Proc. of the 3rd Int. Display Workshops*, p.97 (1996).
10. T.Minami, Y.Sakagami and T.Miyata, *Extended Abstract of the 3rd Int. Conf. on the Sci. and Tech. of Display Phosphor*, p.37 (1997).
11. T.Minami, T.Miyata and Y.Sakagami, *Surface and Coatings Tech.*, **108-109**, p.594 (1999).
12. T.Minami, T.Nakatani and T.Miyata, *Proc. of the 8th Int. Symp. on Phys. and Chem. of Lumin. Mat.*, **99**, p.9 (2000).
13. T.Minami, T.Shirai, T.Nakatani and T.Miyata, *Jpn. J. Appl. Phys.*, **39**, L524 (2000).
14. T.Minami, T.Shirai and T.Miyata, *Proc. of the 7th Int. Display Workshops*, p.897 (2000).
15. T.Minami, T.Shirai and T.Miyata, *Proc. of the SPIE Smart Structure and Devices*, **4235**, p.476 (2000).
16. T.Minami, T.Miyata, T.Shirai and T.Nakatani, *Mat. Res. Soc. Proc.*, **621**, p.Q 4.3.1. (2000).
17. T.Minami, H.Yamada, Y.Kubota, T.Miyata and Y.Sakagami, *Proc. of the 4th Int. Display Workshops*, p.605 (1997).
18. T.Minami, Y.Kubota, T.Miyata and H.Yamada, *SID Int. Symp., Digest of Tech. Papers*, p.953 (1998).
19. A.H.Kitai, T.Xiao, G.Liu and J.H.Li, *SID Int. Symp. Digest od Tech. Papers*, p.419 (1997).
20. D.Stodilka, A.H.Kitai, Z.Huang and K.Cook, *SID Int. Symp., Digest of Tech. Papers*, p.11 (2000).
21. T.Minami, S.Orito, H.Nanto and S.Takata, *Proc. of the 6th Int. Display Research Conf.*, p.140 (1986).

Mat. Res. Soc. Symp. Proc. Vol. 667 © 2001 Materials Research Society

Novel Fluorescent Labels Prepared by Layer-by-Layer Assembly on Colloids for Biodetection Systems

Wenjun Yang[1], Dieter Trau[1], Reinhard Renneberg[1], Nai-Teng Yu[1]* and Frank Caruso[2]
[1]Department of Chemistry, The Hong Kong University of Science and Technology, Clear Water Bay, Kowloon, SAR Hong Kong, People's Republic of China
[2]Max Planck Institute of Colloids and Interfaces, D-14424 Potsdam, Germany

ABSTRACT

Fluorescent polystyrene microparticles with different sizes were produced by the consecutive assembly of fluorescently labeled polyelectrolytes using the layer-by-layer self-assembly method. Film growth was characterized by microelectrophoresis and fluorescence microscopic image (FMI) analysis. Alternating negative and positive ζ-potentials with deposition of each successive polyelectrolyte layer demonstrated that the alternate adsorption of polyelectrolytes was achieved. FMI analysis provided direct measurement of the fluorescence intensity of single microparticles. The subsequent deposition of a protein (immunoglobulin G, IgG) layer onto the fluorescent microparticles was confirmed by a sandwich immunoassay.

INTRODUCTION

During the past ten years, the construction of functional multilayer films using the layer-by-layer (LbL) assembly approach introduced by Decher *et al.* [1] has attracted much interest in many aspects of materials research. Electrostatic interaction between the oppositely charged substrate and assembling species results in the formation of stable multilayers with unique optical, electronic or magnetic properties, depending on the functionality of the assembling species. The substrate widely used for assembling multilayer films was initially charged, flat surfaces. This was later extended to charged colloidal particles. Functional particles with controlled surface modification have recently been produced using the LbL method applied to colloids [2]. Fluorescent microparticles are widely used in biological and immunological studies [3] and different synthetic methods have been attempted for the fabrication of particles with high fluorescence output at desired emission wavelengths. Semiconductor nanocrystal core-shell composites, such as CdSe-CdS nanocrystals [4], CdSe-ZnS quantum dots [5], and hollow spheres of silica filled with fluorescein isothiocyanate (FITC) molecules [6] were all successfully synthesized and used as fluorescent markers in immunotests. Here we report the construction of fluorescent colloidal microparticles by using the LbL approach. The fluorescence intensity from individual microparticles was obtained using fluorescence microscopic image (FMI) analysis. The subsequent passive adsorption of IgG onto polyelectrolyte multilayer-coated fluorescent microparticles provides the biospecific recognition function required for the successful application of the particles in immunodetections.

EXPERIMENTAL DETAILS

Materials. Polystyrene (PS) particles with a negatively charged surface and 640 nm in diameter (PS640) were prepared by the method described by Furusawa et al. [7]. PS particles with diameters of 270 nm (PS270) and 86 nm (PS86) were purchased form Microparticles

GmbH, Germany. Positively charged polyelectrolyte, poly(allylamine hydrochloride) (PAH, M_w 15,000), negatively charged polyelectrolyte, poly(sodium 4-styrenesulfonate) (PSS, M_w 70,000) and β-morpholino-ethansulfonic acid (MES) were from Aldrich. FITC, mouse IgG, goat anti-mouse (Gt α M) IgG, Gt α M IgG-FITC conjugate and bovine serum albumin (BSA) (fraction V) were all from Sigma-Aldrich. The conjugation of FITC with PAH followed the standard method used for protein labeling [8]. The degree of labeling was about one dye molecule to 100 PAH monomers without further mention.

Assembly of Polyelectrolyte and Protein Multilayers onto PS Microparticles. The assembly followed the method reported elsewhere [9]. Briefly, 0.5 ml of PAH-FITC solution or 1 ml of PSS (both containing 1 mg ml^{-1} polyelectrolyte and 0.5 M NaCl) was alternatively added to 0.5 ml of a PS particle suspension (1 wt%, ~ 10^{11} particles ml^{-1}) with a 15min-incubation and three times of centrifugation/washing in between. The incubation/washing steps and the sequential polyelectrolyte depositions were repeated to achieve the desired assembly layer numbers. For protein adsorption, 1 ml of 1 mg ml^{-1} protein solution (0.05 M MES buffer, pH 6.0) was added to 0.3 ml (0.5 wt%, ~5 x 10^{10} particles ml^{-1}) of a coated particle suspension (with PSS as the outer layer). After 45 min of adsorption, the suspension was centrifuged and washed by water.

Microelectrophoresis. The electrophoretic mobility of coated PS particles was measured with a Zetasizer 4 (Malvern Instruments, Worcs, UK) by taking the average of 5 measurements at the stationary level. The mobility u was converted into a ζ-potential by using the relation $\zeta = u\eta/\varepsilon$, where η is the viscosity of the solution and ε is the permittivity.

Fluorescence Microscopic Image Analysis. Fluorescence microscopic images were obtained using a monochromatic CCD camera (Cohu) attached to a fluorescence microscope with a mercury arc as excitation source (CX-40, Olympus). Images were captured via a frame grabber (LG-3, Scion) using a commercial imaging software package (Scion Image, Scion). The gain and offset of the frame grabber were adjusted to match the dynamic ranges of the frame grabber and the CCD camera.

Sandwich Immunoassay using 96-well microplate as substrate. Polystyrene 96-well microplate (MaxiSorp, Nunc) was coated with 200 ng/well of Gt α M IgG in 0.1 M carbonate buffer at 4 °C overnight and washed with washing buffer (0.1 M PBS, 0.1% BSA, 0.5 % Tween 20). The plate was then incubated with 100 μl/well of mouse IgG solutions (1~500 ng ml^{-1}) at RT for 2 hr. After washed and incubated with 1:100 dilution of Gt α M IgG-PS86 particle suspension at RT for 2hr, the plate was again washed and examined by a fluorescence microplate reader (Bio-Rad) with excitation/emission at 485/535 nm.

RESULTS AND DISCUSSION

Microelectrophoresis measurement provides the information of surface charge on colloidal particles, known as the ζ-potential. The sign of the ζ-potential value reflects the charge on the surface, while the magnitude indicates the stability of the charged colloidal particles. The higher the absolute magnitude of the ζ-potential, the more stable the particles are in the suspension. Fig. 1 summarizes the ζ-potential changes after deposition of each polyelectrolyte layer under three different assembly conditions. The alternating ζ-potentials show the change of sign of surface

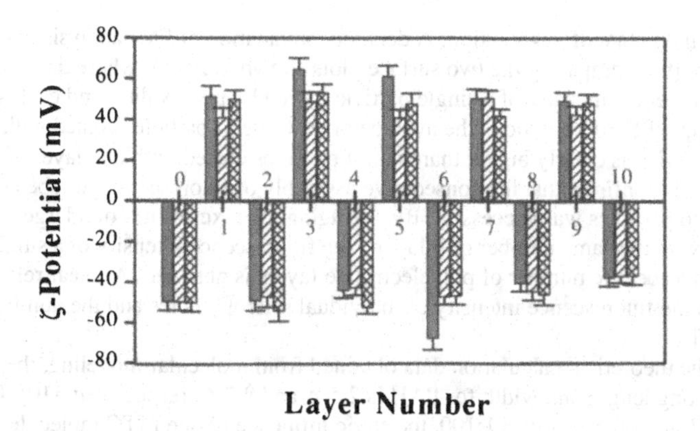

Fig.1 ζ-Potential as a function of layer number for PS640/(PAH/PSS)$_n$ (grey block), PS640/(PAH-FITC/PSS)$_n$ (oblique line) and PS270/(PAH-FITC/PSS)$_n$ (crossed line)

charge after each deposition step, which suggests the stepwise, LbL deposition of oppositely charged polyelectrolyte layers onto the PS particles. Comparing the three curves in Fig. 1, it can be seen that the size of the PS particle does not influence the assembly of polyelectrolyte. In addition, the conjugated FITC molecules on the PAH-FITC chain had little effect on the assembly and the resulting ζ-potentials of the microparticles.

As it is not possible to measure the fluorescence intensity of single microparticles by steady-state fluorescence using a conventional spectrometer, a more direct and straightforward method was used. FMI analysis was employed to examine the fluorescence intensity from individual

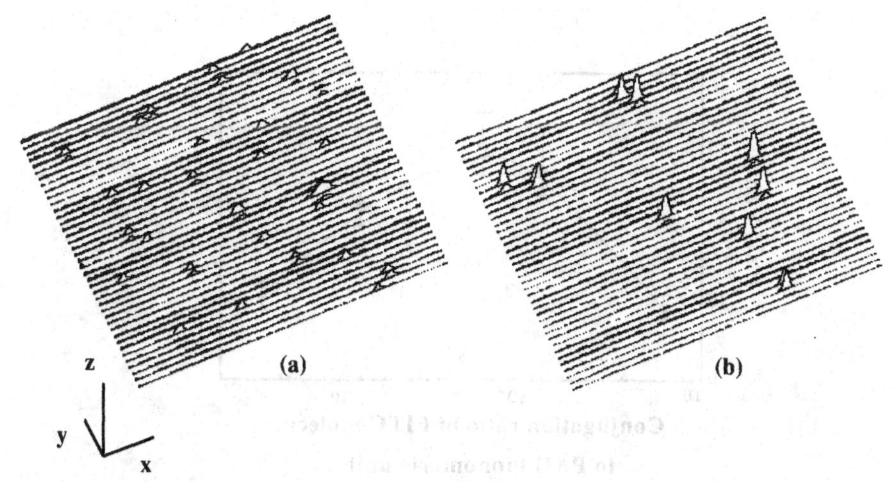

Figure 2. FMI surface plots of PS640 particles coated with one (a) and five (b) layers of PAH-FITC. The z-axis represents the fluorescence intensity of the particles, given as a pixel value.

particles as well as their state of aggregation. A detailed explanation of FMI analysis was published elsewhere [9]. Comparing the two surface plots shown in Fig. 2, where the z-axis represents the fluorescence intensity of a single particle (given by pixel value) and each spot corresponds to a single PS microparticle, the average pixel value of particles coated with 5 layers of PAH-FITC (Fig. 2 (b)) is clearly bigger than that of particles coated with one layer of PAH-FITC (Fig. 2 (a)). This confirms that the consecutive assembly of fluorescent dye-labeled polyelectrolytes onto particles was successful. By averaging the pixel values of a large number of particles coated with the same number of bilayers, the fluorescence intensity of a single particle coated with a specific number of polyelectrolyte layers is obtained. A linear relationship was found between the fluorescence intensity of individual microparticles and the number of deposited layers [9].

According to the theoretical calculation data obtained from molecular modeling, the characteristic repeating length and width for PAH is 2.5 Å and 8.7 Å, respectively [10]. In the case of FITC : PAH monomeric unit = 1:100, the steric influence of one FITC molecule to one hundred monomeric units of PAH is negligible. Thus, for a close-packed adsorbed layer of PAH-FITC molecules on PS640 microparticles, the number of FITC molecules (FITC equivalent, *F.E.*) loaded on one particle can be calculated as follows:

$$F.E. = N \times \frac{A}{A'} = N \times \frac{4\pi R^2}{ab} = \frac{1}{100} \times \frac{4 \times 3.14 \times (320nm)^2}{0.25nm \times 0.87nm} \approx 6 \times 10^4$$

where *A* represents the surface area of one PS640 particle, *A'* represents the area of one PAH monomeric unit, *N* is the ratio of number of FITC molecule to number of PAH monomeric unit upon conjugation. When the particles are coated with more PAH-FITC layers, the fluorescent dye loading would also increase to an even higher degree, which suggests an easy way to control the fluorescence output of single microparticles.

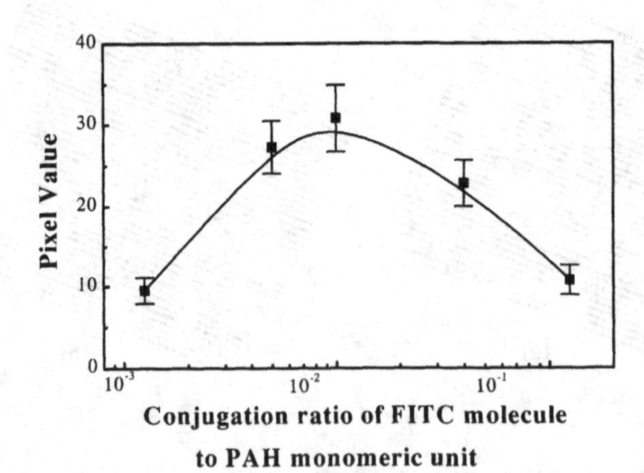

Figure 3 Relationship between fluorescence intensity of PS640/(PAH-FITC/PSS)$_4$ particles, represented as pixel value, and the ratio of FITC molecule to PAH monomeric unit.

According to the above equation, there are two ways to increase the *F.E.* value. One is to use bigger particles which have a bigger surface area *A*; the other is to decrease *N*. The first method is often not applicable for solid-phase immunodetection using particulate labels, because smaller-sized labels are preferred. To test the second method, five different conjugation ratios ($N = 0.125$, 0.04, 0.01, 0.005, 0.00125) were examined. The influence of the conjugation ratio to fluorescence output of PS640 particles coated with 4 bilayer of PAH-FITC/PSS was shown in Fig. 3. When the ratio increases from 0.00125 to 0.01, the conjugation of more FITC molecules with the PAH polymer chains yields a higher fluorescence output from the PAH-FITC coated particles. However, the more FITC molecules are conjugated with the PAH chains, the closer the average distance is between two FITC molecules. When a certain ratio is reached, internal quenching of the fluorescent dye molecules will be observed. As the PAH-FITC polymer chains may loop on the latex surface or interpenetrate with each other, the quenching may be caused by FITC molecules conjugated in the same or different PAH chains. So when the ratio is greater than 0.01, the fluorescence output decreases due to the quenching effect. Thus, to obtain optimal fluorescence output from the coated particles, a medium-range ratio (N~0.01) is required.

Passive adsorption of IgG, a widely used immunoreagent, onto the fluorescent microparticles was achieved and characterized by microelectrophoresis measurements [9, 11]. The suitability of the Gt α M IgG-coated fluorescent PS microparticles for immunodetections was investigated by a solid-phase sandwich immunoassay for mouse IgG. PS latex particles with 86 nm in diameter were used as a template to prepare the fluorescent particulate label. The microparticles were coated with three PAH-FITC/PSS bilayers before the passive adsorption of Gt α M IgG. The immunoresponse from the prepared fluorescent particulate label was compared with that obtained from Gt α M-FITC conjugate, the conventional soluble fluorescent label.

Fig. 4 shows that the prepared fluorescent PS label gave a comparable fluorescent signal response to the soluble fluorescent label. The detection limits for both tests were 2 ng ml^{-1}. The

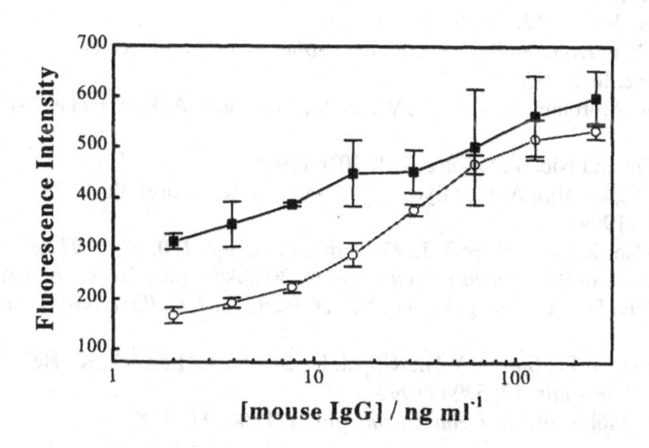

Figure 4 Sandwich fluorescence immunoassay of mouse IgG using Gt anti-M IgG conjugated PS86/(PAH-FITC/PSS)$_3$ microparticles (squares) and Gt anti-M-FITC (open circles) as detectors.

higher background signal and less reproducibility produced by the particulate label may be due to the insufficient coverage of IgG molecules on the particles, which resulted in non-specific binding between the labeled particles and the solid-phase. The preliminary result of the sandwich immunoassay showed that the fluorescent microparticles are capable of acting as immunolabels in fluorescent immunoassays.

CONCLUSIONS

Alternating layers of oppositely charged polyelectrolytes of PAH-FITC and PSS were deposited onto PS microparticles via the LbL self-assembly approach. Microelectrophoresis measurements showed reversal of the surface charge on the microparticles after each deposition step, suggesting the successful assembly of positively and negatively charged polyelectrolyte. The fluorescence intensity of the coated microparticles was studied by fluorescence microscopic image analysis. The influence of conjugation ratio of FITC molecule to PAH monomeric unit on the fluorescence output of the microparticles was examined. Medium conjugation ratios (N~0.01) were found to lead to the highest fluorescence output for the PAH-FITC coated microparticles. The IgG molecules adsorbed on the fluorescent microparticles were found to be immunoactive in the solid-phase sandwich assay. Due to the simplicity of the fluorescent particle preparation, the accessibility of a variety of conjugated fluorescent dyes, the ability to deposit immunospecies, and the high fluorescence output of the prepared particles, the biofunctional fluorescent microparticles constructed via the LbL approach are promising for application in immunodetection systems.

REFERENCES

1. Decher, G. and Hong, J. -D., *Makromol. Chem. Macromol. Symp.*, **46**, 321-327, 1991
2. Caruso, F., *Adv. Mater.*, **13,** 11 (2001).
3. Haugland, R. P., in *Handbook of Fluorescent Probes and Research Chemicals*, sixth edition, Molecular Probe, Inc., 1996.
4. Bruchez Jr., M., Moronne, M., Gin, P., Weiss, S., Alivisatos, A. P., *Science*, **281**, 2013 (1998).
5. Chan, W. C. W., and Nie, S., *Science*, **281**, 2016 (1998).
6. Makarova, O. V., Ostafin, A. E., Miyoshi, H., Norris, J. R., Meisel, D., *J. Phys. Chem. B*, **103** (43), 9080 (1999).
7. Furusawa, K., Norde, W., Lyklema, J., *Kolloid Z. Z. Polym.*, **250**, 908 (1972).
8. Hermanson, G. T., in *Bioconjugate Techniques*, p. 304. Academic Press, London, 1996.
9. Yang, W.J., Trau, D., Renneberg, R., Yu, N.T., Caruso, F., *J. Colloid. Interf. Sci.*, **234**, 356 (2001).
10. Donath, E., Walther, D., Shilov, V.N., Kippel, E., Budde, A., Lowack, K., Helm, C.A. and Moehwald, H., *Langmuir*, **13**, 5294 (1997).
11. Caruso, F., and Möhwald, H., *J. Am. Chem. Soc.*, **121**, 6039 (1999).

Mat. Res. Soc. Symp. Proc. Vol. 667 © 2001 Materials Research Society

Synthesis and optical characteristics of $ZnGa_2O_{4-x}S_x:Mn^{2+}$

J.S. Kim , H.L. Park[1] and G.C. Kim[2]
Department of Physics, Yonsei University, Seoul 120-749, Korea
[2]School of Liberal Arts, Korea University of Technology and Education,
Cheonan 330-708, Korea
[1] Corresponding author E-mail address : hipark@yonsei.ac.kr

ABSTRACT

The first zinc gallium oxysulfides ($ZnGa_2O_{4-x}S_x$) are synthesized and emission properties of $ZnGa_2O_{4-x}S_x$ and $ZnGa_2O_{4-x}S_x:Mn^{2+}$ have been investigated. The sulfur in $ZnGa_2O_{4-x}S_x$ forms tetrahedrons and the cations in $ZnGa_2O_{4-x}S_x$ are randomly distributed between tetrahedrons and octahedrons such that $ZnGa_2O_4$ structure can be viewed as a mixed spinel. The four-fold enhancement of the brightness of Mn^{2+} emission can be the significant effect of forming oxysulfide in $ZnGa_2O_{4-x}S_x:Mn^{2+}$

INTRODUCTION

Recently, zinc gallate ($ZnGa_2O_4$) and Mn-doped zinc gallate have attracted enormous attention for applications in vacuum fluorescent display (VFD) and field emission display (FED) as a low-voltage cathodoluminescence phosphor [1-4]. $ZnGa_2O_4$ has the spinel structure with space group of Fd3m. Buscherdorf [5] described $ZnGa_2O_4$ as a normal spinel. In normal spinel, Zn^{2+} ions occupy the tetrahedrally coordinated A-sites, whereas Ga^{3+} ions occupy the B-sites which are octahedrally coordinated. However, Hoffman and Brown [6] reported that $ZnGa_2O_4$ has the inverse spinel structure with half of Ga^{3+} ions on the tetrahedral site and half of Ga^{3+} ions along with Zn^{2+} ions on octahedral sites.

Jeang et al. [7] did report the new self-activated optical center in $ZnGa_2O_4$ related to tetrahedrally coordinated Ga-O groups. The green emitting Mn^{2+} site in $ZnGa_2O_4:Mn^{2+}$ is generally accepted as tetrahedrally coordinated [3], but there are two kinds of ions which are tetrahedrally coordinated where Mn^{2+} ions can be replaced in $ZnGa_2O_4$ [7]. Thus, the exact arrangements of cations in $ZnGa_2O_4$ structure are not clarified up to date. A given optical center in different host lattice will exhibit different optical properties due to the changes of the direct surroundings of the center. When we substitute oxygens with sulfurs in $ZnGa_2O_4$, i.e., $ZnGa_2O_{4-x}S_x$, we can expect some changes on its optical properties.

The typical examples are Y_2O_2S and yttrium oxysulfide incorporated with Eu^{3+} widely used as a red phosphor for color monitors because of its bright luminescence and high energy efficiency [8-12].

The purpose of our study is two-fold, i.e., clarifying the exact arrangements of cations in $ZnGa_2O_4$ and investigating the changes on emission properties of $ZnGa_2O_{4-x}S_x$ along with Mn^{2+} emission characteristics in $ZnGa_2O_{4-x}S_x:Mn^{2+}$. Our study would reveal the exact nature of cation arrangements in $ZnGa_2O_4$, the synthesis of zinc gallium oxysulfide ($ZnGa_2O_{4-x}S_x$), and the emission characteristics of $ZnGa_2O_{4-x}S_x$: Mn^{2+}

EXPERIMENTAL DETAILS

The $ZnGa_2O_{4-x}S_x$ powder samples were synthesised through solid state reactions of ZnO, Ga_2O_3 and elemental sulfur in the evacuated quartz tube at 1000 °C for 12 hours in an electric furnace. The concentrations of sulfur in $ZnGa_2O_{4-x}S_x$ are varied from x=0 to x=0.05. The phase of $ZnGa_2O_{4-x}S_x$ was identified by the conventional X-ray diffraction technique(XRD) and the valence state of sulfur was determined by a X-ray photoelectron spectroscopy(XPS). The cathodoluminescence(CL) spectra were recorded by a homemade spectrometer with the acceleration voltage of 10 kV and the vacuum of the sample chamber was kept at 10^{-6} torr during the CL measurements. A 75 watt Xe-lamp was employed for an excitation source in photoluminescence excitation(PLE) measurements.

RESULTS AND DISCUSSION

The XRD patterns of $ZnGa_2O_{4-x}S_x$ are shown in Figure. 1. One phase i.e., [$ZnGa_2O_4$], exists for three x-values in $ZnGa_2O_{4-x}S_x$, Also a very small change in lattice constant is observed and the lattice constant is shortened with increasing sulfur concentration in $ZnGa_2O_{4-x}S_x$. The (220) diffraction of spinel is originated solely from cations at tetrahedral site, whereas the (440) diffraction peak originated from tetrahedrally and octahedrally coordinated cations [13]. Therefore we can get the information about the substituted sulfurs and their environments when we get the integrated intensity ratio of (220)/(440). The (220)/(440) integrated intensity ratios are given at following: x=0.001, 0.8828, x=0.01, 0.8864 and x=0.05, 0.9284. Thus the more oxygen ions are replaced by sulfur ions in $ZnGa_2O_{4-x}S_x$, the more oxygen ions forming tetrahedrons in $ZnGa_2O_4$ are substituted with sulfur ions in the case of $ZnGa_2O_{4-x}S_x$. In other words, the sulfur ions, replacing the oxygen in $ZnGa_2O_{4-x}S_x$ have a tendency to form a tetrahedrally coordinated polyhedrons in $ZnGa_2O_{4-x}S_x$.

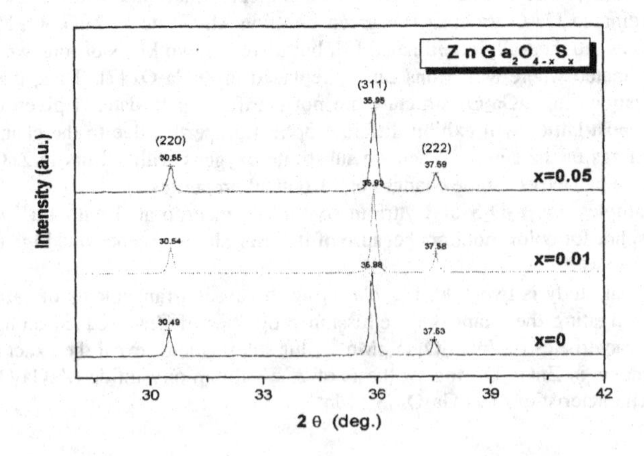

Figure 1. XRD patterns of $ZnGa_2O_4$ and $ZnGa_2O_{4-x}S_x$

The CL spectra of $ZnGa_2O_{4-x}S_x$ are shown in Figure. 2. As one can see in Figure. 2, the self-activated emission band in $ZnGa_2O_4$ is shifted to higher energy with increasing sulfur contents in $ZnGa_2O_{4-x}S_x$. Also the emission intensity is enhanced with increasing sulfur concentration in $ZnGa_2O_{4-x}S_x$.

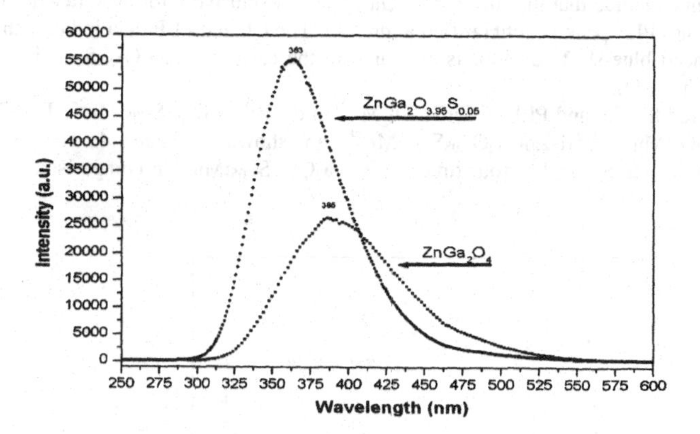

Figure 2. CL spectra of $ZnGa_2O_4$ and $ZnGa_2O_{3.95}S_{0.05}$

The photoluminescence excitation spectra(PLE) of $ZnGa_2O_4$ and $ZnGa_2O_{3.95}S_{0.05}$ are shown in Figure. 3. For PLE measurements, the emission peaks are fixed at 385 nm ($ZnGa_2O_4$) and 360 nm ($ZnGa_2O_{3.95}S_{0.05}$). The 244 nm and 342 nm PLE peaks of $ZnGa_2O_4$ are assigned as tetrahedrally coordinated Ga^{3+} [7, 14]. Jeang et al. [7] assigned 255 nm PLE peak at Ga^{3+} ion in octahedrally coordinated in $ZnGa_2O_4$. The PLE peaks showed a blue-shift behavior when oxygen ions are replaced with sulfur ions in $ZnGa_2O_{3.95}S_{0.05}$.

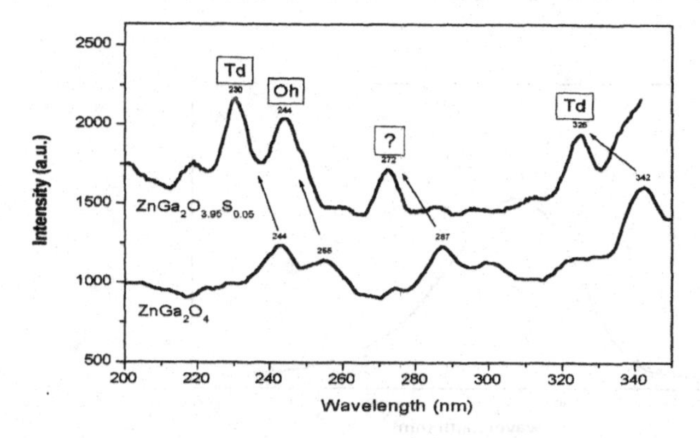

Figure 3. PLE spectra of $ZnGa_2O_4$ and $ZnGa_2O_{3.95}S_{0.05}$ monitored at 385 and 360 nm.

The relationship between the PLE peaks of $ZnGa_2O_4$ and $ZnGa_2O_{3.95}S_{0.05}$ can be interpreted as followings; The 244 and 342 nm PLE peaks related to Ga^{3+} in the tetrahedrally coordinated in $ZnGa_2O_4$ shifted to 230 and 325 nm in $ZnGa_2O_{3.95}S_{0.05}$, whereas the 255 nm

PLE peak originated from Ga^{3+} in octahedrally coordinated in $ZnGa_2O_4$ moved to 244 nm in $ZnGa_2O_{3.95}S_{0.05}$. The blue-shift behavior at 244 and 342 nm PLE peaks is more pronounced in the tetrahedrally coordinated Ga^{3+} in the case of $ZnGa_2O_{3.95}S_{0.05}$. From the integrated intensity ratio of (220)/(440) at XRD data and the blue-shift behavior at 244 and 342 nm PLE peaks, we can conclude that the sulfur in $ZnGa_2O_{4-x}S_x$ exclusively forms tetrahedrons. The origin of 287 nm PLE peak is not clarified at present. The CL and PLE data indicate that a markedly enhanced blue-shift behavior is exhibited in the case of $ZnGa_2O_{3.95}S_{0.05}$ in the comparison with $ZnGa_2O_4$.

We also compared the CL and PLE of $ZnGa_2O_4:Mn^{2+}$ and $ZnGa_2O_{3.95}S_{0.05}:Mn^{2+}$. The CL spectra of $ZnGa_2O_4:Mn^{2+}$ and $ZnGa_2O_{3.95}S_{0.05}:Mn^{2+}$ are shown in Figure. 4. The Mn^{2+} emission intensity increases roughly four times in $ZnGa_2O_{3.95}S_{0.05}:Mn^{2+}$ in comparison with $ZnGa_2O_4:Mn^{2+}$.

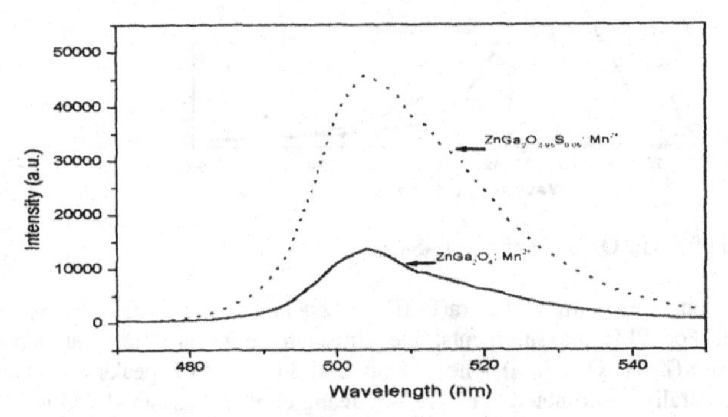

Figure 4. CL spectra of $ZnGa_2O_4$: Mn^{2+} and $ZnGa_2O_{3.95}S_{0.05}$: Mn^{2+}

The PLE signatures are almost the same between $ZnGa_2O_4$: Mn^{2+} and $ZnGa_2O_{3.95}S_{0.05}$: Mn^{2+} (see Figure. 5)

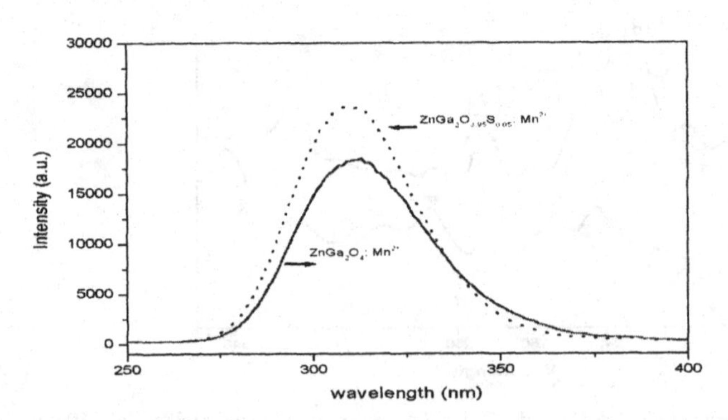

Figure 5. PLE spectra of $ZnGa_2O_4$: Mn^{2+} and $ZnGa_2O_{3.95}S_{0.05}$: Mn^{2+} monitored at 505 nm.

The XPS spectrum is depicted in Figure. 6. One can see clearly S^{2-} signature.

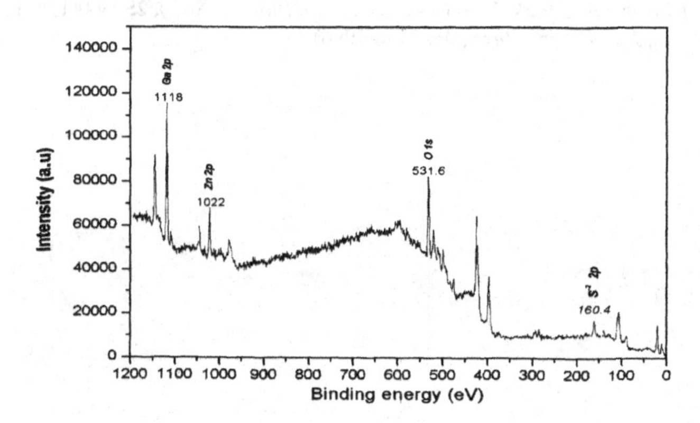

Figure 6. XPS spectra of $ZnGa_2O_{3.95}S_{0.05}$

CONCLUSIONS

The first zinc gallium oxysulfides ($ZnGa_2O_{4-x}S_x$) are synthesized and emission properties of $ZnGa_2O_{4-x}S_x$ and $ZnGa_2O_{4-x}S_x:Mn^{2+}$ have been studied in our investigation. From XRD data and PLE data, the sulfur in $ZnGa_2O_{4-x}S_x$ forms tetrahedrons and the cation in $ZnGa_2O_{4-x}S_x$ are randomly distributed between tetrahedrons and octahedrons such that $ZnGa_2O_4$ structure can be viewed as a mixed spinel. The four-fold enhancement of the brightness of Mn^{2+} emission can be the significant effect of forming oxysulfide in $ZnGa_2O_{4-x}S_x$: Mn^{2+}. $ZnGa_2O_{4-x}S_x$: Mn^{2+} can be the promising candidate for FED and VFD phosphors.

ACKNOWLEDGMENTS

This work has been supported by KOSEF in 2000(#2000-2-0636)

REFERENCES

1. I.K. Jeong, H.L. Park and S.I. Mho, *Solid State Comm.*, **108**, 823(1998)
2. S.K. Choi, H.S. Moon, S.I. Mho, T.W. Kim and H.L. Park, *Mat. Res. Bull.*, **33**, 693(1998)
3. L.E. Shea, R.K. Datta and J.J. Brown. Jr., *J. Electrochem. Soc.*, **141**, 2198(1994)
4. S.H.M. Poort, D. Cetin, A. Meijerink and G. Blasse, *J. Electrochem. Soc.*, **144**, 2179(1994)
5. F. Buschendorf, Z. Phys. Chem., **14**, 297(1931)
6. C.W.W. Hoffman and J.J. Brown, *J. Inorg. Nucl. Chem.*, **30**, 63(1968)
7. I.K. Jeong, H.L. Park and S.I. Mho, *Solid State Comm.*, **105**, 179(1998)
8. M.R. Royce, *U.S. Patent* 3, **418**, 246(1968)
9. P.N. Yocom, *U.S. Patent* 3, **418**, 247(1968)
10. L. Ozawa, *J. Electrochem. Soc.*, **124**, 413(1997)
11. O. KaneHisa, T. Kano and H. Yamamoto, *J. Electrochem. Soc.*, **132**, 2023(1985)

12. S.Y. Chang, C.I. Jeon, C.H. Pyon, Q.W. Choi, C.H. Kim and S.I. Mho, *Bull. Korea. Chem, Soc.*, **11**, 386(1990)
13. J.S. Lee, S.H. Kim and H.L. Park, *New Physics (Korean Physical Soc.)*, **25**, 460(1985)
14. C.F. Yu and P. Lin, *Jpn, J. Appl. Phys.*, **35**, 5726(1996)

Mat. Res. Soc. Symp. Proc. Vol. 667 © 2001 Materials Research Society

Blue Room-Temperature Photoluminescence of AlN Films, Prepared by RF Magnetron Sputtering

V. Ligatchev, S.F. Yoon, J. Ahn, Q. Zhang, Rusli, K. Chew, S. Zhgoon
School of Electrical and Electronic Engineering, Nanyang Technological University, Nanyang Avenue, Singapore 639798, Republic of Singapore

ABSTRACT

Photo luminescence (PL) signal from the aluminum nitride (AlN) films, excited by near UV (363.8 nm) laser has been measured at the room temperature. The AlN films are deposited by radio frequency (RF) sputtering of aluminum in argon-nitrogen-hydrogen gas mixture. Positions of the PL peaks maximums are influenced by the AlN preparation regimes. The analysis of the PL data is based on the results of the structural studies and electron spectrum investigations.

INTROUDUCTION

Light emitting in the blue area is important for the applications and has been extensively studied during the last decade [1 - 3]. Gallium and aluminum nitrides as well as GaN-based alloys are applied frequently for the blue-light devices producing, however preparation of the high quality layers is still quite expensive that limits of application areas of such devices [2 - 4]. Results of studies of AlN films, deposited by a radio frequency (RF) sputtering of aluminum (Al) target are presented in the paper. This technology allows to increase sufficiently the rate of the films growing with respect to the standard molecular-beam epitaxy or chemical vapor deposition processes.

EXPERIMENTAL DETAILS

Two series of aluminium nitride (AlN) films are obtained by RF (13.56 MHz, 300 W) magnetron sputtering of pure Al target in argon (Ar) and nitrogen-hydrogen (N_2-H_2) mixture on the 'Denton'-18 sputtering system. Nominal substrate temperature (T_s) of 275 °C and Ar flow rate (FR) of 20 sccm are invariable for 'a' series whereas the N_2+H_2 mixture (N_2:H_2=95:5) FR is varied in the 4 – 40 sccm range. For the 'b' series the Ar and N_2+H_2 flow rates are equal to 10 and 5 sccm (respectively), N_2:H_2=80:20, the T_s value is changed from 600±50 to 900±50 °C. Standard substrate heater and substrate temperature control unit are used at the "a' series deposition.

A new heater was specially designed for the AlN films deposition at the relatively high substrate temperature. The heater current is only considered as a reliably controlled deposition parameter, while the T_s value is estimated indirectly for the sample of the "b" series. Nominally undoped (ρ > 20 Om*cm), single-side polished and <110> oriented silicon plates are used as substrates. Typical films thickness is 0.3 – 0.5 μm for the "a" series and of 1.0 – 2.0 μm for the "b" series films.

Spectra-Physics Beamlok 2065 Ar ion laser (363.8 nm wavelength, 40±2 mW) is used for the AlN photoluminescence (PL) excitation at the room temperature (293 K). Schematic diagram of the measurement system is shown on the figure 1.

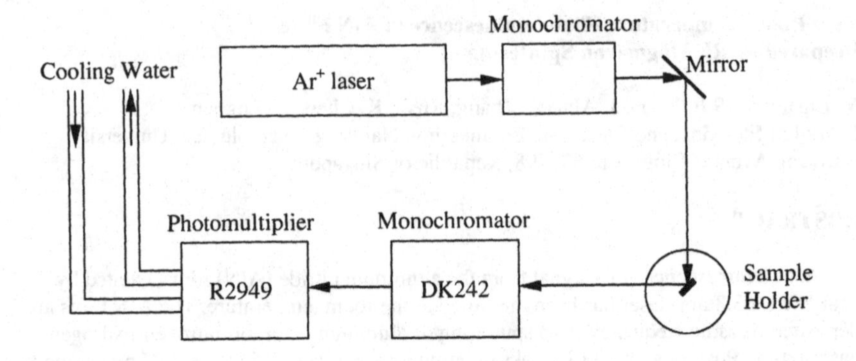

Figurel. Schematic diagram of PL measurement system.

The beam has a diameter of 1.5 mm. The PL of the AlN sample is excited at an angle of incidence of about 60°. This large angle of incidence is chosen to increase the absorption of the laser radiation in the nearly transparent AlN film by increasing its path, and also to reduce the scattered laser radiation being collected by the monochromator.

The monochromator used for discriminating the PL radiation is a CVI DIGIKROM DK242 double grating ¼ meter. The R2949 (Hamamatsu) photomultiplier is water-cooled to 10°C during PL measurement to reduce the background noise level. As usual, the PL spectra are corrected by taking into account the effect of the combined response of the R2949 and the monochromator DK242. All the PL spectrums have been measured at the room temperature of the samples.

The spectra contain main peaks at wavelengths of 420 – 440 nm for the all samples of the 'b' series (figure 2) and for a part of samples from the 'a' series (figure 3a), deposited at the nitrogen-hydrogen FR values below 25 sccm. The PL peaks positions for the other samples from the 'a' series are in the 450 – 510 nm wavelength range (figure 3b).

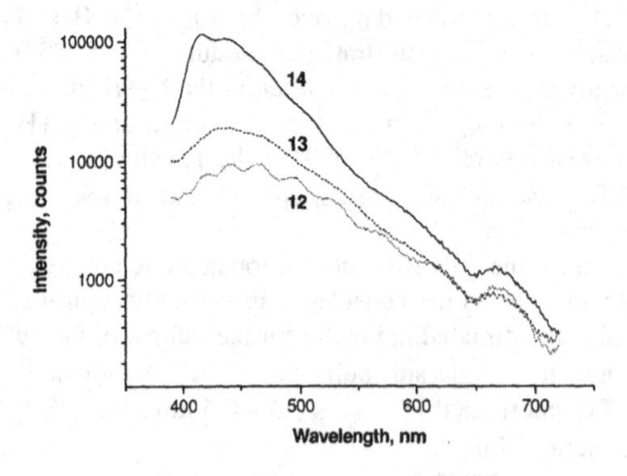

Figure 2. Typical normalized PL spectra of the AlN samples from the series "b", obtained at the different values of the heater current (shown on the figure in amperes).

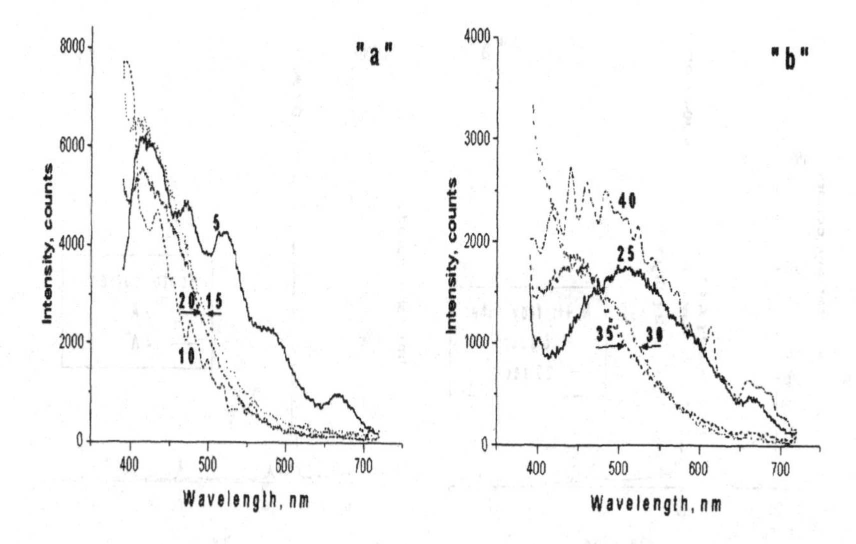

Figures 3 a, b. Normalized PL spectra of the AlN samples from the series "a", obtained at the different values of the nitrogen-hydrogen flow rate (shown on the figure in sccm).

The typical peaks fullwidth at the halfmaximum (FWHM) is 20 – 50 nm for the all studied AlN films. The PL signal of the 'b' series samples is higher then the signal from any sample of the 'a' series. Blue PL output for the 'b' series films enhanced more than ten times while the heater current is increased from 12 A to 14 A.

Alteration of the PL signal intensity from the AlN samples of the two series can be attributed to the differences of the films atomic structure and the films thickness. The X-ray diffraction (XRD) measurements are used for the structural characterization. The measurements are carried out in a 'detector scan' mode on the 'Siemens' D5005 equip ment at the room temperature. The equipment employs Cu K_α X-ray source (1.54 Å) operated at 40 kV and 40 mA. Low ($\theta = 6^o$) angle of X–ray incidence is applied in order to obtain a high output signal from the relatively thin AlN films.

The XRD data analysis reveals that the signal intensity is higher for the samples of the "b" series (figures 4 a, b). The XRD peaks are identified in according to [5 - 7]. The peaks intensity alteration can be originated both by the films thickness variation and by the changes of the crystalline phase fraction in the material. But variations of the peaks structure have to be attributed to the structural differences only. The results clearly show-that polycrystalline phase within the films of the "b" series are oriented more uniformly and it leads to the PL intensity gain.

The PL peaks positions on the films of the "b" series are practically not dependent on the substrate temperature (figure 2), while for the "a" series the peaks energies are essentially influenced by the nitrogen-hydrogen flow rate (figures 3 a, b). Possible alterations of the PL peaks red shift in the part of films from the "a" series (see figure 3 b) can be interpreted on the base of electron density of states N(E) deconvolution data for these

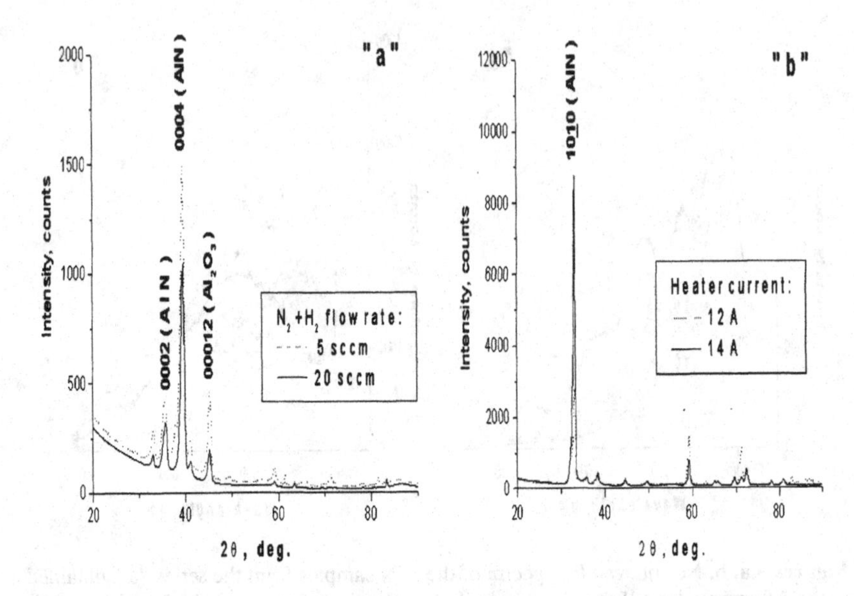

Figures 4 a, b. Normalized XRD spectra of the typical AlN samples from the "a" (figure a) and the "b" (figure b) series.

AlN samples. The N(E) spectrums for the films were experimentally investigated by the deep level transient spectroscopy (DLTS) and constant photocurrent method (CPM). The CPM shows two relatively wide peaks with maximums at approximately 2.3 eV and at 3.0 – 3.5 eV below conduction band (CB) bottom [8]. Dark Fermi level position in the samples is placed at the 0.2 – 0.4 eV below CB [9], hence without illumination these states are filled with electrons. Results of the DLTS method are given on the figure 5. All the peaks belong to nitrogen vacancy donor-like triplet [9 and references therein]. The photon flux under the laser illumination is $2.5*10^{17}$ particles per second, while the number of the electrons in the illuminated film volume are of $2.5*10^{14}$ in the peak at 3.0 – 3.5 eV below CB and of $2.5*10^{12}$ in the donor-like triplet peaks, respectively. It means that the electron levels temporary become partially unfilled due to the optical electrons excitation from these localised electron states to the free levels above the CB, and radiative electron recombination is permitted between the triplet peaks and the peak at 3.0 - 3.5 eV below CB. The suggestion is supported by the comparative analysis of the PL (figure 3 a, b) and the N(E) (figure 5) peaks widths and positions. Indeed, the FWHM of the PL peaks are of 0.14 - 0.35 eV, while the FWHM for the N(E) peaks are of 0.1 – 0.3 eV. In addition, energy reducing of a strongest visible PL peak is of 0.4 ± 0.03 eV at the N_2+H_2 FR increment from 10 to 25 sccm (figures 3). The strongest N(E) peak shift to midgap is of 0.4 ± 0.05 eV for the same AlN samples (figure 5). Thus origin of the visible PL in the studied AlN samples is the radiative electron transitions from the partially filled triplet levels to the temporary unfilled deep electron states at the 3.0 – 3.5 eV below CB.

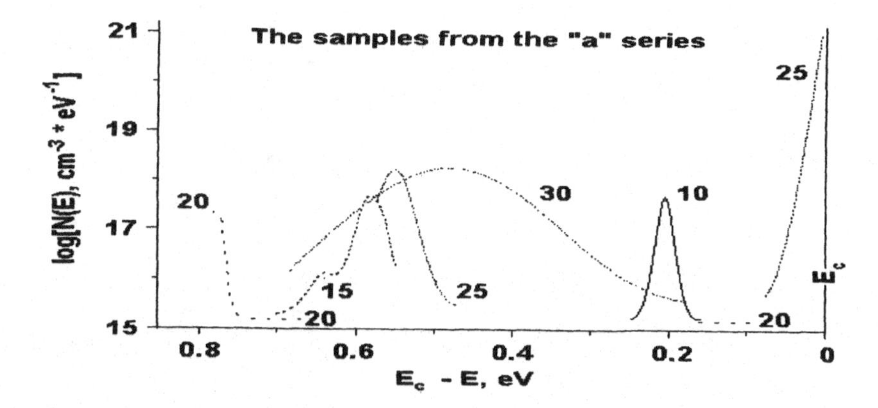

Figure 5. The N(E) spectrums measured by DLTS technique on the Al/AlN/Si structures, deposited at the different values of the nitrogen-hydrogen flow rate (given in sccm at the curves).

CONCLUSION

Room-temperature visible photoluminescence of the AlN films, deposited by magnetron sputtering of Al target in argon-nitrogen-hydrogen mixture on the silicon plates is investigated in the paper. It was found out that intensity of the PL signal is sufficiently influenced by the variations of the films deposition conditions (substrate temperature and nitrogen-hydrogen flow rate). Experimentally detected changes of the atomic structure and the electron spectrum of the AlN films are supposed to be responsible for the photoluminescence spectrum alteration.

REFERENCES

1. I. Akasaki et al., GaN-based UV/blue light emitting devices, Inst. Phys. Conf. Series No. 129, Chapter 10, Int. Symp. *GaAs and related compounds*, (Karuizawa 1992), pp. 851-856.
2. J. H. Harris, R. A. Youngman, Photoluminescence and Cathodoluminescence of AlN, *Properties of Group III Nitrides*, ed. J. H. Edgar, (INSPEC, 1994) pp. 203-221.
3. S. Nakamura et al., *J. Appl. Phys.*, **76**, 8189 (1994).
4. S. Nakamura et al., *Appl. Phys. Lett.* **58**, 2021 (1991).
5. Jang Jie, Wanf Chen, Tao Kun et al., *Appl. Phys. Lett.*, **62,** 2790 (1993).
6. K. Kaya, Y. Kanno, H. Takahashi, *Jpn. J. Appl. Phys.* **35**, 2782 (1996).
7. A. Saxler, P. Kung, C.J. Sun et al., *Appl. Phys. Lett.*, **64**, 339 (1994).
8. V. Ligatchev, S.F. Yoon, J. Ahn, Q. Zhang, Rusli, S. Zhgoon, K.L. Chew, *Diamond and related materials*, **10**, 1335 (2001).
9. V. Ligatchev, S.F. Yoon, J. Ahn, Q. Zhang, Rusli. Abstracts of COMMAD 2000, Melbourne, Australia, 6-8 December 2000, p.138.

Figure 3. The Mössbauer absorption spectra of Fe in medium-doped $Na_3V_2(PO_4)_3$ samples taken at the different annealing temperatures. The sample was measured at room temperature.

CONCLUSION

Room temperature Mössbauer spectra copies of the $Na_3V_2(PO_4)_3$ samples by annealing.

REFERENCES

Quantum Wells and Quantum Dots

Mat. Res. Soc. Symp. Proc. Vol. 667 © 2001 Materials Research Society

Nanocrystal Quantum Dots: Building Blocks for Tunable Optical Amplifiers and Lasers

[1]Jennifer A. Hollingsworth, [1]Alexander A. Mikhailovsky, [1]Anton Malko, [1]Victor I. Klimov, [2]Catherine A. Leatherdale, [2]Hans –J. Eisler, and [2]Moungi G. Bawendi,

[1]Physical Chemistry and Applied Spectroscopy, Chemistry Division, Los Alamos National Laboratory, Los Alamos, NM 87545, USA

[2]Department of Chemistry and Center for Materials Science and Engineering, Massachusetts Institute of Technology, Cambridge, MA 02139, USA

ABSTRACT

We study optical processes relevant to optical amplification and lasing in CdSe nanocrystal quantum dots (NQD). NQDs are freestanding nanoparticles prepared using solution-based organometallic reactions originally developed for the Cd chalcogenides, CdS, CdSe and CdTe [*J. Am. Chem. Soc.* **115**, 8706 (1993)]. We investigate NQDs with diameters ranging from 2 to 8 nm. Due to strong quantum confinement, they exhibit size-dependent spectral tunability over an energy range as wide as several hundred meV. We observe a strong effect of the matrix/solvent on optical gain properties of CdSe NQDs. In most of the commonly used solvents (such as hexane and toluene), gain is suppressed due to strong photoinduced absorption associated with carriers trapped at solvent-related interface states. In contrast, matrix-free close packed NQD films (NQD solids) exhibit large optical gain with a magnitude that is sufficiently high for the optical gain to successfully compete with multiparticle Auger recombination [*Science* **287**, 10117 (2000)]. These films exhibit narrowband stimulated emission at both cryogenic and room temperature, and the emission color is tunable with dot size [*Science* **290**, 314 (2000)]. Moreover, the NQD films can be incorporated into microcavities of different geometries (micro-spheres, wires, tubes) that produce lasing in whispering gallery modes. The facile preparation, chemical flexibility and wide-range spectral tunability due to strong quantum confinement are the key advantages that should motivate research into NQD applications in optical amplifiers and lasers.

INTRODUCTION

It was realized almost two decades ago that semiconductor nanocrystal quantum dots (NQDs) should provide superior performance in lasing applications in comparison with other semiconductor bulk and low dimensional materials. In very small dots, the spacing of the electronic states is much greater than the available thermal energy (strong confinement), inhibiting thermal depopulation of the lowest electronic states. This effect should result in a lasing threshold that is temperature insensitive at an excitation level of only one electron-hole (e-h) pair per dot on average [1]. Additionally, NQDs in the strong confinement regime have an emission wavelength that is a pronounced function of size, adding the advantage of continuous spectral tunability over a wide energy range simply by changing the size of the dots. The prospect of realizing lasers for which the output color can be controlled by facile manipulation of the dot size and semiconductor composition has been a driving force in NQD research for more than a decade.

Lasing has previously been demonstrated for epitaxially grown III-V NQDs [2-4]. These dots have relatively large lateral sizes (typically greater than 10 nm); therefore, the spacing between their electronic states is smaller than room-temperature carrier energies (weak confinement) and the lasing threshold is temperature sensitive. Further, large lateral dimensions and difficulties in size control limit their spectral tunability using quantum confinement effects. As a result, the emission wavelengths in epitaxial dots are usually controlled by a material's composition rather than QD size [4].

Direct colloidal chemical synthesis provides routine preparations of freestanding semiconductor nanoparticles with sizes that correspond to the regime of extremely strong confinement. For the studies reported here, we synthesized CdSe NQDs via the pyrolysis of dimethylcadmium ($CdMe_2$) and trioctylphosphine selenide (TOPSe) in a coordinating solvent mixture of trioctylphosphine (TOP) and trioctylphosphine oxide (TOPO), as schematically shown in figure 1 [5]. In addition to providing the medium for the solution-based reactions, the coordinating solvent molecules covalently bind to the NQD surface atoms, forming an organic layer, or cap, on the surfaces of the individual NQDs. This capping provides electronic and chemical passivation of surface dangling bonds and imparts solubilities and reactivities that are characteristic of the organic ligand. As-prepared NQDs are further size selected using size-selective precipitation that allows isolation of ensembles with size dispersions of <5 %. Particles of ~2 to 8 nm in diameter were prepared, spanning the range from strong quantum confinement to weak confinement.

In this size range electronic interlevel spacings can exceed hundreds of meV, and size-controlled spectral tunability over an energy range as wide as 1 eV can be achieved. Furthermore, improved schemes of surface passivation allow significant suppression of surface trapping and produce room-temperature photoluminescence (PL) quantum efficiencies as high as 50% with emission wavelengths tunable across the entire visible spectrum. However, despite their excellent emission properties, favorable for optical amplification, colloidal nanostructures have failed to yield lasing in numerous efforts. In this paper we analyze the underlying physics of processes relevant to optical gain and lasing in strongly confined NQDs. We show that there are solvent/interface effects that prevent the observation of gain in most common NQD/solvent systems. Further, we show that there are intrinsic Auger nonradiative recombination processes that complicate the development of stimulated emission in strongly confined NQDs, but do not inherently prevent it. *Finally, we demonstrate stimulated emission in close-packed films of CdSe nanocrystal NQDs in the strong confinement regime that provides for the first time proof-of-principle for NQD lasing.*

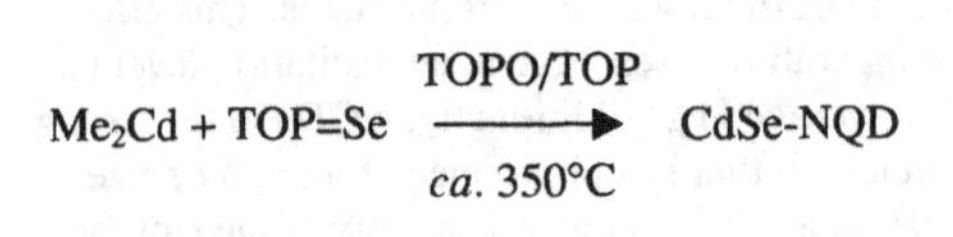

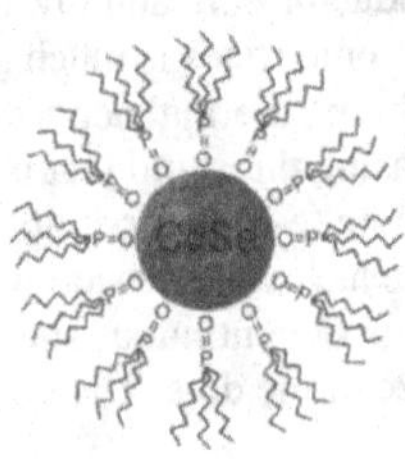

Figure 1. An organometallic reaction used to fabricate CdSe NQDs surface passivated with an organic capping layer; TOPO molecules of cap are shown.

EXPERIMENTAL DETAILS

CdSe NQDs were fabricated as shown in figure 1 [5]. NQD solids were prepared by drop casting films from hexane/octane solutions; solvent evaporation yielded amorphous dry solid films of close-packed NQDs. Transient absorption (TA) measurements in the visible were performed using a femtosecond (fs) pump-probe experiment as previously described [6]. Briefly, the samples were pumped at 3.1 eV by frequency doubled 100 fs pulses from an amplified Ti-sapphire laser. Pump-induced absorption changes were measured using time-delayed broad-band probe pulses of a fs white light continuum generated in a sapphire window.

RESULTS/DISCUSSION

Solvent/interface dependent optical gain dynamics in CdSe NQDs

In an attempt to observe gain in NQD CdSe solutions, we studied the TA pump dependence for NQD CdSe of various sizes in hexane (figure 2). The gain threshold corresponds to $-\Delta\alpha/\alpha_0 = 1$, where α_0 is a linear absorption coefficient and $\Delta\alpha$ a pump-induced absorption change. None of the samples in figure 2 shows crossover from absorption to gain even at very high pump densities corresponding to $N_{eh} > 10$, where N_{eh} is the number of electron-hole pairs per dot. Rather, a "universal" size-independent TA pump-intensity dependence, showing a saturation of the 1S absorption bleaching before crossover to gain, is observed. The universal curve shows a linear growth below $N_{eh} = 1$ and saturation at a level $-\Delta\alpha/\alpha_0 \approx 1$ above $N_{eh} = 1$.

These data are compared with the pump dependence expected for state filling in a system for which the 1S optical transition couples a populated electron state to either a populated (dashed line) or an unpopulated (solid line) hole state. The data are consistent with the latter, suggesting fast hole relaxation to a state that is lower in energy than that involved in the 1S transition [7]. We recently studied this relaxation process using an ultrafast PL experiment [8]. Fs PL data indicate a sub-ps hole relaxation from the state responsible for the 1S absorption ("absorbing" state) to a lower energy state involved in

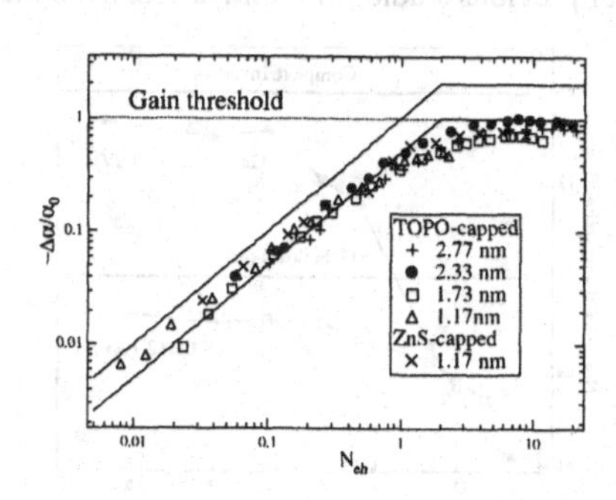

Figure 2. Pump dependence of the normalized 1*S* absorption changes in a NQD-CdSe/hexane sample.

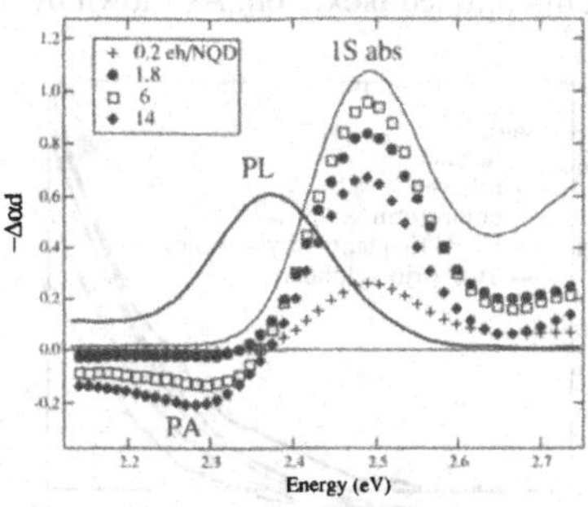

Figure 3. Pump dependence of 1*S* bleaching ($\Delta\alpha<0$) and photoinduced absorption ($\Delta\alpha>0$) in a NQD-CdSe/hexane sample.

the band-edge PL ("emitting" state). The formation of these "fine-structure" hole states is due to splitting of the lowest $1S_{3/2}$ hole level induced by electron-hole exchange coupling, interactions with crystalline field and NQD shape asymmetry [9]. The transition involving the low-energy hole state is not well pronounced in absorption due to its relatively small oscillator strength, but since it is the lowest in energy it is responsible for processes involving light emission and, hence, optical gain.

Instead of gain, however, CdSe NQD/hexane samples show photoinduced absorption (PA) in the region of the lowest "emitting" transition (figure 3). In contrast to 1S bleaching which saturates at high pump intensities, PA does not show saturation and, therefore, cannot be circumvented by simply increasing the excitation density. Analysis of TA data for NQDs in different liquid and solid-state matrices indicate that the PA is dependent on the matrix material, suggesting that it is due to excited-state absorption involving carriers trapped at solvent-related interface states. Such commonly used solvents as toluene, chloroform and HMN show a PA band comparable in intensity with that in hexane (figure 4). However, in the case of NQDs dispersed in polyvinyl butyral, the PA is reduced in magnitude and is red shifted with respect to the PL band. The strongest PA suppression was observed for TOP, one of the NQD growth solvents. CdSe NQDs/TOP samples show gain at the position the PL band (figure 5). The transition from absorption to gain occurred at carrier densities between 1 and 2 electron-hole (e-h) pairs per dot on average, consistent with the 1 e-h pair theoretical threshold expected for 3D strongly confined systems. The gain saturated at a level of $\sim 0.7\alpha_0$, corresponding to ~70% population inversion in terms of a simple two-level model.

Despite the fact that CdSe NQD/TOP samples show optical gain, they do not exhibit stimulated emission. In this case, as well as in other cases of relatively dilute NQD systems (e.g., NQD-doped glasses fabricated by high-temperature precipitation), the development of stimulated emission is inhibited by ultrafast decay of the optical gain, as analyzed below.

Multiparticle Auger recombination and optical gain dynamics in NQDs

In NQDs the band-edge optical gain is dominated by two e-h pair states, i.e. by a quantum-confined biexciton. As shown by our previous studies of multiparticle dynamics

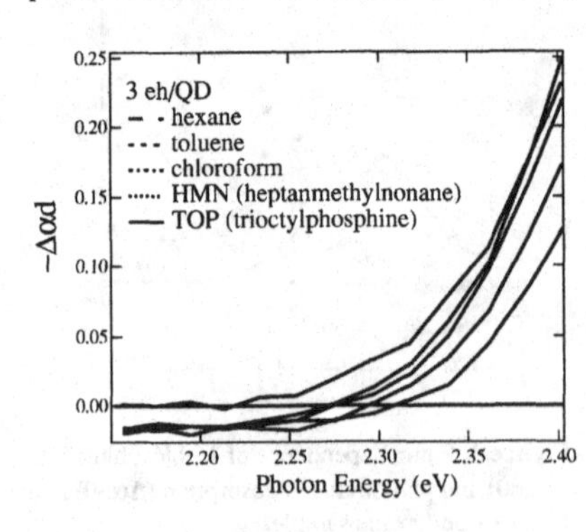

Figure 4. Solvent-dependent PA.

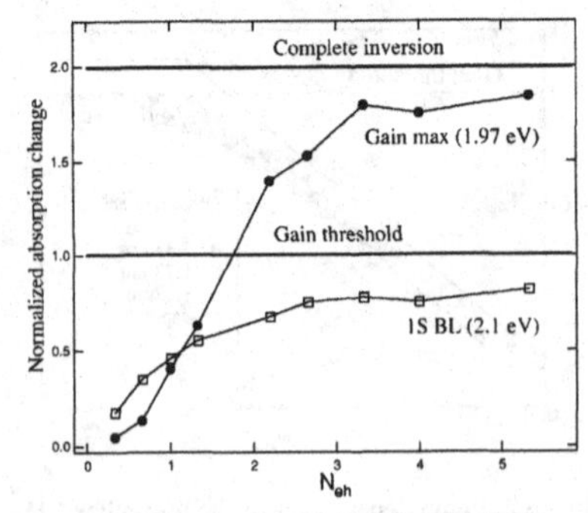

Figure 5. Gain observed in NQD-CdSe/TOP sample is red-shifted from the 1S bleach band, where no gain is observed.

[10], the biexciton decay in CdSe NQDs is dominated by Auger recombination. Auger recombination is a nonradiative process that leads to the recombination of e-h pairs via energy transfer to a third particle (an electron or a hole) that is re-excited to a higher energy state within the dot or even outside it (Auger ionization). Auger recombination has a relatively low efficiency in bulk semiconductors, for which significant thermal energies are required to activate the effect. However, Auger decay is greatly enhanced in quantum confined systems for which the relaxation in momentum conservation removes the activation barrier [11].

In figure 6 we show the size-dependence of the biexciton decay. These data indicate that the biexciton lifetime (τ_2) rapidly decreases with decreasing NQD size following a cubic size dependence ($\tau_2 \propto R^3$); $\tau_2 = 363$ ps in NQDs with $R = 4.1$ and reduces to only 6 ps for $R = 1.2$ nm. Due to its very short time constants, nonradiative Auger relaxation of doubly excited nanoparticles dominates over surface trapping even in samples with incomplete surface passivation. This implies that the fast optical gain decay due to Auger recombination is the principal factor inhibiting lasing in small-size colloidal dots.

Stimulated emission in NQDs: Proof of principle for NQD lasing

To overcome the problem of ultrafast gain decay, development of stimulated emission must occur on time scales that are shorter than those for Auger decay. One approach to enhancing the rate of stimulated emission is to increase the gain magnitude by increasing the dot density in the sample. Our estimations show that the stimulated emission rate exceeds that for the Auger recombination at NQD filling factors, ξ, of ~1% or higher. Because of the limited NQD solubility in TOP, such densities are difficult to achieve in TOP solutions. However, they are readily achieved in solid-state NQD films (NQD solids). In particular, films made of TOPO-capped 1.3-nm NQDs have filling factors as high as ca. 20%, assuming random close packing.

As expected, NQD films show stimulated emission (figure 7). We observed development of stimulated emission at both room and liquid nitrogen temperatures, and the emission color was tunable by simply changing the size of the NQDs. We have also demonstrated true cavity-mode lasing by depositing films on the inner walls of micro capillary tubes and promoting stimulated emission both along the length of the tube (waveguide mode) and around the inner circumference of the tube

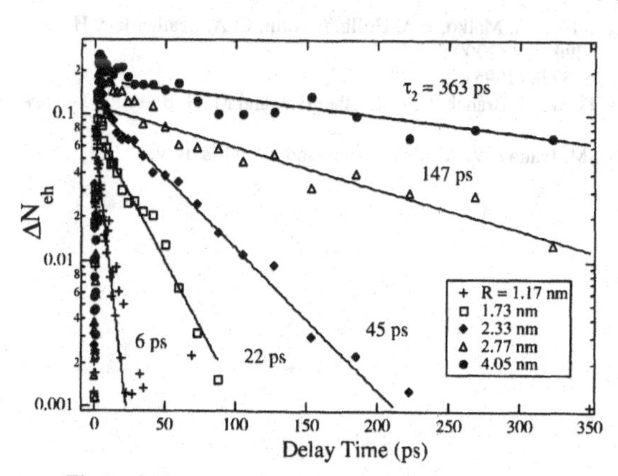

Figure 6. Size-dependent Auger rates: biexciton dynamics as a function of the NQD radius.

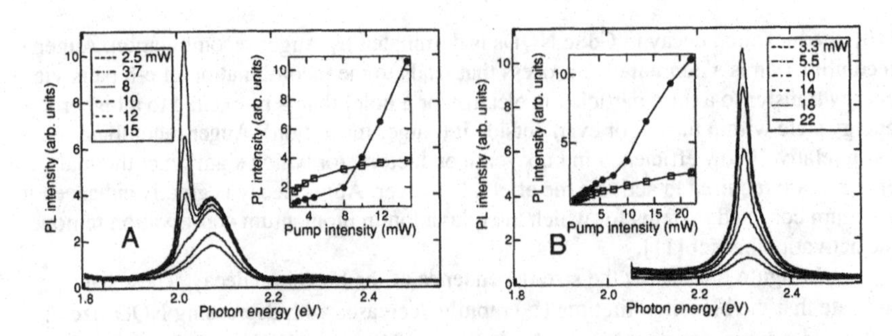

Figure 7. Development of a sharp stimulated emission band as a function of pump intensity in emission spectra of films ($T = 80$ K) fabricated from TOPO-capped NQDs with $R = 2.1$ nm (A) and ZnS-capped NQDs with $R = 1.35$ nm (B). The films are optically pumped perpendicular to the sample plane and the stimulated emission is detected at the edge of the films that act as optical waveguides. Insets: Superlinear intensity dependence of the stimulated emission (circles) showing a clear threshold compared to the sublinear dependence of the PL intensity outside the sharp stimulated emission peak (squares).

(whispering gallery mode) [12]. These results provide a proof-of-principle for lasing in strongly confined NQDs and should motivate the development of tunable NQD-based lasers and amplifiers operating over a broad spectral range.

REFERENCES

1. M. Asada, Y. Miyamoto, and Y. Suematsu, *IEEE J. Quantum Electron.* **QE-22**, 1915 (1986).
2. N. N. Ledentsov, *et al.*, *Semiconductors* **28**, 832 (1994).
3. N. Kistaedter *et al.*, *Electron. Lett.* **30**, 1416 (1994).
4. M. Grundman, *Physica E* **5**, 167 (2000).
5. C. B. Murray, D. J. Norris, and M. G. Bawendi, *J. Am. Chem. Soc.* **115**, 8706 (1993).
6. V. Klimov and D. McBranch, *Opt. Lett.* **23**, 277 (1998).
7. V. I. Klimov, Ch. J. Schwarz, D. W. McBranch, C. A. Leatherdale, and M. G. Bawendi, *Phys. Rev. B* **60**, R2177 (1999).
8. V. I. Klimov, A. A. Mikhailovsky, Su Xu, A. Malko, J. A. Hollingsworth, C. A. Leatherdale, H. –J. Eisler, and M. G. Bawendi, *Science* **290**, 314 (2000).
9. M. Nirmal *et al.*, Phys. Rev. Lett. **75**, 3728 (1995).
10. V. I. Klimov, A. A. Mikhailovsky, D. W. McBranch, C. A. Leatherdale, and M. G. Bawendi, *Science* **287**, 10117 (2000).
11. D. Chepic, A. L. Efros, A. Ekimov, M. Ivanov, V. A. Kharchenko, and I. Kudriavtsev, *J. Luminescence* **47**, 113 (1990).
12. In progress.

Mat. Res. Soc. Symp. Proc. Vol. 667 © 2001 Materials Research Society

Excited State Relaxation Mechanisms in InP colloidal Quantum Dots.

Garry Rumbles, Don Selmarten, Randy E. Ellingson, Jeff Blackburn, Pingrong Yu, Barton B. Smith, Olga I. Micic and Arthur J. Nozik.

National Renewable Energy Laboratory
Center for Basic Sciences
Golden, CO 80401

ABSTRACT

We report photoluminescence, linear absorption and femto-second transient bleaching spectra for a colloidal solution of indium phosphide (InP) quantum dots at ambient temperatures. The photoluminescence quantum yield is shown to depend not only upon the size of the quantum dots, with larger dots exhibiting higher quantum yields, but also upon the excitation wavelength. At short wavelengths, photoluminescence excitation spectra deviate markedly from the absorption spectra indicating that an efficient non-radiative deactivation pathway becomes prominent at these higher photon energies. We interpret this observation in terms of an inefficient relaxation mechanism between the second excited state and the lowest energy excited state from which the emission emanates. The results are consistent with the existence of a phonon bottleneck.

INTRODUCTION

The use of photoluminescence excitation spectroscopy (PLE) for measuring indirectly the absorption spectrum of species that cannot be measured directly is a valuable and reliable tool. For the approach to be valid, however, the intensity of the PL must be directly proportional to the amount of light absorbed and the assumptions made must be recognised in order to avoid errors. For molecular systems in the condensed phase, either solid or solution, it is assumed that relaxation of the initially excited state to the emitting state occurs with an efficiency of 100%. Both internal conversion and vibrational relaxation combine to relax the initially excited state fast and efficiently to achieve this figure and this is often referred to as Kasha's rule [1]. For systems that contain more than one emitting chromophore, the differing photoluminescence quantum yields (PLQY) must be taken into consideration, as PLE is biased towards the more emissive species. It must also be recognised that PLE spectra are a measure of %Absorption (%A), or 1 – Transmission (1 – T) and this only approximates to the absorption spectrum for weakly absorbing samples, such as dilute solutions and thin films.

To correlate the electronic absorption spectra of colloidal quantum dots with the PLE spectra requires careful experimentation. The QD sample is a distribution of dot sizes, each with its own PLQY and emission position. In addition, the relaxation mechanism of the initially excited state to the emitting state may not be as efficient as for molecular systems, as the relaxation of the excited electrons may be hampered by the low density of both electron and phonon states close to the conduction band minimum (CBM). This process, referred to as the phonon bottleneck [2], provides an opportunity to create hot electrons that may be useful in photovoltaic device applications. The phonon bottleneck also provides an opportunity for a non-radiative decay channel to compete with the cooling.

A comparison of the PLE spectra with the absorption spectra of colloidal CdSe dispersed in a polymer matrix has demonstrated that the two spectra are not equivalent and the difference has been attributed to absorption into a high energy absorption continuum that acts as a non-radiative decay channel for the excitation energy [3]. In this article, we report a comparison of the electronic absorption and PLE spectra for InP colloidal quantum dots dissolved in a hexane solvent at ambient temperature. The goal of these studies has been to examine the efficiency of electron cooling and compare the results with those derived for the same sample using femtosecond transient absorption spectroscopy.

EXPERIMENTAL

Colloidal InP was synthesised, purified and characterised according to the procedure published previously [4]. Hexane (Aldrich) was tested for extraneous luminescence and was used without further purification. Solutions that also contained excess TOPO (trioctylphosphine oxide) were purged with helium to remove dissolved oxygen and transferred to 1 cm pathlength, fused silica cuvettes for absorption and emission measurements. Electronic absorption spectra were recorded using a double beam spectrometer (Cary 500) at a spectral resolution of 1 nm. Photoluminescence emission (PL) and excitation (PLE) spectra were recorded using a spectrometer with a photon-counting detection system and a 450 W xenon discharge lamp as an excitation source (Fluorolog 2). Spectral resolution for both PL and PLE was fixed at 1 nm using two additive dispersion double monochromators (1.8 nm/mm dispersion). All emission and excitation spectra were corrected for the wavelength dependence of the detection and excitation systems, respectively. Transient bleaching data were recorded with excitation from a visible OPO pumped by the output of an amplified, mode-locked Ti:Sapphire laser (Clark CPA 2001); a system that provides wavelength tunable, 125 femtosecond pulses at a repetition rate of 1 kHz. White light (440 - 950 nm) pulses were generated by focussing the 775 nm output of the pump laser into a sapphire crystal: compensation for the spectral chirp provided a temporal resolution <200 fs. Transient bleach spectra were measured on solutions contained in 2 mm pathlength cells by recording the white light spectrum with and without alternate pump laser pulses.

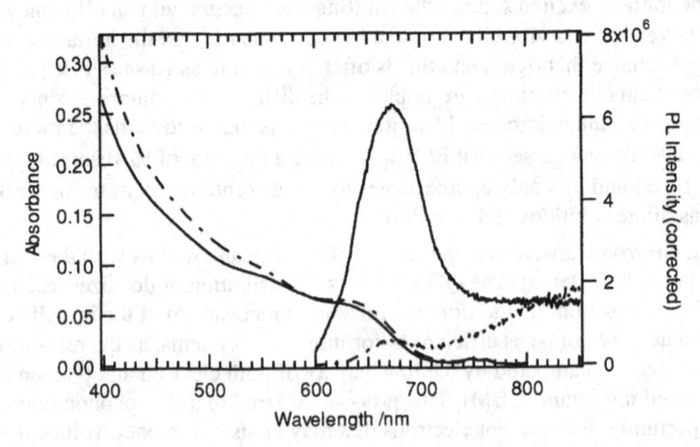

Figure 1 Absorption and PL spectra for etched (solid line) and unetched (dashed line) solutions of InP QDs in hexane at ambient temperature. (The weak spectral feature at 750 nm corresponds to a vibrational overtone of the solvent).

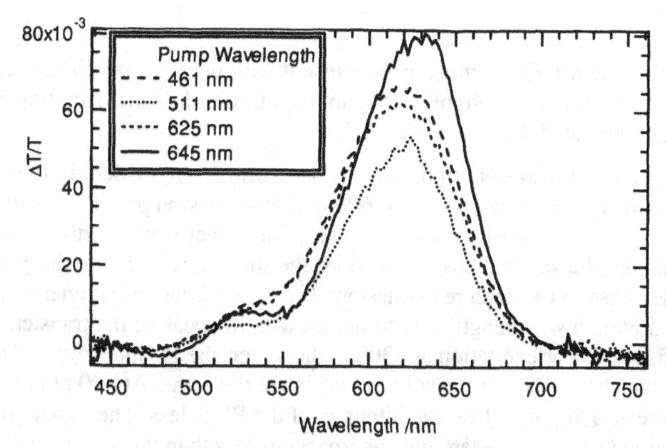

Figure 2 Transient bleaching data for etched InP QDs in hexane recorded 50 ps after excitation at 461 nm, 511 nm, 625 nm and 645 nm.

RESULTS

The electronic absorption data for etched and unetched InP quantum dot samples in hexane at 295K are shown in figure 1. The lowest energy absorption peak, or exciton peak, occurs at ca. 630 nm and is indicative of a dot diameter of 38±4 Å [2,4]. The slight spectral blue shift of the etched sample with respect to the unetched sample is attributed to a modest reduction in the average dot size due to the etching process. The shoulder at 550 nm corresponds to a transition to the second excited state.

Photoluminescence spectra for the two samples exciting at 640 nm are also shown in figure 1. The effect of etching can be seen to increase the emission at 675 nm by an order of magnitude, while the deep red emission at 850 nm decreases marginally.

Transient bleach spectra for the etched sample are shown in figure 2, with excitation at 461, 511, 625 and 645 nm and at a time delay of 50 ps. The excitation density was adjusted to ensure that the excitation density was kept to approximately 0.2 excitations per dot, using a procedure reported elsewhere [5]. There is little difference between these spectra, with the strong signal at 630 nm corresponding to the bleach signal of the first excited state, and the small peak at 530 nm attributed to the second excited state.

Photoluminescence spectra (PL) of the etched sample as a function of excitation wavelength are shown in figure 3 for four representative excitation wavelengths of 420 nm, 540 nm, 620 nm and 660 nm. These data have been corrected for the sensitivity of the detection system, but they have not been normalised to the absorbance at the excitation wavelength.

Two photoluminescence excitation spectra (PLE) for the etched sample were recorded by detecting (i) wavelength-integrated emission and (ii) emission at a fixed wavelength of 675 nm, which corresponds to the peak of the global emission [2]. These data are shown in figure 4 and they have been corrected for the wavelength dependent response of the Si photodetector used to monitor the lamp intensity. This figure also contains a %Absorption spectrum derived from figure 1, that corresponds to the effective absorption at the centre of the 1 cm cuvette, as described later.

DISCUSSION

The effects of etching the InP QD sample are consistent with previous published results [4], with the sharp exciton emission growing significantly in magnitude while the deep trap emission at longer wavelengths is reduced, but to a lesser extent.

The effects of the excitation wavelength on the emission profile are evident in figure 3. For excitation wavelengths in the range 400 nm to 600 nm, the emission profile remains broad, with a bandwidth of 80 nm, with a weak shoulder at 600 nm. For longer wavelength excitation ranging from 600 nm to 720 nm the spectral profile narrows, the shoulder at 600 nm disappears and there is a concomitant decrease in the deep red emission. The peak of the lowest energy absorption feature occurs at 630 nm, a wavelength that coincides with the peak of the transient bleach signals shown in figure 2. For excitation at 630 nm, however, the PL intensity is far greater than at 600 nm , where the absorbance of the solution is almost the same. At 400 nm the absorbance is three times the value at 600 nm and yet the intensity of the PL is less. The maximum PLQY occurs with excitation at 660 nm, where the absorbance is less than the value at 640 nm. These data suggest that the PLQY depends significantly upon the excitation wavelength, with the absorption into the very low energy tail of the absorption yielding more PL than for shorter excitation wavelengths.

This observation is more apparent when the PLE spectra are compared to the absorption data shown in figure 4. As discussed earlier, the PLE spectra are a measure of %Absorption, which approximates to the absorbance for samples of low optical density. The geometry deployed in the PLE experiments uses 1 cm cuvette with emission detected perpendicular to excitation. To avoid inner filter effects, the optical density of the sample must be kept below 0.1, where %Absorption and absorbance are very similar. From figure 1, the absorbance at wavelengths shorter than 500 nm rises beyond a value of 0.1 up to 0.25 at 400nm. The spectrum depicted in figure 4 compensates for this effect, by calculating the %Absorption for the central 2 mm of the sample cuvette, where the emission is detected. Such a procedure thus eliminates any errors that may arise for the absorption of the excitation light at the shorter wavelengths. The spectrum in figure 4

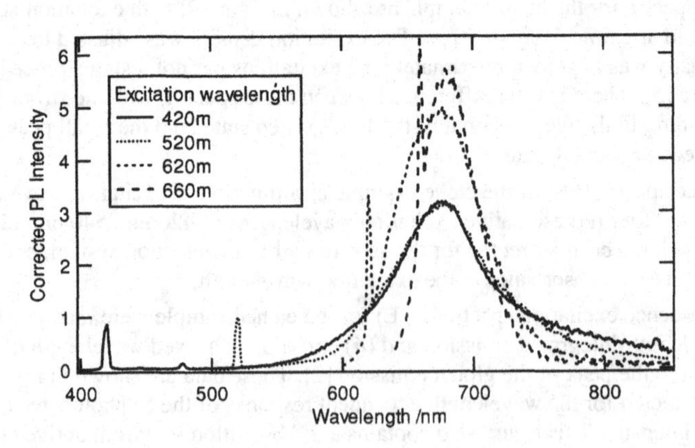

Figure 3 PL spectra recorded with excitation at 420 nm, 520 nm, 620 nm and 660 nm.

is very similar to that in figure 1, which indicates that the effect of the inner filter effect is only a minor one. The PLE data exhibit the same phenomenon as noted of the PL data of figure 3, with short excitation wavelengths yielding less PL than those around the bandedge. The PLE spectrum recorded with detection of wavelength-integrated emission peaks at 650 nm, while detecting emission at 675 nm occurs at a slightly longer wavelength of 660nm. Of these two wavelengths, neither coincide with the absorption maximum or the peak of the transient bleach signal.

The dramatic drop in PLQY at shorter excitation wavelengths and the non-coincidence of the peak PL signal with the transient bleach signal combine to yield a PLE spectrum that deviates significantly from the %Absorption (or absorbance) spectrum. The differences observed are therefore intrinsic features of the QD sample. The 10 nm difference in the position of the PLE spectra recorded with detection at 675 nm and total emission can be attributed to size-selective spectroscopy, where the 675 nm emission is biased towards QDs that emit at this wavelength. Relative to the peak of the first absorption feature, however, both these spectra are shifted noticeably to longer wavelengths. This shift may be explained in terms of a higher PLQY for the larger QDs than the smaller dots. If a major non-radiative decay mechanism is attributed to surface states, then the higher surface-to-volume ratio of the smaller QDs would lead to more non-radiative decay and less PL. The PLE spectra, therefore, should be a good representation of the absorption spectra of the large QDs, with the differing PLQY serving as a tool for refining the spectral resolution. The PLE spectrum recorded for emission detected at 675 nm, is a further enhancement of this effect and should represent the absorption of the QDs that emit preferentially at this wavelength. The difference between the spectral profile and the one recorded with the detection of total emission is, surprisingly, minor. The absorption at short wavelengths for most colloidal quantum dots is both measured [4,6] and calculated [7] to be larger than at the band edge. Excitation at these short absorption wavelengths, however, clearly does not result in the amount of PL that would be predicted for the amount of absorption.

The loss of PLQY at these short wavelengths is a significant observation that requires understanding. It is possible that the optical density at short wavelengths may be due to scattering and not absorption and would therefore not contribute to the PLQY. The solutions appear to be

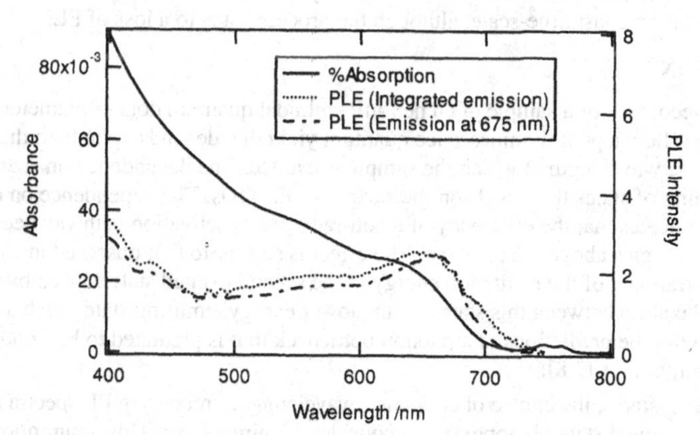

Figure 4 Comparison of %Absorption with PLE spectra recorded with detection of integrated emission and at 675 nm.

optically translucent and there is no direct evidence of aggregation, which would increase particle sizes and enhance the scattering cross-section. An excess of TOPO is used to ensure that the QDs remain in solution. TOPO exhibits a weak absorbance at wavelengths shorter than 400 nm, but the magnitude is far from sufficient to explain the large absorbance observed for the QD solutions. The effect must therefore be due to a non-radiative decay route for the excited state that can only accessed by excitation at short wavelengths. The efficiency of relaxation to the emitting state is therefore not 100%, and depends upon the excess energy of the exciting photon above that of the bandgap. A similar conclusion has been drawn for CdSe QDs dispersed in a polymer matrix, where the PLQY was shown dependent on the excitation wavelength [3]. These authors concluded that there existed a threshold above the bandgap, where a continuum of states acts as sink for the excitation energy. It is unclear what the origin of this continuum of states represents and unlike here, the PLE spectra and absorption spectra coincide perfectly at the bandedge.

An alternative explanation may be derived from the theory of the phonon bottleneck [2], where the transition from the second excited state to the emitting state is hindered by the low density of states that can couple the two together. This would have the effect of increasing the lifetime of the second excited state, allowing non-radiative decay mechanisms to compete more effectively with the relaxation to the emitting state thus lowering the PLQY. This explanation would also be applicable to the case of CdSe.

The efficiency of the relaxation mechanism to the lowest excited state in CdSe has been determined using femto-second transient bleaching data to occur on a sub-picosecond time scale [7]. Our own measurements on InP suggest that similar efficiencies occur here too. These observations appear, at first glance, to conflict with the results from the PL measurements. This discrepancy may be a result of one or two differences between the two experiments. The PL data yield a signal only when the electron is present in the conduction band and the hole in the valence band. The transient bleach signal can also be observed, however, with just the hole or the electron present [5]. The kinetics derived from following the time evolution of the bleaching signal may therefore not represent just the production and decay of the emitting state. The transient bleaching data are measured on a fast time-scale, while the PL data are measured under steady-state conditions. If the non-radiative decay process is very slow, then the transient bleach experiment will not detect it on a fast time-scale, although the process leads to a loss of PL.

CONCLUSION

The PL data recorded for a sample of etched InP colloidal quantum dots of diameter 38 Å in dilute solution show a photoluminescence quantum yield that depends upon both the diameter of the QDs and the wavelength at which the sample is excited. The dependence on size is attributed to the quenching of states that reside on the surface of the QDs. The dependence on excitation wavelength indicates that the efficiency of a non-radiative deactivation pathway becomes more prominent at energies above the bandgap. The effect is similar to that observed in CdSe and may be a result of trapping of the excitation energy in the second excited state caused by the inefficient relaxation between this state and the lower energy, emitting state. Such a mechanism is consistent with the predictions of a phonon bottleneck that is predicted to be important in colloidal quantum dots [2,8].

For molecular systems, the choice of excitation wavelength in recording PL spectra and understanding excited state photophysics is considered unimportant. This assumption is invalid for colloidal quantum dots and caution must be exercised when performing more sophisticated experiments, such as transient absorption and single-dot spectroscopy.

ACKNOWLEDGEMENTS

We would like to that the US Department of Energy, Office of Science, Office of Basic Energy Sciences, Division of Chemical Sciences for generous financial support.

REFERENCES

1. M. Kasha, Radiation Research Suppl., **2**, 243 (1960).
2. A.J. Nozik, Annu. Rev. Phys. Chem. **52**, 193 (2001).
3. W. Hoheisel, V.L. Colvin, C.S. Johnson and A.P. Alivisatos, **101**, 8455 (1994).
4. O.I. Micic, J. Sprague, Z. Lu and A.J. Nozik, Appl. Phys. Lett, **68**, 3150 (1996).
5. V.I. Klimov, J. Phys. Chem. B. **104**, 6112 (2000).
6. A.A. Guzelian, J.E.B. Katari, A.V. Kadavanich, U. Banin, K. Hamad, E. Jubin, A.P. Alivisatos, R.H. Wolters, C.C. Arnold and J.R. Heath, J. Phys. Chem., **100**, 7212 (1996).
7. H. Fu and A. Zunger, Phys. Rev. B., **57**, R15064 (1998).
8. H. Benisty, C.M. Sotomayor-Torrs and C, Weisbuch, Phys. Rev. B., **44**, 10945 (1991).

Mat. Res. Soc. Symp. Proc. Vol. 667 © 2001 Materials Research Society

Photoluminescence of multi-layer GeSi dots grown on Si (001)

J. Wan, Y. H. Luo, G. L. Jin, Z. M. Jiang, J. L. Liu, X. Z. Liao,[1] J. Zou,[1] and Kang L. Wang
Device Research Laboratory, Electrical Engineering Department
University of California at Los Angeles, Los Angeles, CA 90095-1594, U.S.A.
[1]Australian Key Center for Microscopy & Microanalysis, The University of Sydney, Sydney NSW 2006, Australia

ABSTRACT

Temperature and power dependent photoluminescence measurements were carried out on the multi-layer structure of GeSi dots grown on Si(001) substrate by gas-source molecular beam epitaxy. The transfer of photon-induced carriers from wetting layers into the dots and the region near the dots was evidenced. Different power dependent behaviors of the photoluminescence peak position were observed for the dots and the wetting layer. Accordingly, type-II and type-I band alignments were proposed for the dots and the wetting layers, respectively. After annealing, the photoluminescence peaks from the dots and the wetting layers showed blueshift due to the atomic intermixing. For the samples annealed at temperature above 850°C for 5min, the band alignment of the dots changes from type-II to type-I.

INTRODUCTION

Recently many studies have been done on the growth mechanism, electrical and optical properties of GeSi dots embedded in Si [1-3] due to its potential applications, such as optoelectronics. Although photoluminescence (PL) from GeSi dots has been widely studied, little has been done on the band alignment and the correlation between the PLs from the dots and the wetting layers, which is an inevitable result of the Stranski-Krastanow growth mode upon reaching a strain-defined critical thickness [3]. In addition, from a practical point of view, it is desirable to confine both electrons and holes within the dots for many potential applications, such as quantum dot based laser, single electron transistor and quantum computer [4]. However, the band alignment between the GeSi dots and Si matrix is type-II structure and the electron could not be trapped into the dots directly. In this paper, both excitation power and temperature dependent PL spectra of the GeSi dots and the wetting layers were measured. The band alignment of the GeSi dots and the wetting layer was studied. In addition, the annealing effect on the band alignment of the GeSi dots was also investigated.

EXPERIMENTAL DETAILS

The sample was grown on Si (001) substrate by gas-source molecular beam epitaxy (GS-MBE) with a Si_2H_6 gas source and a Ge effusion cell. Ten layers of GeSi dots seperated by Si spacer layers were grown at the temperature of 575°C. The thickness of the Ge dot layer and the Si spacer layer was 1.6 and 40 nm, respectively. Although pure Ge atoms were deposited during the dots growth, the dots are GeSi alloy, as discussed later. Pieces of the sample were annealed for 5 min with temperatures varied between 650 and 900°C by a step of 50°C using rapid thermal annealing in nitrogen gas ambient. The PL measurements were performed by excitation of an Ar^+ laser and with liquid nitrogen cooled Ge detector.

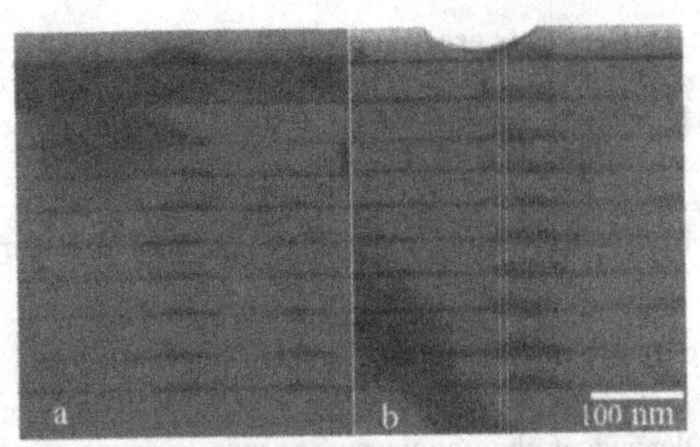

Figure 1. Bright-field cross-sectional TEM image of the 10 stacked layers of GeSi dots grown on Si (001) substrate: a) as-grown (sample A), b) annealed at 900°C for 5 min (sample B). TEM images showed that the dots are dislocation-free and the aspect ratio of height to base diameter of the dots increase from 0.18 to 0.30 after 900°C annealing.

DISCUSSION

Figure 1 shows two typical bright-field cross-sectional transmission electron microscopy (TEM) images of sample A (as-grown) and the sample B (annealed at 900°C for 5 min). As shown in the imgaes, the GeSi dots in both cases were dislocation free. Comparison of the two images indicated that (i) the size and the shape of quantum dots were changed and (ii) more importantly, the aspect ratio of height to base diameter of the quantum dots increased dramaticlly from 0.18 to 0.30 after the annealing process. For the as-grown sample, the average base and height of the dots were 80 and 14 nm, respectively. After annealing in sample B, the average base and height of the dots changed to be 100 and 30nm, respectively.

Typical PL spectra of the as-grown sample measured at 4.5K for different excitation power levels were shown in Fig. 2(a). Apart from the peaks of the Si, the specta consist of two separate components which are characteristic of the wetting layer and the GeSi dots, respectively. Two main peaks located at 1.007 and 0.949 eV are attributed to the NP_{WL} transition and its TO_{WL} phonon-assisted transitions of upper pseudomorphic wetting layers, while two peaks as denoted by NP_{WL1} and TO_{WL1} lines can be assigned to the NP transition and its TO replica in the first grown GeSi wetting layer [1-3]. The reason is that the first layer grows on unstrained Si and the following layers grow on prestrained Si, which induce the change of the thickness of the wetting layers. As a results, the PL from the upper thinner wetting layers show blueshift compared with the first wetting layer [1]. The broad PL peak located around 0.80eV is assigned to the dot and it could be deconvoluted into two Gaussian-line-shaped peaks (TO_{dot} and its NP_{dot} replica).

Figure 2(b) shows the excitation power dependence of PL peak energies of the wetting layers (NP_{WL}) and the GeSi dots (NP_{dot}). The TO_{dot} and NP_{dot} peaks of the dots show a large blueshift of 34 meV with increasing excitation power from 1 W/cm^2 to 40 W/cm^2. This large blueshift of the PL band with the increase of excitation power suggests a type-II band alignment [3,5]. As the electrons and holes occupy two separate regions, a dipole layer is then formed due to the fact that the holes are confined in the GeSi dots while electrons are confined outside when the sample were photon-excited. Then band bending will occur at the interfaces due to the Hartree potential. At high excitation power intensities, more photon-induced electrons and holes will result in a higher Hartree potential, which will cause the electrons and holes to be in higher

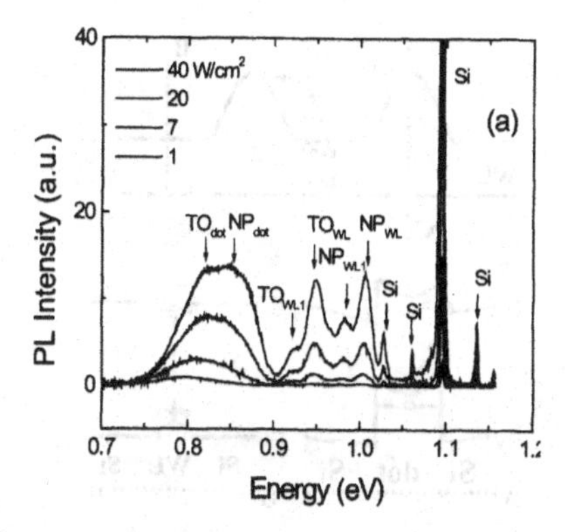

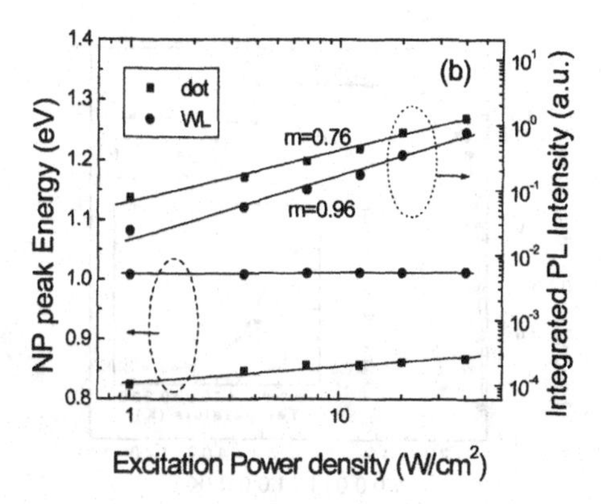

Figure 2. (a) 4.5 K PL spectra of the GeSi/Si (001) dots under different excitation powers. (b) Power dependence of PL peak energies and integrated PL intensity from the dots and the wetting layer. The peak energies from the dots show a blueshift with increasing excitation power while the peak energy from the wetting layer almost remains unchanged. This means that the band alignment of the dots and the wetting layer are type-II and type-I, respectively.

energies states, resulting in the blueshift of PL band [2,3,5]. It is known that the band filling effect could also cause the PL blueshift with the increase of excitation power. However, it was calculated that in a type-II structure the line shift due to band filling is only ~ 2 meV for general PL measurements and the band filling effect is almost an order of magnitude smaller than the shifts induced by band bending [5]. In contrast to the PL from GeSi dots, the NP_{WL} peak positions of the wetting layers only increase about 2 meV when the excitation power increases. For this case, a type-I band alignment at the wetting layer/Si interface is more reasonable and this small blueshift of wetting layers is due to band filling effect.

Different band alignments of the dots and the wetting layers could also be verified by their different power-dependent behaviors of intergrated PL inetensities. Figure 2(b) shows the power dependence of the integrated PL intensity from the GeSi dots and the wetting layers. According to the formula $I \propto P^m$, where I represents the PL intensity and P the excitation power [6], the coefficients m are found to be 0.97 and 0.78 for the PL from the dots and the wetting layers, respectively. The sublinear power dependence of PL intensity of the dots is due to a typical saturation effect for the type-II alignment. In a type-II alignment, the indirect excitons are firstly localized at the interfaces, and then recombine. As the interface state density is limited, the PL intensity quickly saturates [6]. For the wetting layer, there is no saturation as it is type-I structure.

It is known that the band alignment of strained $Si_{1-x}Ge_x$/Si heterostructure with higher Ge composition is type-II [7]. If there was no atomic intermixing at the Ge/Si interface, the wetting layer should be pure Ge, and then type-II band alignment would be expected. However, first, atomic intermixing occurs during MBE growth even at low temperatures. Second, the formation of dots will introduce a strain into Si substrate, and thus resulting in Si atoms in high energy

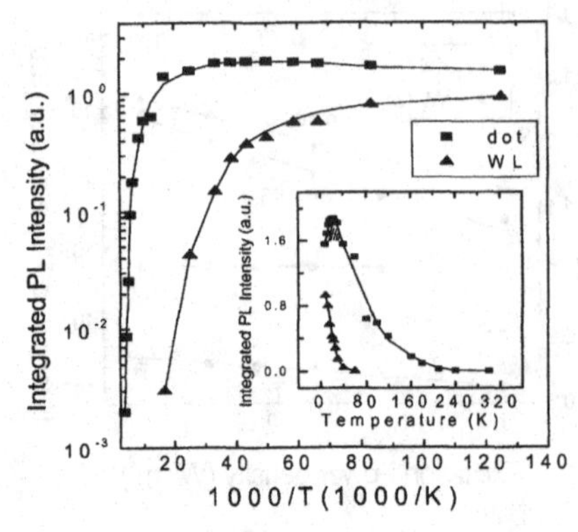

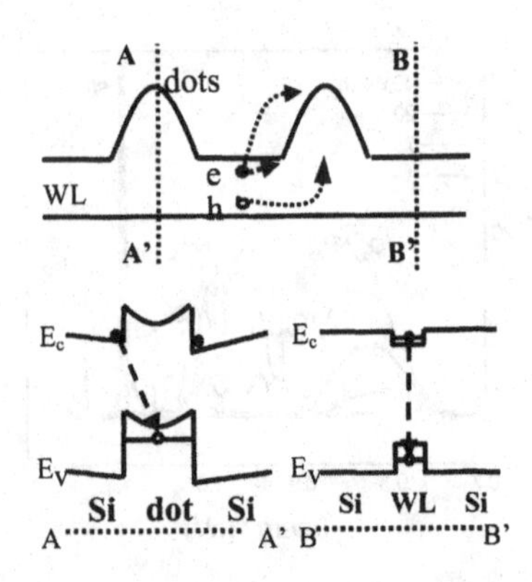

Figure 3. (a) Temperature dependences of integrated PL intensities from the dots and the wetting layers. It is clearly shown that the anomalous increase of the dots PL intensity at the temperature of 8~20K is accompanied with a rapid decrease of the wetting layer PL intensity. (b) Schematic diagram of the photon-induced carrier transfer process from the wetting layers to the dots and the band alignments for the dots and the wetting layer. Type-II and type-I band alignments are shown for the dots and the wetting layers, respectively.

states and reducing the diffusion barrier of Ge into Si, where both of them will enhance the interdiffusion of Si and Ge [8]. For our samples grown by GS-MBE, the dots were found to be $Si_{0.58}Ge_{0.42}$ alloy dots [9]. Compared with the Ge content in the dots, the Ge content in the wetting layers is usually lower according to the theoretical calculation [10]. So the type-I band alignment at the wetting layer/Si interface is reasonable because the Ge content is lower.

Figure 3(a) shows the integrated PL intensities of the dots and the wetting layer as a function of the measurement temperature. Interestingly, as the temperature increases, the PL intensity from the dots increases slightly first in the temperature range of 8~20K and then decreases. At the same time, the PL from the wetting layer drops quickly and disappears at 40K. This implied that a large portion photon-induced holes in the wetting layers could be transferred into the dots. As the band alignment of the dot is type-II, the photon-induced electron will be transferred to the region near the dots. Figure 3(b) illustrates the photon-induced carriers transfer process and the band alignments of the wetting layer and the GeSi dots sandwiched by two Si layers. Dashed arrows show the PL transition process in the wetting layer and the dots, respectively. For the wetting layer, when the temperature increases, holes and electrons which are previously both trapped in the wetting layers at low temperatures could thermally be activated to move to the dots and the region near the dots, respectively, as shown in Fig. 3(b) (top). With the increase of measuring temperature, more holes are trapped in the dots to attract more electrons to near the dots, resulting in an increase of PL intensity of the dots.

After thermal annealing, the PL peaks related to GeSi dots and the wetting layer shifted to higher energies. At the same time, the shape of the PL peak changed when the annealing temperatures increased and the broad dots luminescence peak evolved into two well separted peaks after annealing. The detailed annealing effect on the PL spectra was discussed in Ref. [11], where one can find that the PL from wetting layers become more and more weak with the

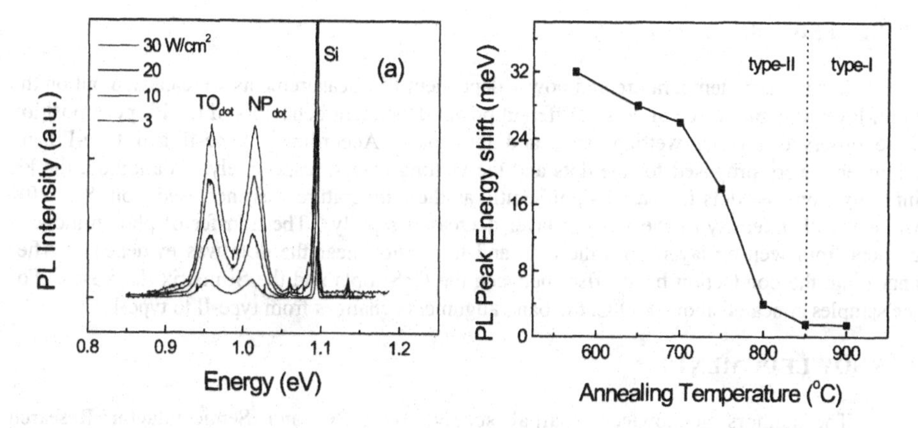

Figure 4 (a) Typical power dependence PL spectra of the sample annealed at 900°C for 5 min. (b) PL peak blueshift due to the increase of excitation power density for the samples annealed at different temperatures. After 850°C annealing, the band alignment of the GeSi dots changes to type-I from type-II.

increase of annealing temperature. And for the sample annealed at 900°C, only PL peaks from the GeSi dots and Si substrate were observed, as seen in Fig. 4(a). In contrast to the power dependence PL from as-rown sample, the PL peak blueshift is only 1.5 meV when the excitation power increases from 1 to 30 W/cm^2. Figure 4(b) shows the power-dependence PL peak blueshift for the samples annealed at different temperatures when the excitation power increased from 1 to 30 W/cm^2. It was seen that the blueshift decreases with the increase of the annealing temperature. For a SiGe/Si type-II quantum well structure, the band-bending effect depends on the band offset, well width, and the change in the sheet carrier density. The blueshift increases with the increase of the well width and the increase of conduction band offset [5]. So the decrease of the blueshift is not induced by the increase of the dot height after annealing but the decrease of the conduction band offset between the GeSi dots and the Si matrix. For the samples annealed at the temperatures above 850°C, the blueshift is only ~1.5 meV. This blueshift is much smaller than the value usually observed in SiGe/Si type-II system. For example, even when the excitation power density increase from 1 to 4 W/cm^2, the blue shift was as large as ~3.6 meV in a 3-nm type-II $Si_{0.7}Ge_{0.3}$/Si quantum well [12]. So the small blueshift of the samples annealed up to 850°C can be reasonably explained by a type-I band alignment and the blueshift is due to the band filling effect. This means the band alignment of the dots changes from type-II to type-I after high temperature treatment.

The band alignment of GeSi/Si dots is complicated and is influenced by the Ge content, the shape of the quantum dots, and strain [13]. The as-grown dots grown by our gas-source MBE were determined to be $Si_{0.58}Ge_{0.42}$ alloy (average) [9]. After annealing, the Ge content in the GeSi dots would decrease due to the Ge/Si interdiffusion. For the sample annealed at 900°C (sample B), the volume of the dots increased about 3.3 times, as seen in Fig. 1, so the average Ge content in the dots is estimated to be ~ 0.12. It is suggested that the decrease of Ge content in the dots plays a main role to change the type-II band alignment to type-I after annealing.

CONCLUSIONS

In summary, temperature and power dependent PL measurements were carried out on the multi-layer structure of GeSi dots. Different power dependent behaviors of the PL peak position were observed for the wetting layer and the dots. Accordingly, type-II and type-I band alignments were proposed for the dots and the wetting layers, respectively. In addition, the PL intensity from the dots increased significantly as the temperature was increased from 8 to 20K while the PL intensity of the wetting layer decreased rapidly. The transfer of photon-induced carriers from wetting layers into the dots and the region near the dots was evidenced. After annealing, the conduction band offset between the GeSi dots and the Si matrix decreases. For the samples annealed above 850°C, the band alignments changes from type-II to type-I.

ACKNOWLEDGMENTS

The authors acknowledge partial support by ARO and Semiconductor Research Corporation. Australian Research council is also acknowledged for their financial support.

REFERENCES

1. O. G. Schmidt, O. Kienzle, Y. Hao, K. Eberl, and F. Ernst, Appl. Phys. Lett. **74**, 1272 (1999).
2. O. G. Schmidt, C. Lange, and K. Eberl, Phys. Stat. Sol. (b) 215, 319 (1999).
3. J. Wan, G. L. Jin, Z. M. Jiang, Y. H. Luo, J. L. Liu, and Kang L. Wang, Appl. Phys. Lett. **78**, 1763 (2001).
4. K. L. Wang, J. L. Liu, and G. Jin in *First International workshop on new group IV semiconductors*, I03, (Japan, 2001).
5. T. Baier, U. Mantz, K. Thonke, R. Sauer, F. Schaffler, and H.–J. Herzog, Phys. Rev. B **50**, 15191 (1994).
6. G. Bremond, M. Serpentini, A. Souifi, G. Guillot, B. Jacquier, M. Abdallah, I. Berbezier, and B. Joyce, Microelectronics Journal **30**, 357 (1999).
7. Chris G. Van de Walle, and Richard M. Martin, Phys. Rev. B **34**, 5621 (1986).
8. S. A. Chaparro, Jeff Drucker, Y. Zhang, D. Chandrasekhar, M. R. McCartney, and David J. Smith, Phys. Rev. Lett. **83**, 1199 (1999).
9. Z. M. Jiang, J. Wan, G. L. Jin, Y. H. Luo, J. L. Liu, K. L. Wang, X. M. Jiang, Q. J. Jia, and W. L. Zheng, Appl. Phys. Lett. (submitted).
10. J. Tersoff, Phys. Rev. Lett. **81**, 3183 (1998).
11. J. Wan, Y. H. Luo, Z. M. Jiang, G. Jin, J. L. Liu, and Kang L. Wang, J. Appl. Phys. (to be published).
12. M. L. W. Thewalt, D. A. Harrison, C. F. Reinhart, and J. A. Wolk, Phys. Rev. Lett. 79, 269 (1997).
13. O. G. Schmidt, K. Eberl, and T. Rau, Phys. Rev. B **62**, 16715 (2000).

Mat. Res. Soc. Symp. Proc. Vol. 667 © 2001 Materials Research Society

Chemical mapping of indium rich quantum dots in InGaN/GaN quantum wells

N Sharma, H K Cho[1], J Y Lee[1] and C J Humphreys
Department of Metallurgy and the Science of Materials, University of Cambridge, Pembroke Street, Cambridge, CB2 3QZ, U.K.
[1]Department of Materials Science and Engineering, Korea Advanced Institute of Science and Technology, 373-1 Kusong-dong,Yusong-gu, Taejon, 305-701, South Korea

ABSTRACT

Indium clustering in InGaN/GaN multiple quantum wells (MQWs) is believed to be responsible for the high luminescent efficiency of GaN based light emitting diodes. In this paper we show that substantial clustering can be induced by reducing to zero the interruption time between growth of the GaN barrier layer on the InGaN quantum well. Photoluminescence (PL) shows that this has the effect of increasing the luminescence intensity and decreasing the band gap energy (higher indium concentration). The clusters or quantum dots were examined and quantified by energy filtered transmission electron microscopy (EFTEM), which was used to form chemical distribution maps of indium, gallium and nitrogen. In this paper we will show that this technique can accurately calculate the indium concentration and distribution in the quantum wells. The calculations show that $In_xGa_{1-x}N$ quantum dots (width = 1.3nm) exhibit an In concentration of up to x = 0.5, which are embedded in a quantum well matrix with x = 0.05.

INTRODUCTION

It has been suggested that the high luminescent efficiency of InGaN/GaN MQWs is due to quantum dot type indium clusters in the InGaN layers [1,2]. This has been suggested be due to the low solubility of InN in GaN [3]. An important requirement for the determination of indium clusters is the accurate measurement of the indium concentration. The techniques which have been used to measure this are X-ray diffraction (XRD) [4], high resolution TEM (HRTEM) [5] and energy dispersive X-ray analysis (EDX) [6]. For most cases HRXRD and HRTEM are used to measure the (0002) c-spacing. As thin InGaN layers are grown pseudomorphically ($a_{GaN} = a_{InGaN}$) the c-spacing will be adjusted with the a-spacing by an amount determined by Poisson's ratio, which will be different for GaN and InN. Also in the case of HRTEM, the thin samples deemed necessary change the projected lattice spacing through surface relaxation effects, making quantitative analysis difficult. Direct analysis using EDX in the STEM [6] is an excellent method for acquiring qualitative comparisons of local regions. However the N K peak is strongly absorbed by the detector window and hence cannot be used for calibration of indium concentrations.

Chen *et al.* [7] used a combination of molecular beam epitaxial growth of thin InGaN layers and scanning tunneling microscopy to show that indium segregates to the growth surface to form a 'floating layer' during growth due to weak In-N bonds compared to Ga-N bonds. They also observed N-vacancy nanopits, 5nm apart, which were claimed to relieve the quantum well strain. Total energy calculations showed a substantial energy reduction due to the lateral segregation of indium into these nanopits, leading to indium clusters along the InGaN layer. In our series of experiments we have varied the interruption time between the InGaN and GaN layers, in order to control this floating layer of indium. It was found from EFTEM that the indium segregates into quantum dots, with an indium concentration of up to x = 0.5, for the sample with 0 sec growth interruption. We suggest that a decrease in the growth interruption time leads to less desorption of the indium off the surface. This may lead to an increased driving force for indium clustering.

GROWTH

Five InGaN/GaN MQW samples were grown in a 1 x 2" MOCVD reactor. All samples had the same GaN template consisting of a 25nm nucleation layer grown at 560 °C on a sapphire substrate, followed by a 2μm silicon doped buffer layer at 1130 °C. The templates were grown under an ambient growth pressure of 400 mbar with an H_2 carrier gas. The gallium, nitrogen, silicon and indium were supplied from TMGa, NH_3, SiH_4 and TMI gases respectively. The MQW consisted of at least five InGaN layers and four barrier layers grown at 750 °C under a N_2 carrier gas. HRTEM and XRD showed that the bi-layer thickness was around 10nm, with each InGaN layer measuring 1.5 nm on average. The differences between the five samples is the growth interruption time between the InGaN layers and the GaN barrier layers, which ranged from 0 to 60 sec.

PHOTOLUMINESCENCE

PL studies showed that the emission intensity and wavelength changed significantly with interruption time (figure 1). This included a large drop in the PL intensity from 0 to 5 sec interruption followed by a small increase from 10 to 30 sec interruption. Secondly the measured bandgap was found to increase with increasing interruption, which corresponds to a small decrease in the indium concentration. Unfortunately the indium concentration cannot be calculated from the PL measured band gap as this is a function of strain, piezoelectric field and indium clustering in GaN.

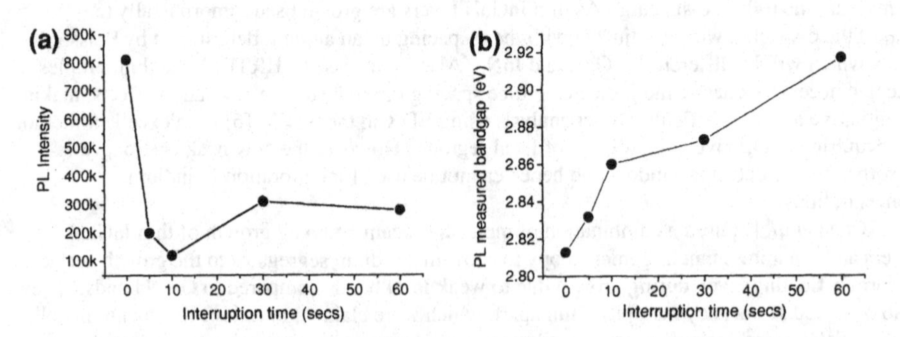

Figure 1. (a) PL intensity and (b) PL bandgap measurements as a function of interruption time

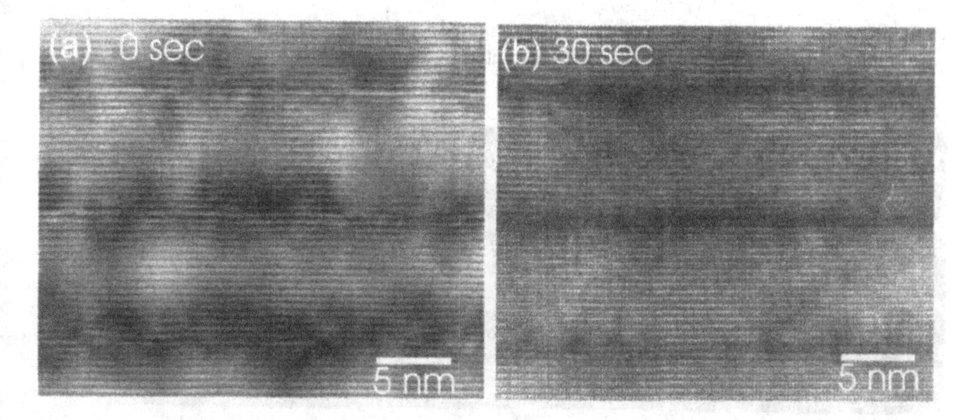

Figure 2. HRTEM images from the <1-210> axis of (a) 0 sec and (b) 30 sec interruption samples. The 0 sec sample shows localized strain contrast, which may be due to quantum dots.

TRANSMISSION ELECTRON MICROSCOPY

HRTEM from the <1-210> GaN axis was performed on the multiple quantum well samples using a JOEL 2000EX transmission electron microscope. The results show that the local strain in the InGaN layers increased with decreasing interruption time (figure 2). The 0 sec interruption sample exhibited strong local strain fluctuation contrast, with a spacing of 4nm, whereas the 5, 30 and 60 sec samples showed little strain contrast (figure 2b). Measurements of the c lattice parameter, assuming Vegard's law, showed the indium concentration within the fluctuations peaked at x = 0.5, however there was a 25% standard deviation in the calculated values.

EFTEM on samples with 0, 5 and 30 sec interruption was performed using a Philips CM300 field emission gun TEM with a Gatan Imaging filter. Previous work has shown that EFTEM can be used to form chemical maps of indium, gallium and nitrogen in InGaN with a high spatial resolution [8,9]. The data were obtained by acquisition of an energy-selected image-series over the electron energy-loss (EELS) range of 0 to 1400 eV, captured with an energy-step and energy-selecting slit width of 20 eV. Quantitative elemental maps were obtained by fitting an AE^{-r} pre-edge background equation and subtracting this from the post-edge region of Nitrogen (K-edge at 401 eV) and Gallium (L_{23}-edge at 1142 eV) by utilising a high confidence of fit parameter (h < 5) and a high signal-to-noise ratio (snr > 50). Dividing the gallium by the nitrogen elemental maps gives an accurate value of the gallium concentration, as this cancels the effects of TEM foil thickness (a problem in HRTEM), inelastic scattering cross-sections and diffraction contrast. This is provided that the nitrogen concentration is constant from the InGaN quantum well to the GaN matrix. Hence a quantitative indium elemental map can be calculated from the inverse of the gallium elemental map.

The EFTEM series were taken from TEM foil thicknesses of less than 20nm, so that plural scattering effects were negligible. The systematic error was reduced by sampling different parts of the 2" wafer. Also the random error was calculated from the noise under the post-edge signal, which gave a low error of x = 0.01 in the calculated indium concentration.

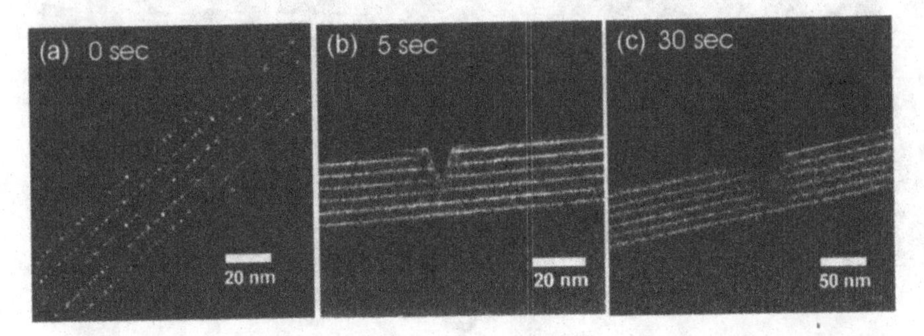

Figure 3. EFTEM indium elemental maps for the (a) 0 sec, (b) 5 sec and (c) 30 sec interruption samples. Images show that the 5 and 30 sec samples have continuous quantum wells, apart from V-defects. However the 0 sec quantum wells are made up of InGaN quantum dots.

For all three samples the nitrogen map showed a flat profile, hence the concentration of indium could be calculated. The 0 sec sample exhibited quantum dots along the quantum wells, which peaked at x = 0.5, with an average width of 1.3nm (which is the width of the quantum well) and a spacing of 2 to 5 nm (figure 3a). Furthermore the 5 and 30 sec interruption samples have wells that are continuous and exhibit a uniform indium concentration, despite having a large density of V-defects (figure 3b,c) [8]. In order to further prove that the indium fluctuations are not artifacts of the EFTEM technique the cumulative EELS spectrum was obtained from all the pixels in the clusters and compared to all the pixels in the matrix (figure 4). This shows that there is a substansive In $M_{4,5}$ core-loss edge (peaks at 500 eV) from the combined indium quantum dots, while the matrix material showed only a sloping N K core-loss edge. There was no trace of sample contaminating oxygen, which would be represented by an O K edge that peaks at 550eV.

Two types of line scans were taken from the quantitative indium elemental maps. The first was a box line-scan across the quantum wells, showing the average indium concentration from well-to-well for the 0 sec interruption sample (figure 5a). The second was a thin line-scan along the quantum wells showing the extent of indium fluctuations from 0 to 30 sec interruption (figure 5b). The results show that even though the 0 sec interruption sample has large fluctuations with up to x = 0.5 in the quantum dots and x = 0.05 between the quantum dots. However the average indium concentration for the 0 sec sample is x = 0.3 (figure 5a). This is very similar to the average indium concentration of the 5 sec (x = 0.31) and 30 sec (x = 0.26) samples (figure 5b). Even considering their relatively high indium concentration the 5 and 30 sec samples shows no sign of quantum dots in both the EFTEM and from the PL intensity measurements, even though InN is very insoluble in GaN [3]. Hence the quantum dots are more likely to be related to the indium floating layer, which is quickly desorped for interruption times higher than 0 sec.

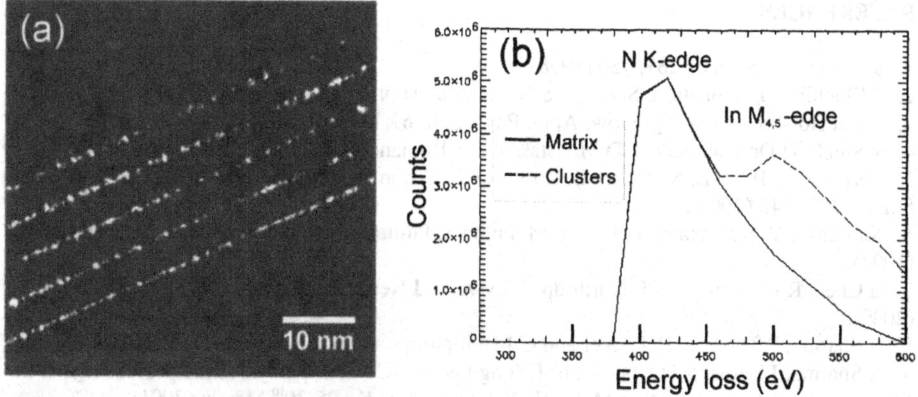

Figure 4. (a) EFTEM indium elemental map of the 0 sec sample. (b) Shows the collective EELS spectra from the quantum dots (clusters) and the matrix

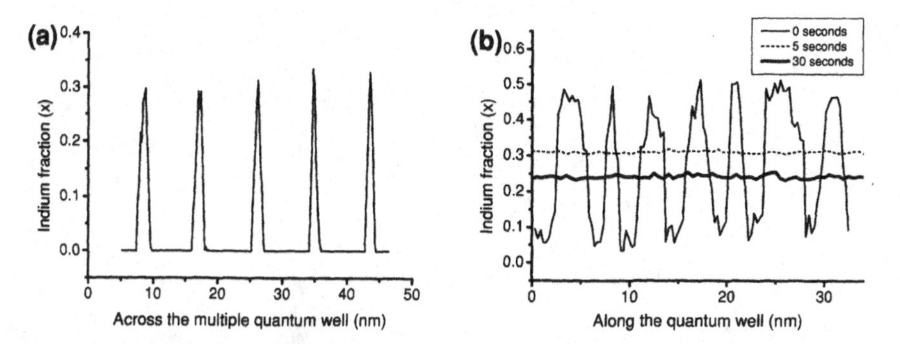

Figure 5. Line scans taken from the indium elemental maps. (a) 0 sec sample box line-scan across the quantum well, (b) comparison of line-scans along the quantum wells.

CONCLUSIONS

In conclusion the results show that having zero growth interruption time between the InGaN quantum well and GaN barrier layer leads to strong indium clustering. The clusters facilitate a substantial increase in the luminescence intensity from the MQW. Quantitative indium distribution maps show indium concentrations as high as x = 0.5 in quantum dots of an average width of 1.3nm, which are 2 to 5 nm apart along the quantum well. HRTEM show the clusters exist in highly strained nanoregions. By reducing the growth interruption time from 30 to 5 sec we have increased the indium incorporation in the layer without formation of quantum dots. However reducing the interruption time to 0 sec shows clear indication of quantum dots suggesting that in the absence of an indium floating layer there is no driving force for indium segregation.

REFERENCES

1. S Nakamura, Science **281**, 956 (1998).
2. S Chichibu, T Azuhata, T Sota and S Nakamura Appl. Phys. Lett. **69**, 4188 (1996).
3. I-hsui Ho and G B Stringfellow, Appl. Phys. Lett. **69**(18) 2701 (1996).
4. R Singh, D Doppalapudi, T D Moustakas, L T Romano, Appl. Phys. Lett. **70**, 1089 (1997).
5. Gerthsen D, Hahn E, Neubauer B, Rosenaur A, Schon O, Heuken M and Rizzi A Phys. Stat. Sol. (a) **177**, 145 (2000).
6. Narukawa Y, Kawakami Y, Funato M, Fujita S, Fujita S, Nakamura S,. Phys. Lett. **70**(8) ,981 (1997).
7. H Chen, R M Feenstra, J E Northrup, T Zyweitz, J Neugebauer, Phys. Rev. Lett. **85**, 1902 (2000).
8. N Sharma, P Thomas, D Tricker and C J Humphreys, Appl. Phys. Lett. **77**, 1274 (2000).
9. N Sharma, J Barnard, H Koun Cho, J Yong Lee and C J Humphreys, in the proceedings of the Microscopy of Semiconducting Materials XII, Oxford, U.K., 26-30th March, (2001).

Mat. Res. Soc. Symp. Proc. Vol. 667 © 2001 Materials Research Society

InGaAs-InP Quantum Wire Stark Effect Modulators: Effect of Wire Width in the Optimization of Changes in Excitonic Absorption and Index of Refraction

M. Xu, Microsoft Corp., 15400 NE 13th Pl, Bellevue, WA 98007
W. Huang, Electrical Engineering and Computer Science Department
United State Military Academy, West Point, NY 10996
F. Jain, Department of Electrical and Computer Engineering
University of Connecticut, Storrs, CT 06269-2157

Abstract

Quantum wire/dot modulators offer superior performance over their quantum well counterpart due to enhanced excitonic binding energy. This paper presents simulations on InGaAs-InP quantum wire Stark effect optical modulators showing a novel trend. While the excitonic binding energies and absorption coefficients increase as the width of the wire is decreased, the refractive index change Δn is maximized at a wire width depending on the magnitude of the applied electric field. For example, Δn is maximized at a width of about 100Å for an external electric field of 120kV/cm in an InGaAs quantum wire. This behavior is explained by considering the opposing effects of the wire width on binding energy and changes in the electron-hole overlap function in the presence of an external electric field. Practical InGaAs-InP modulators using V-groove structures are also presented.

Introduction

Changes in the optical absorption and index of refraction in quantum well layers have been extensively investigated for electroabsorptive and electrorefractive modulators employing quantum confined Stark effect, QCSE [1-2]. There is a significant interest in enhancing optical gain in lasers and electrooptic effects for modulators using quantum wire structures [3-5]. Recently, Arakawa et al. [5] investigated GaAs quantum wires using V-grooves, fabricated via selective epitaxial layers through SiO_2 masks.

This paper computes absorption coefficient and refractive index changes in $In_{0.33}Ga_{0.67}As$-InP quantum wires involving exciton transitions resulting in Stark Effect electrooptic and electrorefractive modulators [6]. The exciton binding energies and electric field induced refractive index changes are found to be significantly higher compare to InGaAs based quantum well system [7]. In general, the narrower the wire width, the higher the exciton binding energy, thus the larger the index change due to electric field. However, our simulation indicates that the index of refraction change is maximized at a wire width of 100Å. This is due to the opposing effects of the wire width on binding energy and on the changes of electron-hole overlap functions due to electric field.

Exciton binding energy InGaAs-InP quantum wire modulators

Fig. 1 shows the cross-sectional schematic of an $In_{0.33}Ga_{0.67}As$-InP multiple quantum wire structure. This device can be grown using MOCVD via selective area epitaxy through a SiO_2 mask, forming quantum wires in the bottom of V-grooves (similar to the work of Arakawa et al. [5]).

Quantum confinement of electrons and holes is achieved using InP or InGaAsP lattice-matched barrier layers as shown in Figure 1. The exciton binding energies as a function of electric field for the system are calculated based on square wire potential approximation, as shown in Fig. 2.

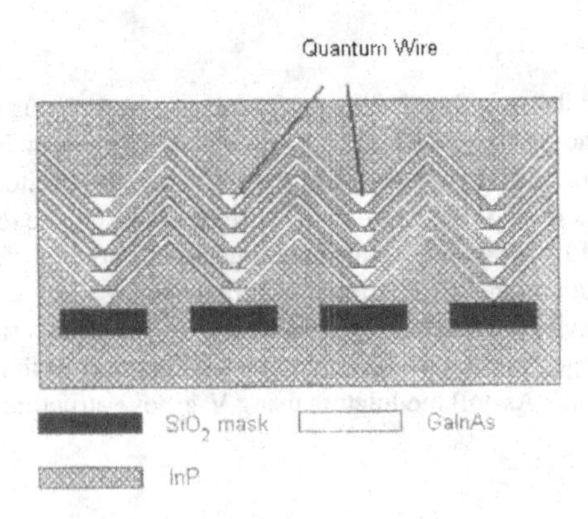

Figure 1. $In_{0.33}Ga_{0.67}As$-InP multiple quantum wires formed in the bottom of V-groove realized using selective area epitaxy.

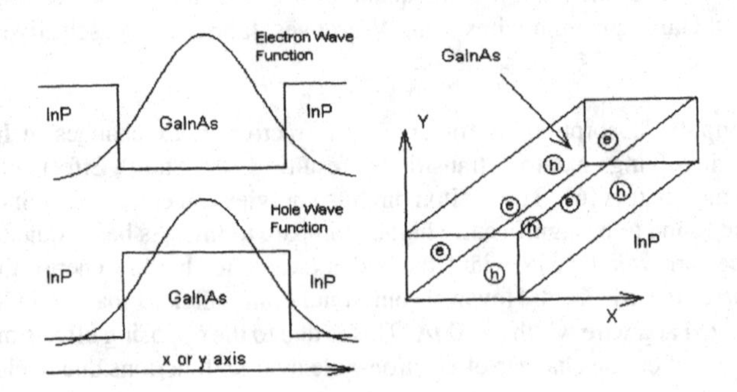

Figure 2 Confined electrons and holes in an InGaAs quantum wire (width L_x and thickness L_y).

The Hamiltonian of the electron and hole system can be written as [3,6]

$$H = H_h + H_e + H_{ex}$$

$$H_h = -\frac{\hbar^2}{2m_h}\left[\frac{\partial^2}{\partial x_h^2} + \frac{\partial^2}{\partial y_h^2}\right] + V_h(x_h, y_h) + eFy_h$$

$$H_e = -\frac{\hbar^2}{2m_e}\left[\frac{\partial^2}{\partial x_e^2} + \frac{\partial^2}{\partial y_e^2}\right] + V_e(x_e, y_e) - eFy_e$$

$$H_{ex} = -\frac{\hbar^2}{2\mu}\frac{\partial^2}{\partial (z_e - z_h)^2} - \frac{e^2}{4\pi\varepsilon\sqrt{(x_e - x_h)^2 + (y_e - y_h)^2 + (z_e - z_h)^2}}$$

The total wavefunction for the above Hamiltonian is assumed to be

$$\Psi = \varphi_e \varphi_h \phi_{ex}$$

Where φ_e and φ_h are electron and hole wavefunctions under the influence of electric field F. ϕ_{ex} is exciton trial function, which is assumed to be

$$\phi_{ex}(r) = \sqrt{\frac{2}{\pi}\frac{1}{\lambda}} e^{-\frac{r}{\lambda}}, r = z_e - z_h$$

Here, λ is variational parameter that denotes exciton radius. It can be determined by minimizing the system energy.

The computed exciton binding energies as a function of wire width under various externally applied electric fields are shown in Fig. 3. The parameters used in the computation are listed in Table I. The exciton binding energy increases as the wire size decreases as a result of increased quantum confinement. The binding energy is as high as 23meV for a wire width of 40Å. The effect of electric field is to separate the electron and hole wavefunctions and decreases the

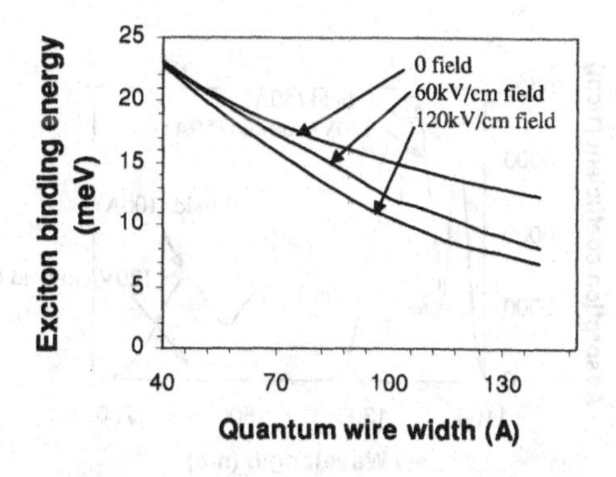

Figure 3. Exciton binding energies as a function of quantum wire width. (Wire thickness and width are assumed to be identical).

exciton binding energy. This in turn changes the absorption coefficient and the index of refraction, which is discussed in the next section.

Absorption coefficient and index of refraction changes due to electric field and their dependence on wire width

$$\alpha = \frac{2\pi\varepsilon_i}{n\lambda}$$

The absorption coefficient α can be calculated from the imaginary part of the permittivity ε_i by: where n is the index of refraction for $In_{0.33}Ga_{0.67}As$ layer, and λ is the operating wavelength. The calculation of ε_i includes both exciton contribution and free carrier contribution terms as following: [3,6]

$$\varepsilon_i(\omega) = \frac{\pi e^2}{m^2 \varepsilon_0 \omega^2} \times D \times M \times Overlap\ function \times L$$

D – Density of States

M – Transition Matrix

L – Line Shape Function (Gaussian)

Fig. 4 show the absorption coefficient for an $In_{0.33}Ga_{0.67}As$-InP 50Å wire and a 100Å wire. The plots indicate the absorption curves without the electrical field as well as with a 120kV/cm electric field. The absorption curves show well-defined peaks for heavy hole and light hole excitonic transitions. As the electric field increases, the absorption coefficient α decreases as the overlap function decreases. Even through α appear to be higher in 50Å wire than 100Å wire, the difference of α

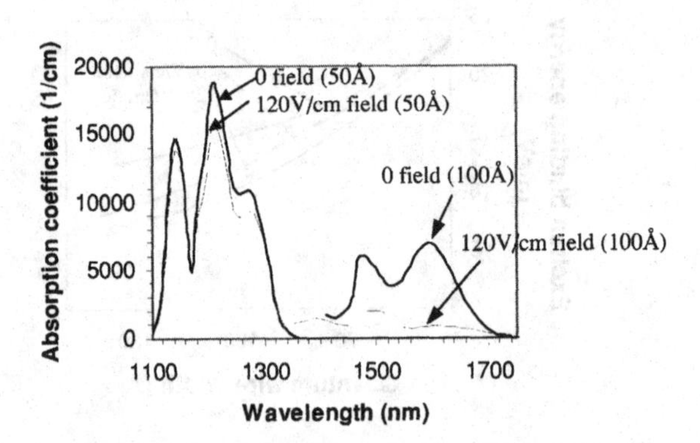

Figure 4. Absorption coefficient for a 50Å and 100Å $In_{0.33}Ga_{0.67}As$-InP quantum wire.

between zero electric field and 120kV/cm electric field is larger in 100A wire. This indicates electric field has a larger effect on 100Å wire than on 50Å wire. The shift of wavelength between the 50Å and 100Å due to the difference of electron and hole quantized energies at different wire width.

Using Kramers-Kronig relations, the refractive index changes (Δn) are computed from absorption changes. Figure 5 illustrates percentage of refractive index change for a 100Å $In_{0.33}Ga_{0.67}As$-InP quantum wire as the electric field increases from 10kV/cm to 120kV/cm. As the field strength increases, Δn increases as expected. It reaches 14.8% at 120kV/cm field. This is a significant improvement in index change over quantum well modulators [7].

Figure 6 summarizes the refractive index change for various wire widths under 120kV/cm electric field. As the wire size varies, it is expected that the smaller the device, the stronger the electro-optic effect due to the increase of exciton binding energy and absorption coefficient. But our computation shows that the index refraction change is maximized at 100Å under the electric field of 120kV/cm. This behavior can be explained by the influence of wire size and electric field on the excitonic binding energy and electron-hole overlap integral, which in turn determine the magnitude of change in absorption and index of refraction. The electric field tends to separate the electron and hole wavefunctions reducing the overlap function between electrons and holes as well as the oscillator strength of the excitons, which leads to reduced absorption and increased index changes. The larger the perturbation the electric field induces, the stronger is the electro-optic effect. But as the wire width is reduced below an optimum value, the electrons and holes are tightly bound/confined inside the wire. The electric field has little influence on the separation of electron and hole wavefunctions, and thus the Stark effect is suppressed in quantum wire sizes below 100Å. Again, this can be seen from Fig. 4, where the change of absorption coefficient between zero field and 120kV/cm field is dramatic for 100Å wire, but a smaller change occurs for 50Å wire.

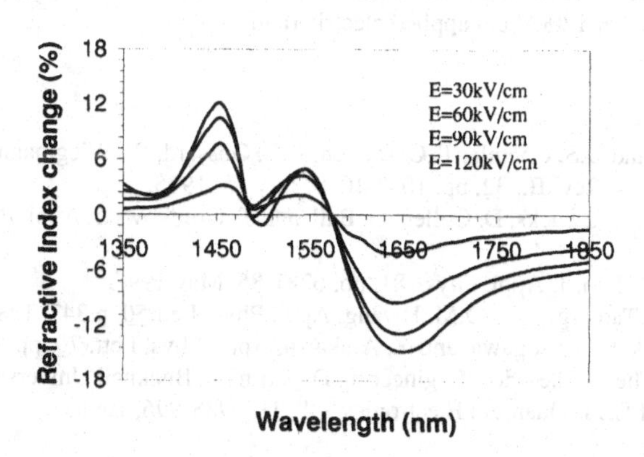

Figure. 5. Refractive index change for an $In_{0.33}Ga_{0.67}As$-InP100Å quantum wire.

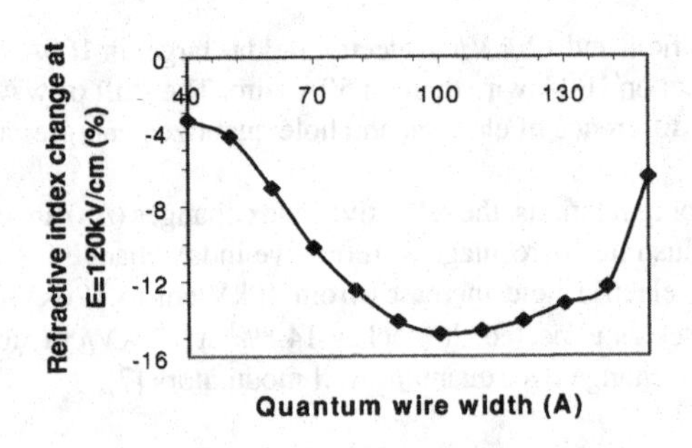

Figure 6. Refractive index change ($\Delta n/n$) as a function of quantum wire width at an electric field of 120kV/cm.

IV Conclusions

Quantum wire structures have shown higher electro-optic effects (change in absorption and index of refraction in the presence of an external electric field) than the quantum well structures. For lattice matched InGaAs quantum wires with width larger than a certain value (e.g. 100Å for electric field E=120V/cm), the reduction in the wire width invariably results in increased exciton binding energy and enhanced electro-optic effects. However, as the wire width drops below 100Å, even though the exciton binding energy and oscillator strength still increase as the quantum wire width is decreased, the magnitude of the refractive index change, Δn, decreases. This is attributed to the effect of the electric field on the separation of electron-hole pair forming the exciton that determines the value of the overlap function. The overlap function is not reduced as much for wire widths under 100Å in the presence of the electric field. Therefore, the index of refraction change Δn peaks around a wire width of 100Å at 120kV/cm applied electric field.

References

1. D. A. B. Miller and D.S. Chemla, T. C. Damen, A.C. Gossard, W. Wiegmann, T. H. Wood, and C. A. Burrus, Phys. Rev. B., 32, pp. 1043-1060, July 15, 1985.
2. S. Cheung, F. Jain, R. Sacks, D. Cullen, G. Ball, and T. Grudkowski, Appl. Phys. Lett., 63, July 19, 1993.
3. W. Huang and F. Jain, J. Appl. Phys., 81, pp. 6781-85, May 1997.
4. E. Kapon, M.C. Tamargo, and D.M. Hwang, Appl. Phys. Lett.50, p.347, 1987.
5. T. Arakawa, Y. kato, F. Sogawa, and Y. Arakawa, Appl. Phys. Lett. 70, pp. 646-648, 1997.
6. X. Ming, M.S. Thesis, Electrical Engineering Department, Bucknell University, April 24, 2000.
7. C. Thirstrup, , IEEE J. Quantum Electronics, QE-31, p988-996, 1995.

Devices and Device Applications

Mat. Res. Soc. Symp. Proc. Vol. 667 © 2001 Materials Research Society

Color Centers in Magnesium Doped Polycrystalline Alumina

L. R. Brock, K. C. Mishra, Madis Raukas, Walter P. Lapatovich and George C. Wei
OSRAM SYLVANIA,
Research and Development,
Beverly, MA 01915, U.S.A

ABSTRACT

We have investigated color centers in MgO-doped polycrystalline alumina (PCA) using absorption, excitation, and emission spectroscopy. Most of the color centers that were reported in earlier studies of the crystalline material have been observed to be present in the polycrystalline material. The absorption spectral features observed in the PCA are attributed to various color centers; however, they are not sufficiently resolved to make unique assignments. Suitable combinations of excitation and emission spectroscopy and also measurements at low temperature were therefore used to identify most of the color centers in this material.

Among the numerous color centers that we have identified in PCA are variations of electron centers including F, F^+, F_2^+, F_2^{2+} and F^+-Mg (($V_o^\bullet$-Mg_{Al}')^x). The most prominent oxygen vacancy related defect observed at room temperature was the F^+-Mg center, with absorption bands located at 217 and 249 nm, and an emission band at 303 nm. This center can be thought of as being formed by association of an F^+ center with a Mg defect. The single crystal sapphire samples containing no Mg show only F^+ ($V_o^\bullet$) centers with 230 and 257 nm absorption bands, and a 328 nm emission band.

Low temperature (22 K) fluorescence excitation measurements of PCA led to emission from F_2^{2+} center at 467 nm. Additionally, there is evidence that the observed 368 nm emission band could be attributed to the zero-phonon line associated with the F_2^+ center.

INTRODUCTION

The new material of choice for High Intensity Discharge (HID) arc tubes is polycrystalline alumina (PCA) [1,2]. For many lighting applications, the optical characteristics of the PCA material are comparable to that of fused silica. PCA can be sintered to translucency with the addition of MgO as a grain-growth inhibiting agent [3]. However, unlike fused silica that devitrifies at 1000 °C, the melting point of PCA exceeds 2000 °C [4]. This allows for a higher operation temperature of the arc tube, leading to improved lamp performance including higher efficacy, better color rendering, and smaller lamp to lamp color spread. Additionally, PCA has been shown to be more resistant to corrosion than fused silica, which may result in longer life for HID lamps.

The envelope of HID lamp arc tubes, whether fused silica or PCA, must retain its translucency throughout the lamp life. One dominant source of lamp degradation is the loss of transmittance of the lamp envelope. This could be partially due to generation of color centers, which absorb visible radiation from the discharge. Here we study the optical properties of MgO-doped PCA using absorption, excitation, and emission spectroscopy, with particular emphasis on color centers.

A color center is a point defect in a crystal lattice. For example, an F center is a color center resulting from an electron or electrons trapped in an anion vacancy. We found that most of the color centers that were reported in earlier studies of the crystalline material are also present in the polycrystalline material.

The crystalline lattice of alumina (α-Al_2O_3) can have a number of point defects such as vacancies, interstitials and chemical impurities [5-16]. The predominant intrinsic defects in alumina involve vacancies at aluminum and oxygen sites. The notation to describe the defects are from Kröger and Vink [17]. For example, V_O or V_{Al} represent a vacancy (V) at an oxygen site (O) or aluminum site (Al) respectively. The superscript **x** represents the neutral charge whereas the superscript • represents a positive charge with respect to the lattice, and an apostrophe (′) represents a negative charge.

The normal oxidation state of an oxygen atom in alumina is O^{2-}, and therefore, when an oxygen atom is missing from the lattice, the effective charge of the oxygen vacancy is 2+. This positive site can trap electrons leading to F centers which are known for their rich absorption and emission spectra. Variations of these F centers (see Table I) have been observed in crystalline alumina.

The antimorphs of F or electron centers are the hole centers. One example of a hole center in our study is Mg_{Al}', which is a Mg^{2+} cation at the Al^{3+} cation site [10, 11]. A summary of color centers in alumina with their characteristic absorption and emission bands is shown in Table I [6-12].

EXPERIMENTAL DETAILS

The MgO-doped PCA samples examined in this study were synthesized from Baikowski International CR-6 alumina powder. The PCA samples were doped with 500 wt. ppm MgO in order to enhance densification of PCA and to inhibit non-uniform or exaggerated grain growth [3]. This enables PCA to be sintered into a translucent ceramic material. The MgO level in sintered PCA was ~240 ppm due to evaporation and decomposition of the MgO dopant during sintering in hydrogen.

Room temperature absorption spectra for each sample were collected with a Perkin-Elmer UV/VIS/NIR Lambda 900 spectrometer equipped with an integrating sphere assembly in the 190 to 800 nm range. Spectra were obtained in 1 nm increments using the transmittance mode of the instrument (%T), and were later converted to absorbance spectra using equation 1.

$$A = -\ln(T) \qquad (1)$$

The absorption spectrometer described above did not take into account any sample fluorescence that could add to the transmittance. However, independent fluorescence spectra were taken for comparison with these spectra.

Room temperature fluorescence excitation and emission measurements of the PCA samples were acquired with a SPEX Fluorolog-2 instrument. This double monochromator instrument employed a xenon lamp for excitation of the sample between 200 and 800 nm. Fluorescence excitation spectra were obtained by using the first monochromator to select the excitation wavelengths to irradiate the sample, and using the second monochromator to collect the sample's fluorescence at a fixed wavelength. Emission spectra were obtained by using the first monochromator to choose a single wavelength to excite the sample, and dispersing the resultant fluorescence with the second monochromator. Each acquired spectrum was corrected for total response of the instrument.

In addition to room temperature measurements, fluorescence excitation and emission spectra of one sample were obtained at 22 K. This was accomplished by using a CTI closed helium cycle cryostat attachment to the SPEX Fluorolog-2 instrument.

Table I. The reported experimental values of some color center absorption and emission features in crystalline alumina (sapphire).

F center Defect	# Electrons	Sapphire (Reported)	
		Absorption (nm)	**Emission (nm)**
F (V_o^x)	2	203[6], 205[7], 207[8]	413[6,8]
F^+ ($V_o^\bullet$)	1	197[8,9], 203[7], 206[6] 229[9], 225[7], 230[6,8] 258[6,7,8]	326[6,8]
F-Mg (V_o^x-Mg_{Al}')'	2	170[10] 200[10] 280[10]	430[10]
F^+-Mg ($V_o^\bullet$-Mg_{Al}')^x	1	217[11] 247[11]	310[10]
F_2 ($V_o^xV_o^x$)^x	4	300[7] 302[12]	322[7] with ZPL* at 313[7] 517[12]
F_2^+ ($V_o^xV_o^\bullet$)$^\bullet$	3	355[7,12]	380[7,12] with ZPL at 369[7]
F_2^{2+} ($V_o^\bullet V_o^\bullet$)$^{\bullet\bullet}$	2	450[7], 459[12]	466[7] with ZPL at 462[7] 558[12], 560[7]

*ZPL = Zero Phonon Line

Table II. The observed F center absorption and emission features in MgO-doped PCA.

F center Defect	PCA (Observed)	
	Absorption (nm)	**Emission (nm)**
F (V_o^x)	200[a] 210[b]	420[b]
F^+ ($V_o^\bullet$)	230 257	328
F^+-Mg ($V_o^\bullet$-Mg_{Al}')^x	249	303
F_2^+ ($V_o^xV_o^\bullet$)$^\bullet$	~357	368 (ZPL)
F_2^{2+} ($V_o^\bullet V_o^\bullet$)$^{\bullet\bullet}$	~459	467

[a] Tentative assignment based on absorption spectra.
[b] Observed in our spectra of sapphire.

DISCUSSION

Absorption Spectra

While there are numerous studies of color centers in crystalline alumina (sapphire) [5-16], and irradiated sapphire [18-26], reports on the optical properties of the ceramic material are limited [18]. For this reason we have studied F centers in pre- and post-irradiated MgO-doped PCA. Table I shows the literature values of some F center absorption and emission features in sapphire, and Table II shows the values observed in this study for PCA. Absorption and emission bands reported in the literature for sapphire range from around 200 nm to approximately 560 nm for F, F^+, F_2, F_2^+, F_2^{2+}, F-Mg and F^+-Mg [6-12].

The absorption spectra (from 190 to 800 nm) of an as-sintered Mg-doped PCA sample (solid trace) and a Mg-doped PCA sample irradiated by 248 nm from an Excimer laser for approximately 3 hours (dotted trace) are shown in Figure 1. Both the as-sintered and irradiated MgO-doped PCA samples show broad, non-zero absorption throughout UV and visible spectral range. Upon irradiation at 248 nm, the PCA exhibited an increase in absorption throughout the entire spectral region studied. The vertical lines superimposed upon the absorption spectra denote the locations of absorption bands of F centers that were reported in the literature for crystalline alumina [6-12]. These bands are also listed in Table I.

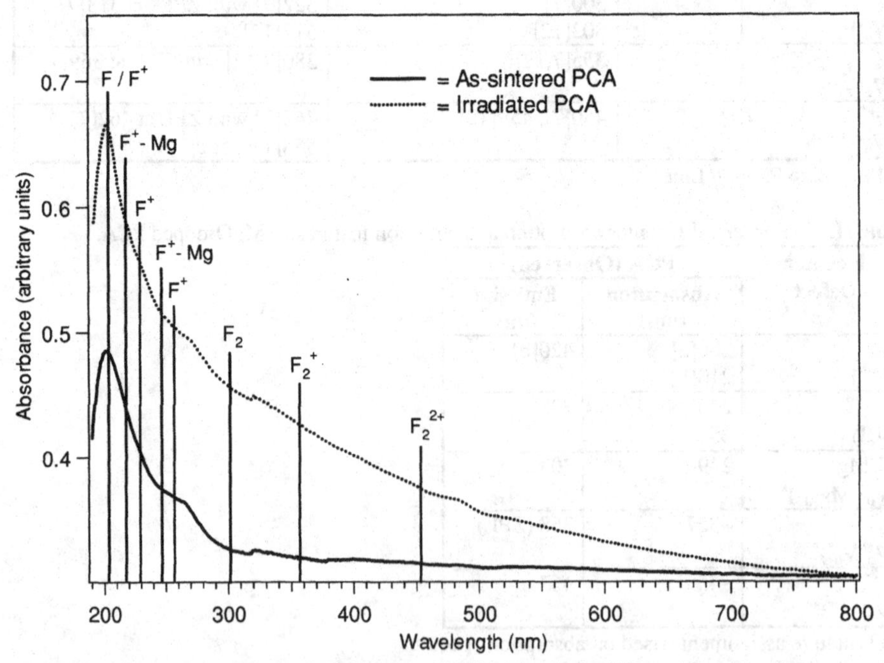

Figure 1. Absorption spectra of a MgO-doped PCA sample (solid trace) compared to an irradiated sample (dotted trace). The vertical lines denote the locations of absorption bands of F centers that were reported in the literature for crystalline alumina (see Table I).

The features that are present in the absorption spectra in Figure 1 are not sufficiently resolved to make unique assignments. For example, the band at approximately 200 nm (in both the as-sintered and irradiated samples) can be assigned to absorption by F, F^+ or F-Mg centers [6-10]. Likewise, the shoulder at approximately 250 nm can be assigned as F^+ or F^+-Mg [6-8,11]. Therefore, suitable combinations of excitation and emission spectroscopy, and also measurements at low temperature were necessary to identify most of the color centers in the PCA.

Emission and Excitation Spectra

A series of emission and excitation spectra were taken of both pre- and post-irradiated MgO-doped PCA samples. For the emission spectra, the excitation wavelengths were carefully chosen to correspond to reported absorption bands in sapphire (see Table I).

The $1s \rightarrow 2p$ F^+ absorption band in sapphire is split into three sub-bands which were reported at approximately 200, 230 and 258 nm [6-9]. The single F^+ emission band was observed at 326 nm [6,8]. The F^+-Mg color center in Mg-doped sapphire has reported absorption bands at 217 and 247 nm, and emission at 310 nm [10]. An F^+-Mg defect is a charge-neutral color center formed by association of an F^+ ($V_o^\bullet$) center with a Mg defect (Mg_{Al}').

We have taken room temperature emission spectra of Mg-doped PCA both before and after irradiation in the 200 to 800 nm range to compare to the reported spectra of sapphire. Figure 2 shows one such emission spectrum in the 280 to 400 nm range. The excitation wavelength of 248 nm was chosen to excite either the F^+-Mg center or the F^+ center. The MgO-doped PCA sample (solid trace) exhibited a luminescence band at 303 nm that we have assigned as the F^+-Mg color center, $(V_o^\bullet\text{-}Mg_{Al}')^x$. However, upon irradiation (dotted trace), the luminescence is shifted to the red to 328 nm, and is assigned as F^+ ($V_o^\bullet$).

An excitation spectrum of sapphire (solid trace), irradiated sapphire (dash-dot trace), PCA (dashed trace), and irradiated PCA (dotted trace) is shown in Figure 3. The PCA and irradiated PCA spectra were multiplied by a factor of 10 so that they could be plotted on the same scale as the sapphire spectra. The samples were excited from 210 to 290 nm, and the emission at 310 nm was collected. The wavelength 310 nm was selected because it is common to the reported F^+ color center in sapphire, with a single emission band of 326 nm, and the F^+-Mg color center in Mg-doped sapphire with a reported emission band around 310 nm [6,8,10]. The Figure 3 excitation spectrum confirms the results observed in the emission spectrum: The MgO-doped PCA sample absorbs 249 nm radiation, and we have assigned this as the F^+-Mg color center, $(V_o^\bullet\text{-}Mg_{Al}')^x$. However, upon irradiation of the PCA, the bands were shifted to 230 and 257 nm, and were assigned as F^+ ($V_o^\bullet$).

Sapphire, which contains no MgO dopant, has F^+ excitation bands at 230 and 257 nm, as shown in Figure 3. The as-sintered PCA only has the F^+-Mg color center at 249 nm. Upon irradiation of the sapphire sample, the intensity of the F^+ excitation bands decreased. This is unlike the $F^+\text{-Mg} \rightarrow F^+$ center conversion in the irradiated PCA where the 249 nm band disappears and the 230 and 257 nm bands appear (see Figure 3). The same $F^+\text{-Mg} \rightarrow F^+$ conversion is manifested in emission in Figure 2.

Although the intensity of the F^+ center decreased upon irradiation of sapphire, the intensity of the F (V_o^x) center increased upon exposure to intense UV radiation, as shown in Figure 4. Similar radiation induced $F^+ \rightarrow F$ center conversion in corundum is discussed in

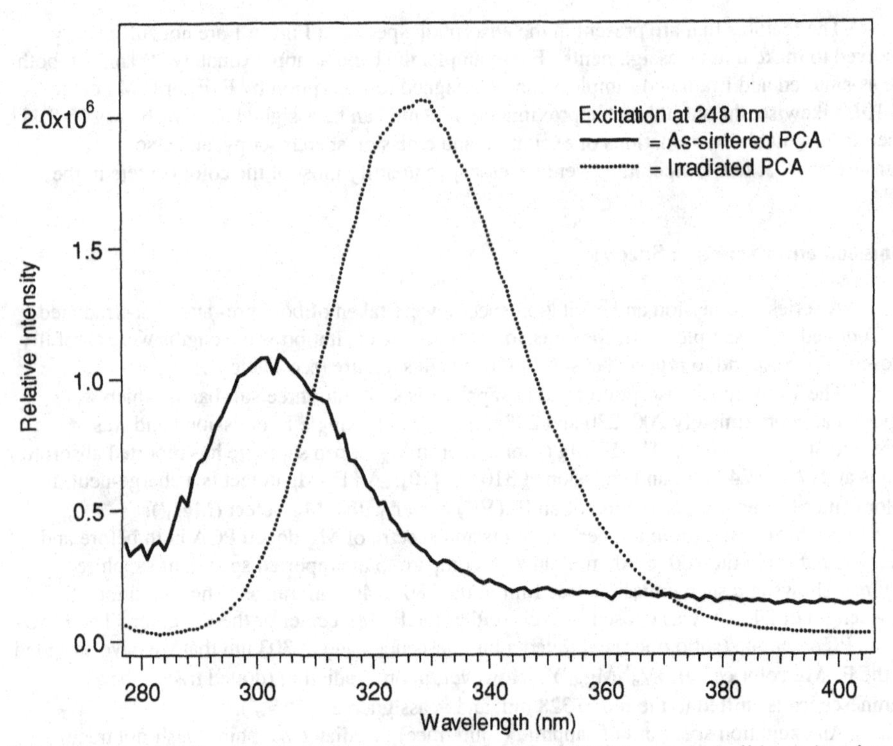

Figure 2. An emission spectrum of an as-sintered Mg-doped PCA sample (solid trace) and an irradiated Mg-doped PCA sample (dotted trace) in the 280 to 400 nm range is shown in Figure 2. The excitation wavelength was 248 nm. The band at 303 nm is assigned as F^+-Mg, and the band at 328 nm is assigned as F^+.

references 18 and 19. In PCA, there is an observed increase in absorption around 200 nm (as shown in Figure 1) that may be attributed to an increase in the F (V_o^x) center.

Optical excitation of other F^+-Mg and F^+ absorption bands was observed. For example, room temperature excitation of an irradiated Mg-doped PCA sample at 225 nm results in an F^+ emission band at 328 nm, and excitation of a sample that was not irradiated yields the expected F^+-Mg peak at approx. 303 nm. The results are summarized in Table II.

The defect center in crystalline Al_2O_3 which consists of two associated F^+ centers is F_2^{2+}. This two electron F center has a measured absorption band around 450 (or 459 nm), a sharp luminescence band at approximately 466 nm, and a broad luminescence band around 560 nm [7,12]. We have also observed this defect center in PCA. Figure 5 shows the emission spectra of a irradiated PCA sample, that was cooled to 22K, with evidence of the F_2^{2+} center. The solid trace in Figure 5 was taken using 459 nm as the excitation wavelength, and the dashed trace serves as the background (excitation at 479 nm where no F_2^{2+} luminescence is expected). Upon subtraction (the dotted trace), the F_2^{2+} band at 467 nm was apparent.

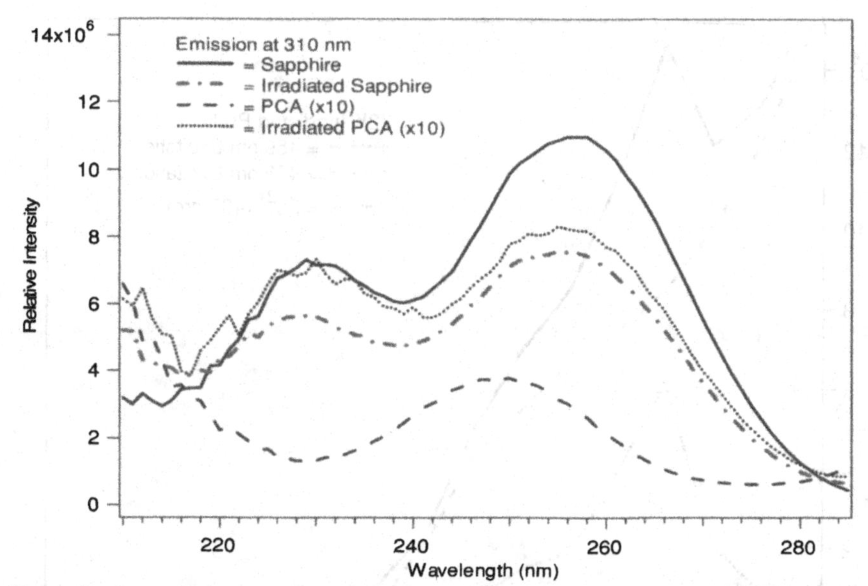

Figure 3. An excitation spectrum of sapphire (solid trace), irradiated sapphire (dashed-dotted trace), PCA (dashed trace), and irradiated PCA (dotted trace). The PCA spectra were multiplied by a factor of 10 so that they could be plotted on the same scale as the sapphire spectrum. The emission at 310 nm was collected.

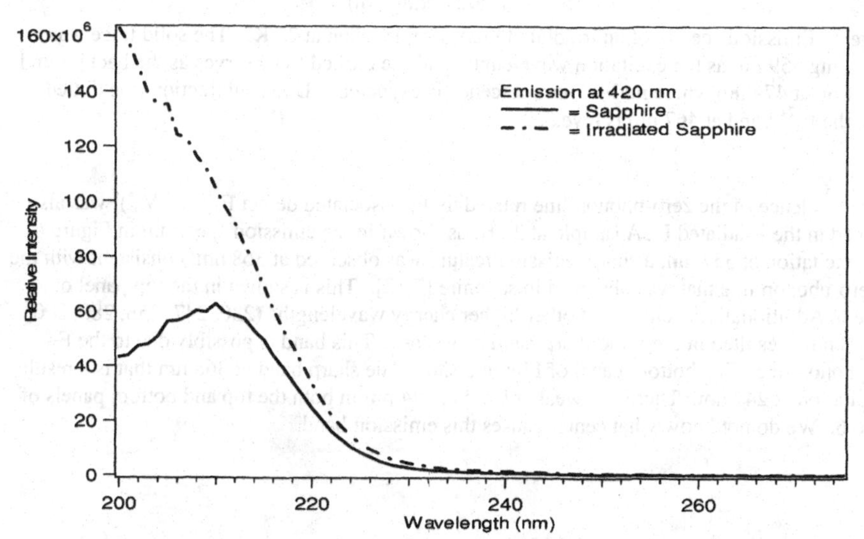

Figure 4. An excitation spectrum of pre- and post-irradiated sapphire showing the increase in F (V_o^x) absorption at 210 nm. The emission at 420 nm was collected.

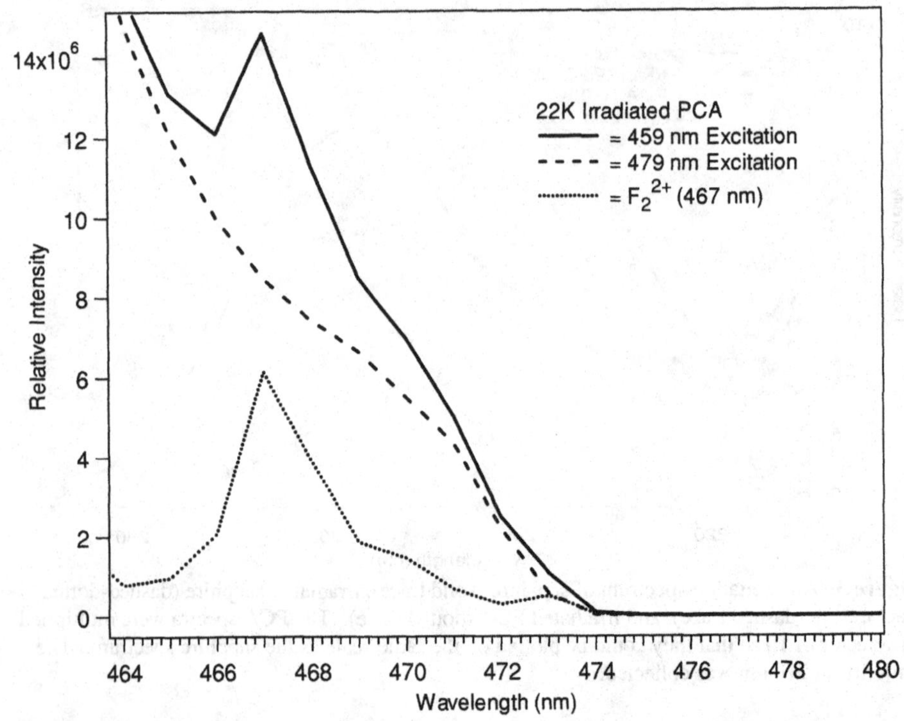

Figure 5. Emission spectra of an irradiated PCA sample taken at 22 K. The solid trace was taken using 459 nm as the excitation wavelength, and the dashed trace serves as the background (excitation at 479 nm where no F_2^{2+} luminescence is expected). Upon subtraction (the dotted trace), the F_2^{2+} band at 467 nm is revealed.

Evidence of the zero-phonon line related to the associated defect F_2^+ $(V_o^x V_o^{\bullet})^{\bullet}$ was also observed in the irradiated PCA sample at 22 K, as shown in the emission spectrum in Figure 6. Upon excitation at 357 nm, a sharp emission feature was observed at 368 nm, consistent with the F_2^+ zero-phonon line that was observed in sapphire [7,12]. This is shown in the top panel of Figure 6. Additionally, excitation at other higher energy wavelengths (230, 247, 256, 280, 300, and 313 nm), resulted in the same sharp band at 368 nm. This band is possibly due to the F_2^+ zero-phonon line. The bottom panel of Figure 6 shows the sharp band at 368 nm that is a result of excitation at 247 nm. There is a weaker band at 374 nm in both the top and bottom panels of Figure 6. We do not know what center causes this emission band.

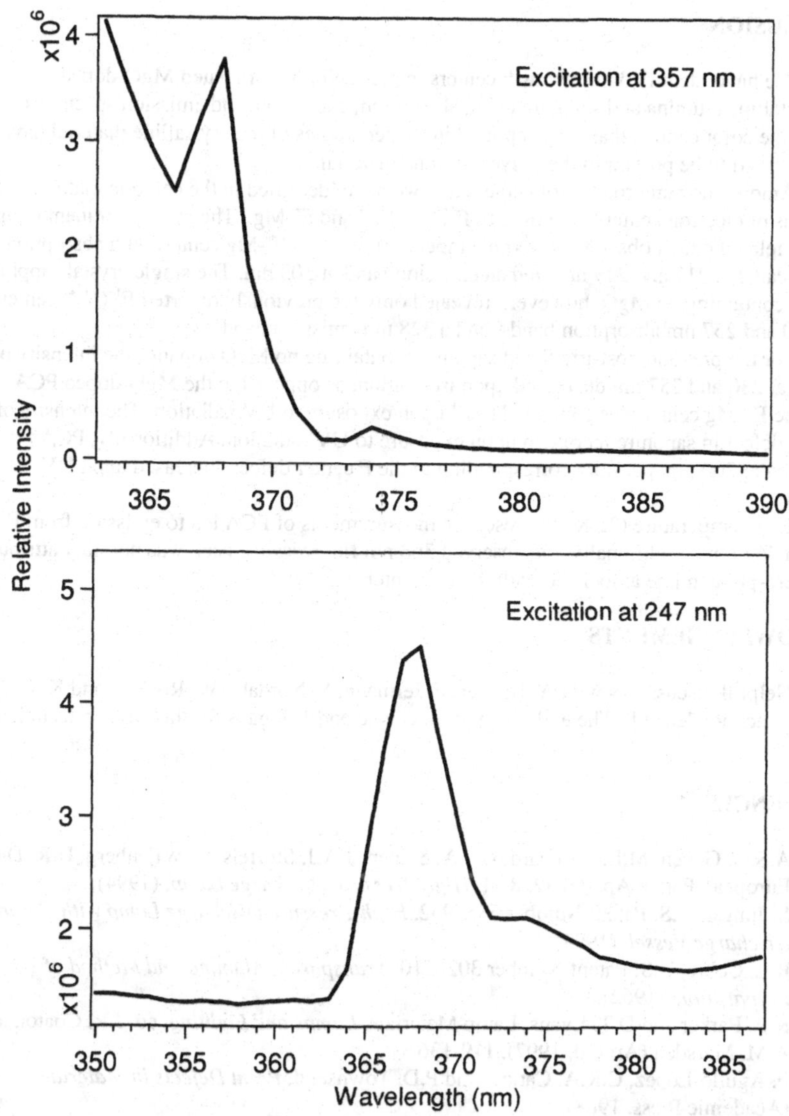

Figure 6. Emission spectra of an irradiated PCA sample taken at 22 K. The top trace was taken using 357 nm as the excitation wavelength, the bottom trace using 247 nm. In both cases, there is a sharp peak at 368 nm corresponding to the F_2^+ zero-phonon line. The band at 374 nm is not assigned.

CONCLUSIONS

We have identified numerous F centers in pre- and post-irradiated MgO-doped polycrystalline alumina and sapphire using absorption, excitation, and emission spectroscopy. Most of the color centers that were reported in earlier studies of the crystalline material have been observed to be present in the polycrystalline material.

Among the numerous color centers that we have identified in the ceramic material were variations of electron centers including F, F^+, F_2^+, F_2^{2+} and F^+-Mg. The most prominent oxygen vacancy related defect observed at room temperature was the F^+-Mg center, with absorption bands located at 217 and 249 nm, and an emission band at 303 nm. The single crystal sapphire samples containing no MgO, however, revealed only the previously reported F^+ ($V_o^{\bullet}$) centers with 230 and 257 nm absorption bands, and a 328 nm emission band.

For the pre- and post-irradiated sapphire (containing no MgO dopant), the intensity of the F^+ band at 230 and 257 nm decreased upon irradiation, as opposed to the MgO-doped PCA where the F^+-Mg center was converted to F^+ upon exposure to UV radiation. The intensity of the F center defect in sapphire increased upon exposure to UV radiation. Additionally, PCA absorption at 200 nm, possibly corresponding to the F center defect, increased upon UV exposure.

Low temperature (22 K) fluorescence measurements of PCA led to emission from F_2^{2+} center at 467 nm. Additionally, the observed 368 nm fluorescence peak was possibly attributed to the zero-phonon line associated with the F_2^+ center.

ACKNOWLEDGEMENTS

Helpful discussions with A. Hecker, F. Jermann, S. Nordahl, W. Rossner and K. Zuk are gratefully acknowledged. The authors thank J. Gagne and J. Ingalls for their expert technical support.

REFERENCES

1. A.S.G. Geven, M.L.P. Renardus, P.A. Seinen, J.A.J. Stoffels, C. Wijenberg, H.R. Dielis, European Patent Ap. 0587238A1, *High Pressure Discharge Lamp*, (1994).
2. S. Jüngst, U.S. Patent Number 5352952, *High-Pressure Discharge Lamp with Ceramic Discharge Vessel*, (1994).
3. R.L. Coble, U.S. Patent Number 3026210, *Transparent Alumina and Method of Preparation*, (1962).
4. K.E. Parker and D.T. Evans, Lamp Materials, *Lamps and Lighting*, ed. J.R. Coaton and A.M. Marsden (Arnold, 1997), 119-136.
5. F. Agullo-Lopez, C.R.A. Catlow and P.D. Townsend, *Point Defects in Materials*, (Academic Press, 1988).
6. J.H. Crawford, Jr., *Semiconductors and Insulators*, **5**, 599 (1983).
7. G.J. Pogatshnik, Y. Chen and B.D. Evans, *IEEE Trans. Nucl. Sci.*, **NS-34**, 1709 (1987).
8. J.H. Crawford, Jr., *Nucl. Inst. And Meth. B*, **1**, 159 (1984).
9. B.D. Evans and M. Stapelbrock, *Phys. Rev. B*, **18**, 7089 (1978).
10. P.A. Kulis, M.J. Springis, I.A. Tale, V.S. Vainer and J.A. Valbis, *Phys. Stat. Sol. B*, **104**, 719 (1981).

11. V.S. Kortov, T.S. Bessonov, M.S. Akselrod, I.I. Milman, *Phys. Stat. Sol. A,* **87**, 629 (1985).
12. B.D. Evans, *Nucl. Instr. And Meth. B*, **91**, 258 (1994).
13. B.D. Evans, *J. Appl. Phys*., **70**, 3995 (1991).
14. G.J. Pogatshnik, Y. Chen, *J. Luminescence*, **40 &41**, 315 (1988).
15. S.K. Mohapatra and F.A. Kröger, *J. Am. Ceramic Soc*., **60**, 141 (1977).
16. K.P.D. Langerlöf and R.W. Grimes, *Acta. Mater*., **46**, 5689 (1998).
17. F.A. Kröger and H.K. Vink, in *Solid State Physics*, edited by F. Seitz and D. Turnbull, (Academic Press, 1956), p. 307.
18. O.A. Plaskin, V.A. Stephanov, P.A. Stephanov and V.M. Chernov, *J. Nucl. Mat*., **271 & 272** (1999).
19. E.A. Kotomin, A. Stashans, and P.W.M. Jacobs, *Radiation Effects and Defects in Solids*, **134,** 87 (1995).
20. F.T. Gamble, R.H. Bartram, C.G. Young and O.R. Gilliam, *Physical Review,* **134**, A589 (1964).
21. T.J. Turner and J.H. Crawford, Jr., *Solid State Comm.,* **17**, 197 (1975).
22. P.W. Levy, *Phys. Rev.,* **123**, 1226 (1961).
23. K.H. Lee, G.E. Holmberg and J.H. Crawford, Jr., *Solid State Comm*., **20**, 183 (1976).
24. K. Atobe and N. Nakagawa, *Cryst. Latt. Def. And Amorph. Mat.,* **17**, 229 (1987).
25. B.D. Evans and M. Stapelbrock, *Solid State Comm*., **33**, 765 (1980).
26. E.A. Kotomin and A.I. Popov, *Nucl. Inst. And Meth. B*, **141**, 1 (1998).

Mat. Res. Soc. Symp. Proc. Vol. 667 © 2001 Materials Research Society

Long-Term Cathodoluminescent Characterization of Thin-Film Oxide Phosphors in a Wide Range of Electron Excitation Densities

Vyacheslav D. Bondar, Thomas E. Felter[1], Charles E. Hunt[2], Yuri G. Dubov, Andrei G. Chakhovskoi[2]
Lviv National University, Department of Physics, 50 Dragomanov Str.,
79005, Lviv, Ukraine.
[1] Lawrence Livermore National Laboratory,
PO Box 808, L - 356, Livermore, CA, 94550
[2] University of California at Davis, Department of Electric and Computer Engineering,
Davis, CA, 95616

ABSTRACT

Long-term processes of cathodoluminescence degradation of thin film phosphors Zn_2SiO_4:Ti and Zn_2GeO_4:Mn were investigated in a wide range of e-beam energies, current and power densities. The time dependencies describing the decreasing of emission intensity have been found. At higher current densities of e-beam irradiation, the specific behavior of long-term degradation processes was observed, which is characterized by rapid initial degradation and a slower long term decrease. The most probable mechanisms are proposed for long-term processes of degradation in the investigated phosphors.

INTRODUCTION

Thin film oxide phosphors are promising for low-voltage field-emission display applications due to their appropriate color coordinates, high efficiency [1,2], and possible increased conductivity [3]. Long-term stability of phosphors is also very important for practical applications, especially for development of field-emission displays with low-voltage excitation (less than 3 keV) where excitation takes place in the near surface layer which is extremely sensitive to degradation. The degradation processes were intensely studied in a series of works [2, 4-10]. Two main mechanisms have been found to be responsible for the decrease of cathodoluminescence (CL) intensity: irreversible loss, caused by physical and chemical reactions, and reversible loss (thermal quenching), caused by thermal heating by the electron beam. Several forms of CL degradation behavior have been found previously [5-8]. Some new forms describing the long-term behavior of a series of thin film oxide phosphors have been found by the authors of the present work [2,10]. Results of more detailed studies of degradation processes are presented in the present work for a wide range of excitation energies, e-beam currents and excitation intensities for the case of phosphors with charged luminescent centers, which are most sensitive to the processes of thermal diffusion quenching.

EXPERIMENTAL

Thin films were deposited by the modified rf-magnetron method [11] with subsequent high-temperature recrystallization at 650-1050°C. CL was measured at continuous electron excitation at 1-3.5 keV, 20-200 μA and beam diameter of 3 mm which corresponds to e-beam current densities of 0.28 - 2.8 mA/cm^2 and power densities from 0.1 to 20 W/cm^2. This mode of measurement during 8 hours corresponds to approximately 1500 hours of excitation of a conventional 21-inch kinescope at 25 kV and 1 mA (assuming equal deposited charges per square cm during irradiation). Measurements were made on the laboratory-made prototype CRT with thin film phosphor samples mounted on the faceplate. The laboratory CRT-prototypes were fabricated using standard technology of CRTs fabrication. It included pumping to 10^{-7} Torr with degassing near 400-450C, sealing and consequent gettering. Final residual gases partial pressure was not higher than 10^{-9} Torr. Therefore the influence of residual gases may be considered as negligible. Two methods were used for avoiding possible errors caused by the instability of the e-beam and the photo detector: periodic deflection of the e-beam to the non-irradiated area and simultaneous measurement of e-beam current and energy with subsequent correction of intensity data. Both methods showed good stability and accuracy in long-term measurements of CL intensity.

RESULTS AND DISCUSSION

It was found that at power densities less than 0.5 W/cm^2 most oxide thin films phosphors do not show substantial decrease of CL intensity within the accuracy of measurements (3-5%). A decrease of CL intensity was observed at power densities greater than 1 W/cm^2 and is in a good agreement with the dependence

$$(Io/I - 1) = K \cdot t^{1/2} \quad (1)$$

that describes a thermal diffusion process [10] of long-term degradation of CL intensity.

At the same time, at higher e-beam power densities in the range of 1-20 W/cm^2 and various current densities and energies, Zn_2SiO_4:Ti phosphors showed distinctive behavior under long-term irradiation which is revealed in a two-stage long-term dependence.

In the case of low excitation energies (< 2 keV) dependence (1) for Zn_2SiO_4:Ti films is linear (figure 1), and quenching (degradation) coefficient K increases from 0.01 to 0.1 with the increasing of current density from 0.28 to 2.8 mA/cm^2.

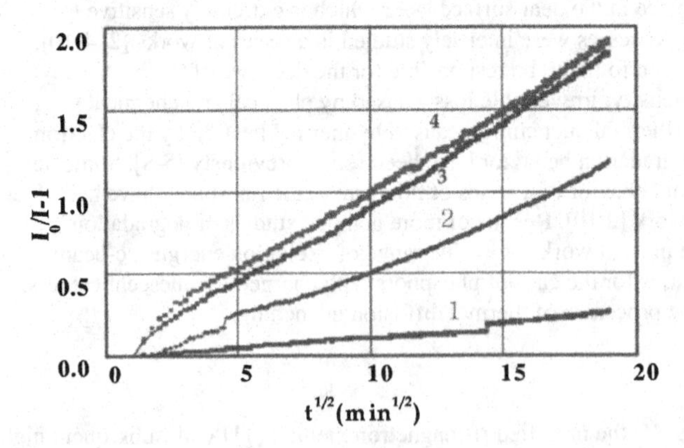

Figure 1. (I_0/I -1) vs. $t^{1/2}$ (min$^{1/2}$) plot of CL intensity of thin-films Zn_2SiO_4:Ti at 2keV and 0.71 mA/cm^2 (1), 1.42 mA/cm^2 (2), 2.12 mA/cm^2(3), 2.83 mA/cm^2(4).

At higher e-beam excitation energies (E > 2.5 keV) long-term behavior differs, though also can be described with dependence (1). It becomes a two-stage structure with a rapid initial decrease with $K = 0.2\text{-}0.4$, and a relatively flat region with $K = 0.01 - 0.06$ during consequent long-term period irradiation dose (figure 2). The rate of decrease at the initial stage increases with an increase of e-beam power density and is accompanied by a relative decrease of degradation rate during the final stage. The value of maximal decrease ($Io/I - 1$) at current density of 2.1 mA/cm^2 changes from 1.85 at 2 keV to 1.45 at 3.5 keV.

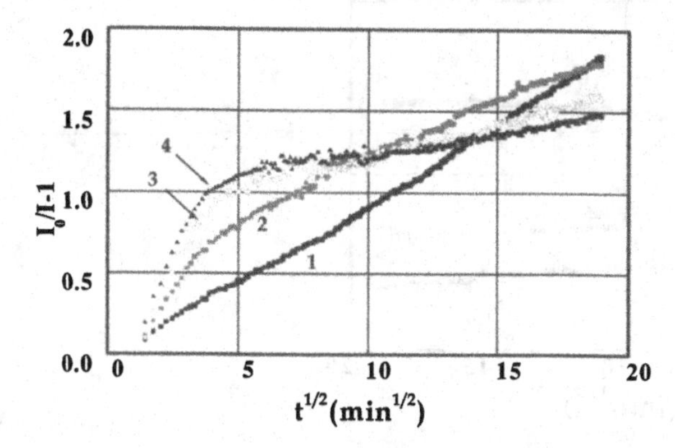

Figure 2. (I_0/I -1) vs. $t^{1/2}$ ($min^{1/2}$) plot of CL intensity of thin-films Zn_2SiO_4:Ti at 2.1 mA/cm^2 and 2 keV (1), 2.5 keV (2), 3 keV (3), 3.5 keV (4).

In the case of Zn_2GeO_4:Mn thin films, the linear behavior of dependence (1) is observed at current densities not greater than 0.28 mA/cm^2 and excitation energies of 1.5-2 keV (figure 3).

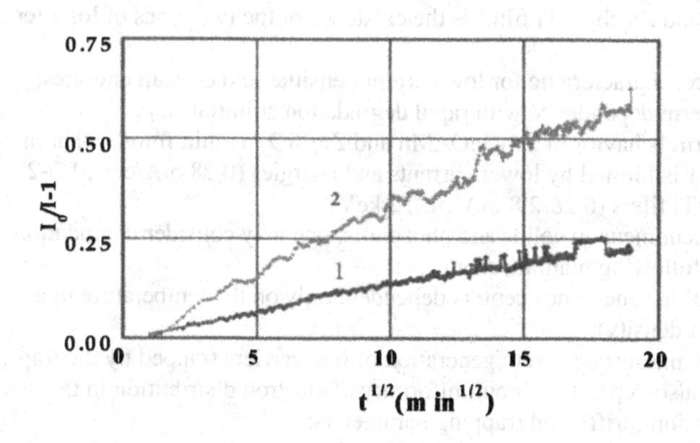

Figure 3. (I_0/I -1) vs. $t^{1/2}$ ($min^{1/2}$) plot of CL intensity of thin-films Zn_2GeO_4:Mn at 0.28 mA/cm^2 and 1.5 keV (1); 2. keV (2).

At higher current densities it becomes a two-stage (figure 4) dependence. At e-beam energy of 1 keV the value of the maximum decrease ($Io/I - 1$) in Zn_2GeO_4:Mn thin films is 0.25, 0.35 and 0.70 at current densities 0.7, 1.4 and 2.1 mA/cm^2, respectively. After the region with rapid decrease, stabilization of intensity or even its partial increase (buildup) is observed. At energies higher than 1.5 keV such stabilization of intensity was not observed.

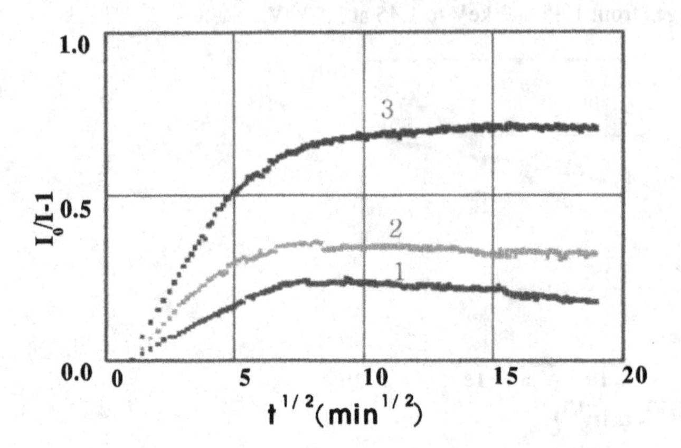

Figure 4. (I_0/I -1) vs. $t^{1/2}$ ($min^{1/2}$) plot of CL intensity of thin-films Zn_2GeO_4:Mn at 1 keV and 0.71 mA/cm^2(1), 1.42 mA/cm^2 (2), 2.12 mA/cm^2 (3).

As can be seen from the results obtained, the long-term dependence of CL of Zn_2GeO_4:Mn and Zn_2SiO_4:Ti thin film phosphors show both similar and different behavior. The common feature for both Zn_2GeO_4:Mn and Zn_2SiO_4:Ti films is the existence of the two types of long-term dependencies:

type 1 – linear dependence, characteristic for low current densities and e-beam energies;

type 2 – two-stage long-term dependence with rapid degradation at initial stage.

The difference in long-term behavior of Zn_2GeO_4:Mn and Zn_2SiO_4:Ti thin films is that in the case of Zn_2GeO_4:Mn type 1 is limited by lower currents and energies (0.28 mA/cm^2, 1.5-2 keV) as compared to Zn_2SiO_4:Ti films (0.28-2.8 mA/cm^2, 2 keV).

From basic analysis of phenomena in solids and phosphors, one may consider degradation processes to be caused by the following main mechanisms:

- temperature quenching of luminescence centers dependent only on the temperature of a sample (power or current density);
- external quenching of luminescence due to generation of free carriers trapped by the trap centers; this component also depends on non-uniformity of electron distribution in the sample, their space diffusion (drift), and trapping parameters;
- generation of defects under electron bombardment which behave as the centers of nonradiative recombination both in the volume or at the surface of phosphor;
- surface reactions (this process may be considered as special case of generation of nonradiative recombination centers on the phosphor's surface).

Based on results obtained, the most probable mechanisms for long-term stability may be:

- intrinsic temperature quenching, which is indicated by increase of degradation coefficient with increase of current density at the same excitation energy;
- generation of nonradiative recombination centers in near-surface area, which is indicated by increase of degradation rate at higher excitation energies;

Let's discuss the possibility of mechanism of nonradiative recombination centers generation. The probability of generating defects by direct impact is small at the e-beam energies used (< 4 keV) because such energies are much less than the defect generation threshold. However, the defects may be generated by means of below-threshold mechanisms as well. In fact, several mechanisms of defect generation at below-threshold energy levels are known: multiple ionization [12] , decomposition of self-trapped holes [13], radiation-enhanced reactions [14]. Indeed, these mechanisms have been found in a series of materials [15, 16], and in particular under electron excitation [17].

When nonradiative recombination centers are generated in the near-surface region, the process of diffusion of free carriers generated in volume to nonradiative recombination centers, as well as the process of radiation-enhanced diffusion of defects take place, and degradation of CL intensity is proportional to $t^{1/2}$ [15]. Stabilization of luminescence degradation observed in Zn_2GeO_4:Mn can be explained as attaining equilibrium between generation of nonradiative recombination centers and their annealing.

Existence of the rapid stage of defects generation in Zn_2GeO_4:Mn at lower currents compared to Zn_2SiO_4:Ti correlates with the fact that the temperature of Zn_2GeO_4:Mn thin films formation (crystallization) is lower than that of Zn_2SiO_4:Ti.

CONCLUSIONS

It was found that at power densities less than 0.5 W/cm^2 most of the oxide thin film phosphors do not show substantial decrease of CL intensity. Decrease of CL intensity was observed at e-beam power energies higher than 1 W/cm^2 and is in a good agreement with the dependence $(Io/I - 1) = K \cdot t^{1/2}$ which describes a thermal diffusion process.

At higher energy densities of e-beam irradiation (in the range of 1 - 20 W/cm^2) and various current or power densities the specific behavior of long-term degradation processes in Zn_2SiO_4:Ti and Zn_2GeO_4:Mn was found which reveals as rapid degradation at the initial stage and a slow consequent decrease of intensity.

The most probable mechanisms responsible for long-term processes of degradation in the investigated phosphors are caused by internal thermal quenching due to heating by the e-beam as well as generation of nonradiative recombination centers in the near-surface area. This mechanism is in agreement with results presented in [7,8] where the chemical modification of the phosphor surface had been observed.

ACKNOWLEDGEMENTS

This work has been partly supported by Ministry of the Science and Education of Ukraine (LSU Project Fz41B). The work of TEF has been performed under the auspices of the US DOE. by LLNL under Contract No. W-7405-ENG-48.

REFERENCES

1. V. Bondar, M. Grytsiv, A. Groodzinsky, M. Vasyliv, A. Chakhovskoi, C. Hunt, M. Malinowsky, and T. Felter, J. Electrochem.Soc. **144**, 704 (1997).
2. V. Bondar, Materials Science and Engineering B. **69,** 505 (2000).
3. V. Bondar, Materials Science and Engineering B. **70**, 510 (2000).
4. A. Pfanhl, Bell. System. Technolog. **42,** 181 (1963).
5. Shmulovich Joseph, Information Display **3**, 17 (1989).
6. H. Widdel, and D. Post, *Color in Electronic Displays*, Plenium Press, NY 1992, p. 242.
7. C. H. Seager, K. R. Zvadil, D. R. Talant, W. L. Warren. *Proc. The Third International Conference on the Science and Technology of Display Phosphors*. Huntington Beach, CA. Nov. 3-5, 1997, pp. 325-326.
8. P.E.Holloway, W.J.Thomes, B. Abrams, S.Jones, L.Williams, The Effectr of Surfaces on FED Phosphors// Proc. The Seventh International Display Workshop (IDW'00). - Kobe (Japan). - November 29-December 1, 2000. - P.837-840. .
9. R. L. Donofrio, *Proc. Electrochemical Society*, May, 1981, pp. 381-382
10. V. Bondar, S. Popovich, Yu. Dubov, T. Felter. *Proc. The Fifth International Conference on the Science and Technology of Display Phosphors.* San Diego, CA. Nov. 8-10, 1999, pp. 329-332.
11. V. Bondar and M. Vasyliv, Ukrainian patent No. 18151 A, 01.07.97.
12. M. I. Klinger, V. V. Emtsev, T. V. Mashovets, S. M. Ryvkin, N. A. Vitovskii, Radiation Effects **56**, 229 (1981).
13. H. N. Hersh, J. Electrochem. Soc. **118**, 144C (1971).
14. D. V. Lang, Ann. Rev. Mater. Sci. **12**, 377 (1982).
15. V. L. Vinnetskii, E. E. Yaskovets, E. V. Kelman, in *Physical processes in crystal with defects*, (Institute of Physics of Ukraine Academic Science Publishers, Kiev, 1972), p. 122 (in Russian).
16. J. Klaer, D. Braugnig, F. Wulf, *Second European Conference on Radiation and its effects on Components amd System*, Saint-Malo, France, Sept. 13-16, 1993, pp.146-153.
17. G. A. Scoggan, T. P. Ma, J. Appl. Phys. **52**, 6231 (1981).

Mat. Res. Soc. Symp. Proc. Vol. 667 © 2001 Materials Research Society

Optical and Electrical Properties of Cr-SiO Thin Films for Flat Panel Displays

Richard Wood, Peter Hofstra and David Johnson
Luxell Technologies Inc.
5170A Timberlea Blvd.
Mississauga, CANADA, L4W 2S5

ABSTRACT

Low temperature fabrication of transparent conducting materials is a key issue in flat panel display production. Though Cr-SiO cermet thin films have predominantly been used as thin film resistors in a variety of microelectronics applications, it is shown in this paper that the material can successfully be used as a transparent to semi-transparent conductor in some applications if the value of the extinction coefficient, k, can be kept low (<0.3). Thus, a detailed study of the interdependence of the resistivity and optical properties of Cr-SiO is presented for the first time within the context of its use in the flat panel display industry. Specifically, Cr-SiO can be employed as part of Luxell's optical interference structure, known as the Black Layer™, US patent 5,049,780, used to increase contrast in TFEL and OLED displays.

INTRODUCTION

Cr-SiO cermet films have been used for various applications in the microelectronics industry [1-6] and the stability of conduction as a function of temperature has been the dominant characteristic investigated. Little consideration, however, has been paid to the optical properties and how they can be used in the fabrication of flat panel displays, specifically TFEL displays. In TFEL technology, which consists of a dielectric/phosphor/dielectric structure deposited on glass and sandwiched between two electrodes, there is a need for transparent and semi-transparent conducting materials for use in contrast enhancing structures, which are integrated into the rear electrode, specifically Luxell Technologies Inc.'s Black Layer™ technology. For this application Cr-SiO is an ideal material as it can be made to be both suitably conducting and transparent in thin film form.

As shown in the following work, the electrical and optical properties of Cr-SiO are highly dependent on the deposition conditions and the composition of the source material. This can be mapped out by examining the interdependence between resistivity and optical absorption, as expressed through the extinction coefficient, k, in various Cr-SiO thin films of thicknesses relevant to the above application, typically 20-150 nm. These relationships can be understood more fully using XPS to examine the composition and bonding states within the films as well as the degree of oxygen incorporation. It is also shown that the material can be used successfully as a semi-transparent conductor when the extinction coefficient is kept low (<0.3). Resistance was also measured as a function of temperature to examine the conduction mechanism. Finally, transmission electron microscopy (TEM) was used to examine crystalline sites in the material.

EXPERIMENTAL

Cr-SiO films were deposited using electron beam evaporation in a Leybold A1100 coating system at a base pressure of 8 x 10-6 Torr. Sintered Cr-SiO pieces, 1-3 mm, in wt% mixtures

(Cr/SiO) of 70/30, and 60/40 were used for evaporation. The films were deposited on Corning 7059 glass, ITO coated glass and Si wafers at rates varying from 0.5 to 10 Å/s. After calibration, all films were grown to a thickness of 1500Å, +/- 10Å. The deposition was conducted at room temperature. The film thickness was measured using a Tencor Alpha Step 500 thickness profiler and was monitored using a quartz crystal.

The extinction coefficient of each film was measured at a wavelength of 546.1 nm using a Rudolph AutoEL IV Ellipsometer. The extinction coefficient as a function of wavelength (visible regime) was also extracted using reflection and transmission measurements from a Filmetrics Reflectometer. The sheet resistances of the samples were measured using a Loresta HP, four point meter. The film structure was observed using a transmission electron microscope (TEM). The bonding structure was observed with XPS using a PHI 5500.

RESULTS and DISCUSSION

Sheet resistivity and optical absorption of the thin films were studied as a function of deposition rate for both the 70/30 and 60/40 mixtures, with results shown in figure 1. Sheet resistance decreased with increasing rate for both Cr-SiO mixtures, while the extinction coefficient increased with rate. This may be explained by two dominant phenomena. First, residual oxygen incorporation in the films is likely more prevalent for lower deposition rates and results in higher resistivity and lower absorption as more of the component metal oxidizes. This is supported by the fact that during calibration the measured film thickness, as compared to the monitored thickness, was found to decrease for higher rate films. This implies that the films deposited at higher rates were denser, with less oxygen incorporation, and relatively more chromium and/or silicon.

Second, the electron beam power required to deposit Cr and SiO are different, with Cr having a lower vapour pressure than SiO at a given temperature. Elemental chromium requires a higher power to deposit at the same rate as SiO. Samples from consecutive runs re-using source material, as well as thicker films, have a higher extinction coefficient, possibly from an increase in Cr concentration. The power required to maintain a given deposition rate increased with thickness. Presumably, as SiO is depleted from the source, more chromium is deposited to maintain the same rate. For consistent results, fresh material was used for each deposition.

From figure 1b, it is seen that for some deposition conditions, using the 60/40 mixture, k is less than 0.3. Thin films with k less than 0.3, at 550 nm, are substantially transparent (more than 50%) up to 50-100 nm thick. Figures 1b and 2 reveal the general trend of increased absorption in the films that were deposited at higher rates and from the higher chromium source concentration. From previous studies [1-6] on Cr-SiO cermets, it is known that Cr exists within a SiO_2 matrix in several forms: elemental Cr; various silicides (CrSi, Cr_3Si, Cr_5Si_3); as well as Cr_2O_3, identified via x-ray diffraction [1]. While figure 2 supports the trend of increased absorption with increasing Cr concentration in the film, it is also seen that the overall k dispersion does not change. The primary conduction mechanism has been previously identified [4,5] as hopping conductivity between components of the conducting phase, Cr_3Si crystals, as well as metallic bridges imbedded in an amorphous SiO_2 matrix. This conducting phase has been shown to partially oxidize [2] during deposition, increasing resistivity. The present results show that this occurs primarily for the films deposited at lower rates (figure 1a). Auger electron spectroscopy has previously revealed a shift in elemental concentration in the films, as compared to the source material, suggesting that SiO dissociates, freeing the components to react with Cr [3]. In films with excess Cr, Si will be replaced by Cr in the crystals, as in the films deposited at a higher rate or from a higher chromium concentration source. Silicon rich centers, Cr_5Si_3 and CrSi, occur in the lower source chromium content depositions. However, silicon rich sites, such as Cr_5Si_3,and CrSi, have been shown to contribute little to conductivity [4].

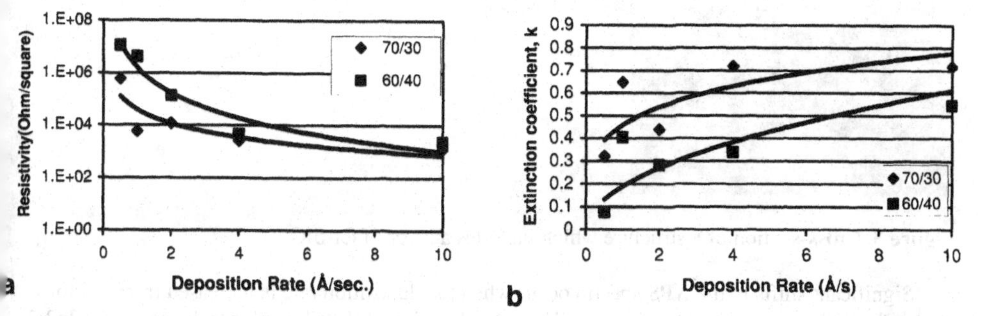

Figure 1. Changes in resistivity and optical extinction coefficient with various e-beam deposition conditions and source material. 1500Å thick samples.

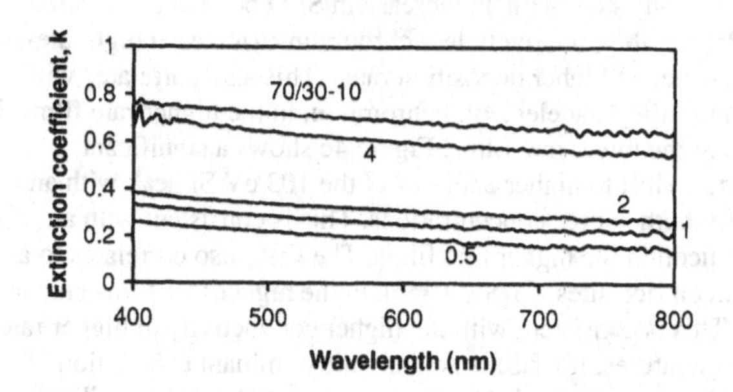

Figure 2. The dispersive absorptive behaviour of various Cr-SiO thin films, 1500Å thick, from 60%/40% Cr/SiO source material at deposition rates ranging from 0.5 to 4 Å/s. The spectrum for a sample deposited from 70/40 Cr/SiO at 10 Å/s is also shown.

Grain size has also been related to conduction, with resistance decreasing with grain size [2] and with grain concentration. TEM of samples of 60/40 deposited at 1.5 Å/s, figure 3, revealed crystals that are 1 to 2.5 nm in size. Other TEM studies have shown grain sizes from 2.5 nm [5] to 10 nm in size (sputtered at 150° C and then annealed) [4]. As our samples were deposited at low temperatures, and were not annealed, the grain size was expected to be smaller.

A Cr-SiO 60/40 sample, deposited at 2 Å/s, was annealed at 350° C and the sheet resistance as a function of temperature was measured. The results revealed a negative temperature coefficient of resistance (TCR) for low temperatures, below 100° C, the normal operating regime of the devices, and a positive TCR for high temperatures. Hence, the films act like semiconductors at low temperature and like metals at high temperature [5]. In order to gain further insight into the chemical composition of the Cr-SiO thin films, XPS was employed to examine bonding states [7,8]. Chromium spectra are shown in figure 4a, (568 to 590 eV Cr 2p1/2 and Cr 2p3/2), oxygen spectra in figure 4b, (528-535 eV, O 1s) and silicon spectra in figure 4c (96 to 107 eV, Si 2p), for

1500 A thick samples fabricated from both the 60/40 (bold lines) and 70/30 source material, at deposition rates of 0.5 and 10 Å/s.

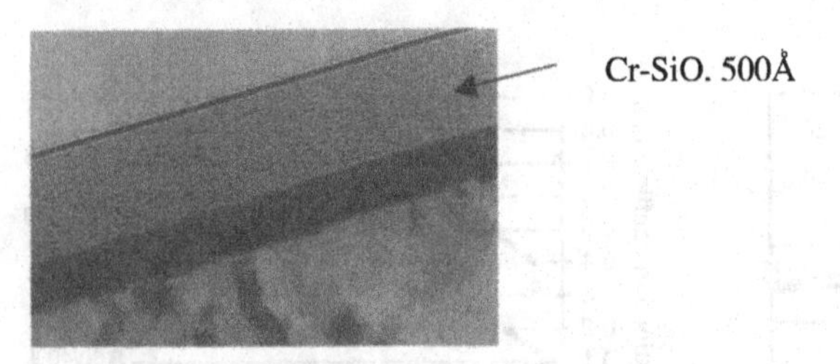

Figure 3. Cross-section of a structure which includes a layer of Cr-SiO.

Significant shifts in the XPS spectra occur when the deposition rate is increased from 0.5 to 10 Å/s for both source concentrations. In figure 4a, there is a shift to lower energies in the Cr 2p3/2 and 2p1/2 peaks, consistent with increases in the concentration of elemental chromium and relative decreases in chromium oxide concentration. There is also a shift in the 533 eV, O1s peak to higher energies for higher rate films (figure 4b), consistent with an increase in Si-O bond concentration, relative to Cr-O bond concentration. Presumably, relatively less chromium oxidizes at high rates, leaving more elemental Cr in films deposited at higher deposition rates. This also correlates with lower absorption in lower rate films suggesting that elemental chromium, in the higher rate films, is more absorbing than chromium oxides in the lower rate films. Figure 4c shows a significant increase in the free Si peak at 99 ev, and a shift to higher energies of the 103 eV Si peak with an increase in deposition rate, occurring for both source concentrations. This is consistent with an increase in the presence of elemental silicon in the higher rate films. The shift also correlates to a decrease in Si^{+}, Si^{2+} concentration, (silicon rich sites, Cr_5Si_3, CrSi), in the higher rate films and an increase in Cr_3Si (Si^{3+}) concentration. This is consistent with the higher conductivity in higher rate films, and the negative TCR at low temperatures, if Cr_3Si is actually the dominant conduction center. In contrast, the lower rate films have lower conductivity and appear to have more silicon rich sites. The relative increase of Si^{4+} in the higher rate films is possibly a result of incomplete dissociation of SiO or insufficient time, due to the high deposition rate, to form silicon rich crystals.

The variations in XPS spectra for shifts between source concentration are more subtle. In figure 4c, there is a small shift to lower energies in the Si 2p peak when the source concentration of Cr increases from 60/40 to 70/30, at a rate of 0.5 Å/s, indicating a possible increase in the relative concentrations of silicon in 3+ ionization states [7], Cr_3Si sites. This would support the Cr_3Si sites as the main conducting mechanism, as the 70/30 mixture deposited at 0.5 Å/s is significantly less resistive than the 60/40 mixture, figure 1a. There is also an increase in the relative intensity of the Si2p 99 eV peak for the higher Cr concentration material, indicating an increase in the presence of free silicon, Si^{o} , in the 70/30 films. The increase in elemental silicon correlates with the formation of less silicon rich crystals. Figure 4a shows a slight shift to higher Cr-O concentrations (relative to Si-O) for the 70/30 mixture. It is likely that there is simply more chromium present, which can be oxidized, in the 70/30 depositions.

The similarities in the XPS results for both source concentrations, (they are virtually identical), deposited at high rates (10 Å/s) is consistent with the similarities in the electrical characteristics of these films, as shown in figure 1a. They appear to reach a saturation point for conduction. There is no obvious XPS evidence for the higher k values for the 70/30 films. However, it is likely that there

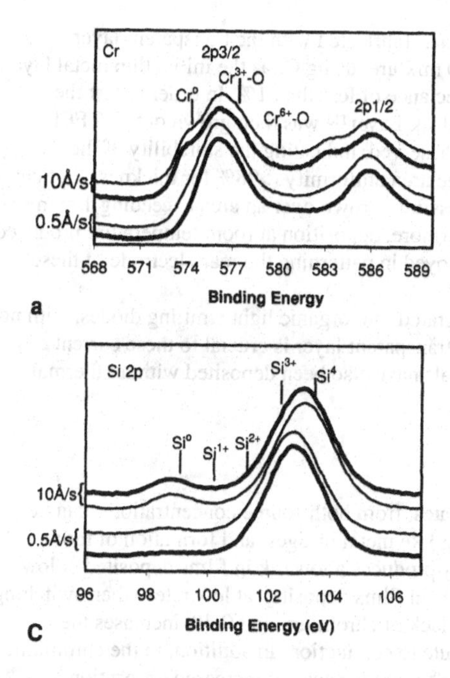

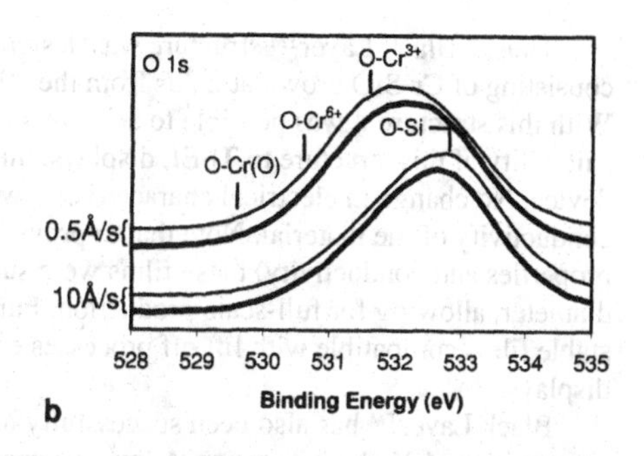

Figure 4. X-ray photoelectron spectroscopy of Cr-SiO thin films, 1500Å thick, produced from 60% (bold lines) and 70% Cr source material, at 10 and 0.5 Å/s.

is additional elemental chromium and chromium oxides in the70/30 films, which are more absorbing than elemental silicon and silicon oxides.

Luxell's contrast enhancing Black Layer™ technology is used in the manufacture of TFEL displays. Incorporated between the rear electrode and the light emitting structure, it consists of a thin metal layer, a transparent to semi-transparent layer, and a thick metal layer using the principle of optical interference to reduce display reflectance. To achieve an acceptable level of light reduction and to be electrically benign within the structure, the transparent layer must have both a suitable k value for the required thickness and have low resistivity for conduction through the film.

Deposition of Cr-SiO under suitable conditions produces thin films, which are relatively transparent up to 50 nm thick, if the extinction coefficient, k, is kept below 0.3. However, it is also necessary to maximize Cr concentration to minimize resistive losses in this film. Note that though sheet resistivity was the measurable quantity, conduction through the thickness of the film is of primary importance in the design and fabrication of devices. From figure 1a it is seen that by simply increasing the rate from 0.5 to 4 Å/s it is possible to decrease sheet resistivity by almost four orders of magnitude for both the 70/30 and 60/40 source material mixtures. At the same time, the k (at 546.1 nm) increases from about 0.1 to about 0.35 for the 60/40 mixture, (figure 1b) which is within the range of suitable transparency for the thicknesses required. However films from the 70/30 mixture showed much higher increases in absorption for little gain in resistivity.

Thus, a Black Layer™ structure was designed and fabricated with the transparent layer consisting of Cr-SiO grown at 2 Å/s from the 60/40 mixture, using Cr as the initial thin metal layer. With this structure it was possible to achieve a reflectance of less than 1%. In order to test the suitability of this structure to TFEL displays, this Black Layer™ was then grown onto a TFEL device. No change in electrical characteristics was observed, indicating the suitability of the conductivity of the material. Note that to prove large area uniformity (>98% for thickness, optical properties and conductivity) these films were successfully grown over an area exceeding 1 m in diameter, allowing for full-scale production. Furthermore, deposition at room temperature produced stable films compatible with lift off processes employed in patterning the rear electrode of these displays.

Black Layer™ has also been successfully integrated into organic light emitting diodes, with no change in the I-V characteristics. A low resistance transparent layer is crucial to these current driven devices. The required Black Layer™ materials have also been deposited without thermal damage to the underlying organic layers.

CONCLUSIONS

Lower resistance in films deposited at higher rates, from both source concentrations, can be attributed to the presence of more chromium, acting like metal bridges, and formation of more Cr_3Si centres. Oxidation of metals, Si and Cr, likely produces a lower k in films deposited at low rates. Resistance decreases by an order of magnitude in films deposited at low rates when switching from the 60/40 to 70/30 mixtures suggesting that a lack of chromium, in 60/40, increases the concentration of silicon rich sites that don't contribute to conduction. In addition, in the chromium rich, 70/30, films, more chromium is present, likely being the cause of increased absorption.

At high rates, both films have excess chromium that appears to remain elemental. Conduction reaches a peak with increasing deposition rate, suggesting that the dominant mechanism is the formation of Cr_xSi_y complexes and not elemental chromium: the Cr_3Si crystals likely reach a concentration and size maximum. A high atom arrival rate at the surface will limit crystal growth.

It has been shown that electron beam evaporation conditions can be changed to alter the material concentrations and states in the films, unlike sputtering and flash evaporation. Results also indicate that deposition conditions could be further adjusted to reduce resistance while keeping the extinction coefficient low. Addition of oxygen during the deposition of 70/30 could reduce the free silicon and chromium content.

REFERENCES

1. L. K. Thomas, W. Pekruhn, T. Chunhe and A. Schroder, Solar Energy Mater. **16**(1-3) 133 (1987).
2. M.Milosavljevic, T. M. Nenadovic, N. Bibic and T. Dimitrijevic, Thin Solid Films **101**(2) 167 (1983).
3. A. G. Taylor, R. E. Thurstans and D. P. Oxley, J. Phys. E: Sci. Instrum. **17** 755 (1984).
4. V. Fronz, B. Rosner and W. Storch, Thin Solid Films **65** 33 (1980).
5. E. Schabowska and R. Scigala, Thin Solid Films **135** 149 (1986).
6. E. Schabowska-Osiowska, R. Scigala and Z. Porada, Vacuum **37** 119 1987.
7. I. Bertoti, A. Toth, M. Mohai, R. Kelly, G. Marletta, M. Farkas-Jahnke, Nucl. Instr. and Meth. B **122** 510 (1997).
8. I. Bertoti, A. Toth, M. Mohai, R. Kelly, G. Marletta, Nucl. Instr. and Meth. B **116** 200 (1996).

Mat. Res. Soc. Symp. Proc. Vol. 667 © 2001 Materials Research Society

Low-temperature technology and physical processes in green thin-film phosphor Zn_2GeO_4-Mn

V.Bondar, S.Popovich, T.Felter[1], and J.Wager[2]
Lviv National University, Department of Physics, 50 Dragomanov Str., 79005, Lviv, Ukraine.
[1] Lawrence Livermore National Laboratory, PO Box 808, L - 356, Livermore, CA, 94550
[2] Department of Electrical and Computer Engineering, Oregon State University, Corvallis, Oregon 97331-3211, U. S. A

ABSTRACT

Thin-film Zn_2GeO_4:Mn phosphors with lower temperature of crystallization and potentially compatible with industrial technologies were investigated. The technology of thin films synthesis has been developed and their structure and crystal parameters have been investigated. Photoluminescence excitation spectra, photoconductivity, temperature dependencies and ESR-spectra of manganese ions were studied. A mechanism for luminescence in this phosphor has been proposed. Results are presented of cathodo- and electro-luminescence of thin film structures of Zn_2GeO_4:Mn.

INTRODUCTION

The manganese dopant is one of the most efficient centers of luminescence in phosphors. Together with high efficiency, one additional advantage is the color of emission, which depends on the crystal lattice: yellow in ZnS:Mn, green in Zn_2SiO_4:Mn, or blue-green in $ZnGa_2O_4$:Mn. Oxide phosphors are more stable in comparison to sulfide ones and therefore are promising for flat panel display applications. However, the use of oxide thin-film phosphors in many cases is limited because of their high crystallization temperature. For this reason, a series of recent works [1, 2, 3] was directed on the development of phosphors with lowered crystallization temperature. In the present work, we developed the technology of synthesis of thin-film Zn_2GeO_4:Mn phosphors with lower crystallization temperature. Thus, these phosphors may be compatible with industrial technologies. We also investigated the crystal structure of thin films and the luminescent processes caused by manganese dopant.

EXPERIMENTAL

Zn_2GeO_4:Mn films were deposited by rf-magnetron sputtering [4] of a pressed target manufactured from the presynthesized powder corresponding to the stoichiometric composition of Zn_2GeO_4:Mn_2O_3. The quality of thin films depends on deposition conditions and was assessed by absorption edge position, refractive index, and transparency of the films in the visible region obtained from optical measurements. The film structure was investigated using an HZG-4A X-ray powder diffractometer ($Cu_{K\alpha}$ radiation, θ-2θ scan mode, step 0.05°, t=10 s per point). The Rietveld profile refinement method of analysis of experimental XRD data was used [5, 6]. Specified parameters were unit cell parameters and a texture parameter [7]. The X-band ($\nu \cong 9.4$ GHz) ESR spectra were recorded using a computer controlled commercial AE-4700 radio frequency spectrometer with 100 kHz magnetic field modulation at room temperature. The microwave frequency was controlled by means of diphenylpicrylhydrazyl (DPPH) g-marker (g=2.0036±0.0001). Cathodoluminescence spectra were measured at 300 K and with e-beam energy excitation between 1 and 6 keV and current density of 1 mA·cm^{-2}. Photoluminescence excitation spectra and photoconductivity were measured in the wavelength range from 200 to 400 nm and the temperature range from 120 to 360 K with MDR-4 monochromator. Hydrogen (LDD-400) or xenon (DKSEL-1000) lamps were used as excitation sources. Measured excitation, emission, and photo-conductivity spectra were corrected according to the spectral distribution of emission and spectral sensitivity of the measuring system.

RESULTS AND DISCUSSION

Films deposited by the rf-magnetron method were initially amorphous. After temperature treatment the structure becomes crystalline corresponding to the structure type of R3b with lattice constants a = 14.206Å, and b= 9.5146Å, (table.1, figure 1). The powder phosphor has a (100) texture axis and texture coefficient of 1.09, whereas the films have a dominant (001) texture axis and texture coefficient of 0.91.

Table 1. Crystal parameters of thin film

Crystal parameters of thin film	Zn_2GeO_4	Zn_2GeO_4-Mn, target	Zn_2GeO_4-Mn, film
Cell dimension (a, Å)	14.2333 ±0.0002	14.2365 ±0.0002	14.2060 ±0.0002
Cell dimension (c, Å)	9.5256 ±0.0002	9.5281 ±0.0001	9.5146 ±0.0001
Cell volume (Å^3)	1671.14 ±0.07	1671.70 ±0.07	1662.22 ±0.07
Texture axis	[100]	[100]	[001]
Texture parameter, τ	1.0	1.09 ±0.03	0.91 ±0.03

It was found that the parameters a and c of the crystal lattice of the film as compared to the powder phosphor decrease from 14.2365Å to 14.206 Å and from 9.5281 Å to 9.514 Å, respectively, and the volume of the unit cell decreases by approximately 9.5 Å^3 (table.1). This decrease of the lattice constants may be caused by two probable reasons:

1. Formation of a defect anionic sublattice due to substitution of germanium atoms in tetrahedral sites for manganese atoms with smaller valence than germanium which results in formation of oxygen vacancies.
2. Increase of covalent component of the Me-O chemical bond by isomorphic substitution of zinc for manganese, which results in decrease of the corresponding atomic distances.

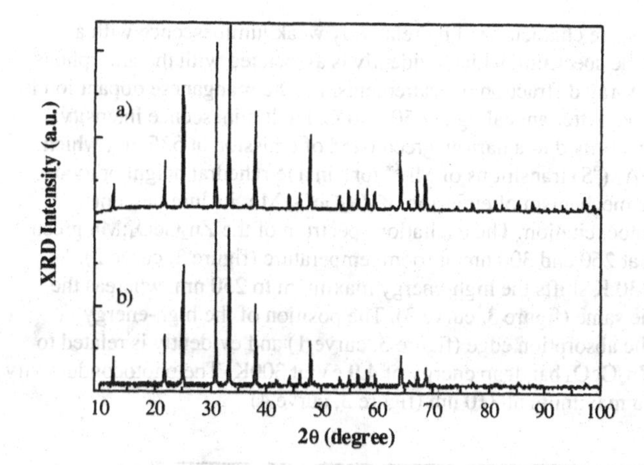

Figure 1. X-rays diffraction patterns of powder (a) and thin-film (b) phosphor Zn_2GeO_4:Mn

The dependence of luminescence of Zn_2GeO_4:Mn phosphor on concentration was investigated. It was found that the maximal luminescence yield is observed at a concentration of manganese 2 wt.%. At higher concentrations, emission intensity decreased because of concentration quenching.

In Zn_2GeO_4:Mn (C_{Mn} = 0.8 wt. %) at 300K, the composite ESR spectrum is observed which consists of the broad ($\Delta H_{pp} \cong 800$ G) line with $g_{eff} = 2.0$ and approximately 30 narrow lines of different intensity (figure 4). The broad line with g_{eff}=2.00 in polycrystalline samples may be interpreted as the unresolved fine structure (FS) of the ESR spectrum in the disordered system of isolated Mn^{2+} ions ($^6S_{5/2}$, $^3d_5$), or as the ESR spectrum from clusters of 2 or more ions bonded with dipole-dipole magnetic interaction. Narrow lines of various intensity around g_{eff}=2.00 (figure 2) may be interpreted as allowed and forbidden transitions of the FS central component of isolated ions Mn^{2+}.

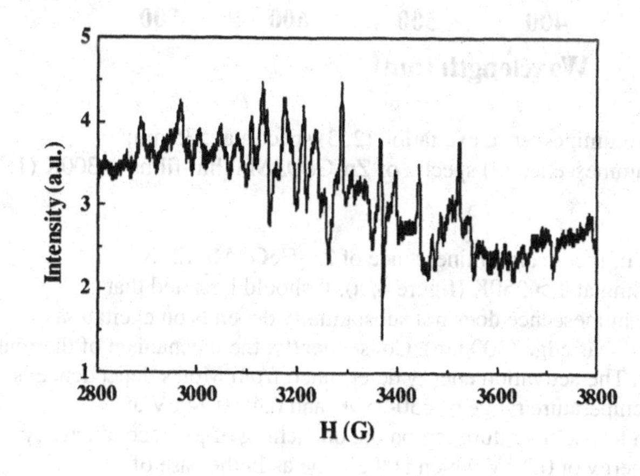

Figure 2. ESR spectrum of Zn_2GeO_4:Mn phosphor

The as-deposited thin films are characterized by relatively weak luminescence with a maximum in the red region of the spectrum, which evidently is associated with the amorphous state of the film confirmed by X-ray diffraction measurements, i.e. the manganese dopant ion is not yet in a crystal neighborhood. After annealing at 650-750ºC, the luminescence intensity grows, and the spectrum is transformed to a narrow green band of emission at 535 nm, which corresponds to the $^4T_1$ (^4G) – $^6A_1$ (^6S) transitions of Mn^{2+} ions in a tetrahedral neighborhood.

In order to investigate the mechanism of emission of Zn_2GeO_4:Mn, its luminescent properties were studied by photoexcitation. The excitation spectrum of the Zn_2GeO_4:Mn green emission band shows maxima at 260 and 300 nm at room temperature (figure 3, curve 2). Lowering the temperature to 130 K shifts the high-energy maximum to 250 nm, whereas the low-energy maximum stays the same (figure 3, curve 3). The position of the high-energy maximum is in the region of the absorption edge (figure 3, curve 1) and evidently is related to band-to-band transition with Zn_2GeO_4 bandgap energy of 4.9 eV at 300K. The photoconductivity of the Zn_2GeO_4:Mn films has a maximum at 310 nm (figure 3, curve 4).

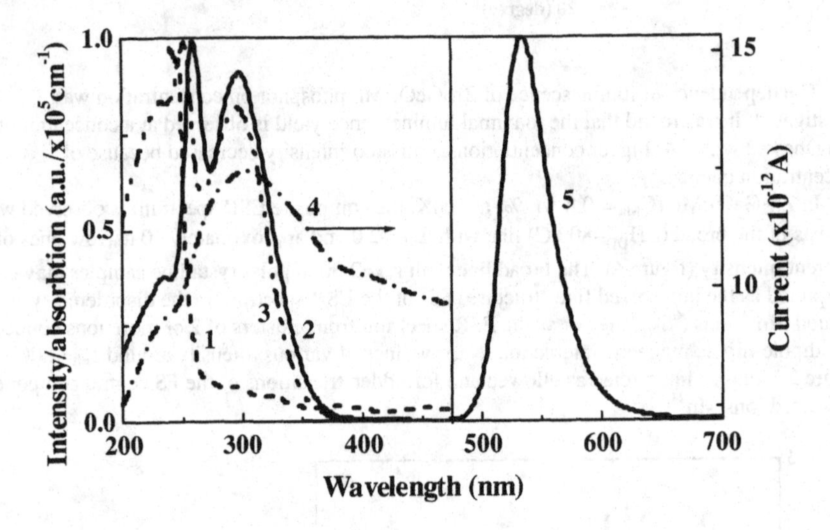

Figure 3. Absorption (1) and photoluminescence excitation (2, 3) spectra at 535 nm; photoconductivity (4) and photoluminescence (5) spectra of Zn_2GeO_4:Mn thin-films at 300K (1, 2, 4, 5) and 130K (3)

The temperature dependence of the green luminescence of Zn_2GeO_4:Mn films is characteristic with thermal quenching at T > 250K (figure 4, a). It should be noted that temperature dependence of photoluminescence does not substantially depends on excitation region – band-to-band (259 nm) or near-edge (300 nm). Consequently, the mechanism of thermal quenching in both cases is similar. The activation energy determined from Mott's dependence is 0.20-0.24 eV (figure 3, b) in the temperature range of 250-280K and 0.40-0.44 eV at temperatures higher than 290K. In low-temperature region the quenching of photoconductivity also is observed with activation energy of 0.2 eV which is the same as in the case of photoluminescence. It indicates on thermal quenching caused by transitions from local level to valence band. The existence of two regions in temperature dependence indicates on the two-component quenching process: low-temperature and high-temperature. The first component can be a result of quenching the excitation, and the second one - quenching of the Mn^{2+} ion.

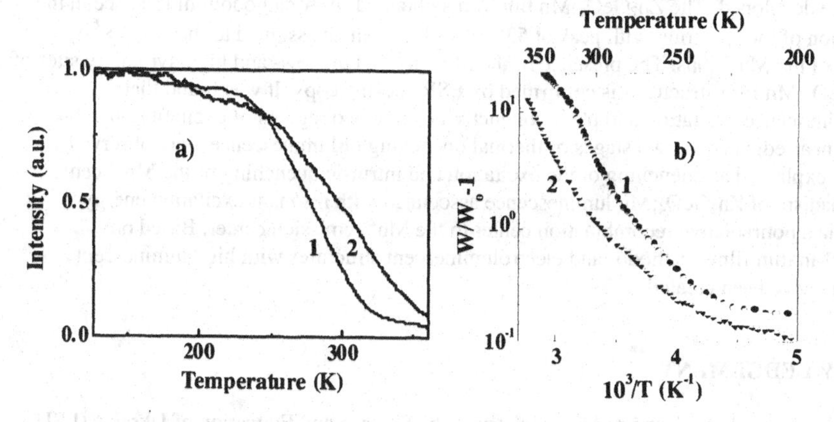

Figure 4. Temperature dependence of luminescence at 535 nm of thin-film phosphor Zn_2GeO_4:Mn under excitation at 250 (curve 1) and 300 nm (curve 2) - (a), and the same dependence in (W/W_0-1) vs. $10^3/T$ coordinates - (b).

Based on the results obtained, the most probable excitation mechanism of Zn_2GeO_4:Mn luminescence at 535 nm may be band-to-band energy transfer to a nonradiative recombination center or excitation of nonradiative recombination center itself, which in turn transfers energy in a resonant way to the Mn^{2+}- center. The intermediate center responsible for resonant energy transfer to the Mn^{2+}- center may be the same center responsible for the maximum at 300 nm in photoconductivity and excitation spectra. Although its optical distance E_{opt} = 0.8 eV from the bottom of the valence band and the thermal depth $E_{therm.}$= 0.2 eV differ, this does not contradict the theoretical and experimental data on the differences between thermal and optical energy levels in phosphors [8]. The optical activation energy can indeed be several times higher than the thermal one, and this difference tends to increase with increase of crystal ionicity level.

We find that Zn_2GeO_4:Mn luminescence shows thermal quenching to start at 250 K, a high temperature compared to that for the widely investigated ZnS:Mn [9] system, 100K. Therefore, the temperature parameters of Zn_2GeO_4:Mn may be considered good for practical use.

For FED applications, the thin-film phosphor Zn_2GeO_4:Mn has a luminescent efficiency of 2.4 lm/W at 2 keV electron excitation. High stability is an additional advantage of this thin-film phosphor for FED applications. In long-term testing with electron excitation, Zn_2GeO_4:Mn showed 5 times higher stability compared to sulfide phosphors – 20-30 Coulombs/cm^2 (dose of electron irradiation resulting in 50% decrease of luminescence intensity).

Using Zn_2GeO_4:Mn the sulfur-free electroluminescent structure "Corning/ ITO/ ATO/ Zn_2GeO_4:Mn/SiON/Al" with green emission have been fabricated. These structures show high brightness (14ft-L at 40 V over threshold at 60 Hz), good efficiency (~0.2 lm/W), excellent color coordinates (CIE*x* = 0.302, CIE*y* = 0.668), robustness (withstand up to 300 V without burning out), and excellent aging characteristics (no measurable aging was observed after continuous excitation). These parameters are better than conventional sulfide electroluminescent displays.

CONCLUSIONS

The technologies for fabrication of the thin film phosphor Zn_2GeO_4:Mn by the rf-magnetron method are developed. The Zn_2GeO_4:Mn thin films obtained show cathodoluminescence in the green region of the spectrum with peak at 535 nm, which can be assigned to the $^4T_{1g} \Rightarrow {}^6A_{1g}$ transition of the Mn^{2+} ions. The presence of the Mn^{2+} ions in the low- and high-symmetry sites of the Zn_2GeO_4:Mn film structures is confirmed by ESR spectroscopy. It was found that photoluminescence excitation and photoconductivity show two regions of excitation: band-to-band and near-edge. Also, two stages of thermal quenching of luminescence were observed which are explained as quenching of the excitation and intrinsic quenching of the Mn^{2+} center. The mechanism of Zn_2GeO_4:Mn luminescence associated with resonant excitation energy transfer via a nonradiative recombination center to the Mn^{2+} emission center. Based on Zn_2GeO_4:Mn thin films, cathodo- and electroluminescent structures with high luminescent parameters have been created.

ACKNOWLEDGEMENT

This work has been supported by Ministry of the Science and Education of Ukraine (LSU Project Fz41B). The work of TEF has been performed under the auspices of the US DOE by LLNL under Contract No. W-7405-ENG-48.

REFERENCES

1. T. Minami, in *Proc. of 4-th Intern. Conf. on the Science and Technology of Display Phosphors* (Bend, Oregon, 1998), pp. 195-198.
2. A. N. Kitaii, in Proc. of *4-th Intern. Conf. on the Science and Technology of Display Phosphors* (Bend, Oregon, 1998), pp. 199-202.
3. J. S. Lewis and P.H. Holloway, *Journal of the Electrochemical Society* **147** (8), 3148-3150 (2000)
4. V. Bondar, M. Vasyliv, Ukrainian patent No. 18151 A, (July 1997).
5. V. Lysoivan, S. Gromilov, Aspects of accuracy in polycrystalline diffractometry (Nauka, Novosibirsk, 1989 (in Russian)), p. 243.
6. V. K. Pecharski, L. D. Axelrood and P. Yu. Zavalii, *Crystallography* **32** (6), 874-877 (1987).
7. A. Segmuller, *J. Vac. Sci. Technol.* **A 9** (4), 2477-2480 (1991).
8. Vu. Kuang, M. V. Fok in *Luminescence centers in crystals* (Trudy FIAN, **79,** Nauka, Moskov, 1974 (in Russian)), pp. 39-63.
9. H. Hanzawa, M. Kobayashi, O. Matsuda, K. Murase and W. Giriat, *Phys. Stat. Sol. (a)* **175,** 715 (1999).

Mat. Res. Soc. Symp. Proc. Vol. 667 © 2001 Materials Research Society

Strong ultraviolet electroluminescence from porous silicon light-emitting diodes

H. L. Tam [1], J. Yuan [1], K. F. Li [1], W. K. Wong [2], and K. W. Cheah [1,3*]
[1] Department of Physics, [3]Center for Surface Analysis and Research,
Hong Kong Baptist University, Kowloon Tong,
[2] Department of Chemistry, Hong Kong Baptist University, Kowloon Tong, Hong Kong, China

ABSTRACT

Porous silicon light-emitting diodes were found to emit strong line-shaped ultraviolet under a forward bias driving voltage of about 20 volts. The intensity was sufficiently strong to pump an organic crystal, Tb-dipicolinic acid, producing clear Tb 4*f* intra-shell transition photoluminescence spectrum. The current-voltage characteristics of the devices also showed negative differential resistance, which was frequency dependent. In addition, purging of the device with various gases could quench the electroluminescence but the intensity recovered partially after each purging, but with no change in emission spectrum. Both results indicate the transport was influenced strongly by local space charge. From the results, the electroluminescence mechanism is tentatively attributed to core recombination in the porous layer, and the spectral characteristics is due to the microcavity effect between the top Au contact and silicon substrate. The present study shows that porous silicon has the potential as UV source in optoelectronics applications.

INTRODUCTION

The development of information and communication technology depends almost exclusively on the advancement of silicon integrated circuits technology. However, the indirect 1.1 eV band nature of silicon restricts its usage as optoelectronics or photonic material. The discovery of efficient visible photoluminescence from electrochemically etched porous silicon (PS) [1] opened a possible way to exploit Si in optoelectronics applications. Intensive investigations had led to many reports on successful fabrication of light emitting diodes (LEDs) from PS [2-5] and other forms of Si materials [6-8]. However, applicable devices are still not available due to the low efficiency, poor durability, and broad emission. In 1993, Kozlowski et al [9] reported ultraviolet light emission from porous silicon LEDs. The emission was sharp and was attributed to N_2 discharging in the microsized cavities inside the PS. The threshold voltage of the LED was in excess of 100V, thus, the device reported by them behaved more like a field emission device. In this report, we demonstrated that it is possible to reproduce the UV emission with lower driving voltage. Our experiment demonstrated that a threshold voltage of as low as 19 volts can be obtained for the devices to work. The emission may not relate to microsized N_2 discharging as proposed but originate from core recombination.

EXPERIMENT

Polished Sb doped CZ n-type (100) silicon wafer with resistivity 0.008~0.02 Ω.cm was used in the experiment. After dipping in HF acid to remove the native oxide, a layer 300 nm of Al was deposited by thermal evaporation on the wafer as back contact. The Al coated Si wafers were annealed at 280 °C in vacuum for 1.5 hours. Electrochemical etching was performed using

HF(48%):ethanol (1:1 in volume) solution under illumination from a 45 W tungsten lamp and a constant etching current density of 4 mA/cm^2. The etching time was about 10-15 minutes. A semitransparent Au thin film with thickness of 60 nm was deposited as top contact. The samples have active areas of 2×2 mm^2. Current-Voltage (I-V) characteristic was measured with the Keithley 2400 sourcemeter. Electroluminescence (EL) was dispersed by an Oriel ¼ m double monochromator equipped with a cooled Hamamatsu photomultipler tube (PMT). Chopped EL signal was collected with a computer controlled Stanford SR830 lock-in amplifier. All the experiments were performed at room temperature and the samples were under d.c driven.

RESULTS

A typical I-V curve is depicted in Fig. 1, in which one can see that the sample shows fair rectification behavior under lower bias. When the Au was positively biased with applied voltage lower than 7 V, the device behaved like a diode in forward bias, and a maximum current of ~90 mA was reached at about 7 V. Further increase in applied voltage led to a negative differential resistance (NDR) region, and the current dropped to ~30 mA at about 15 V. The NDR has a peak to valley current-rate (PVCR) of 3. When the applied voltage was greater than 15 V, the sample showed very high resistance and the current increased very slowly.

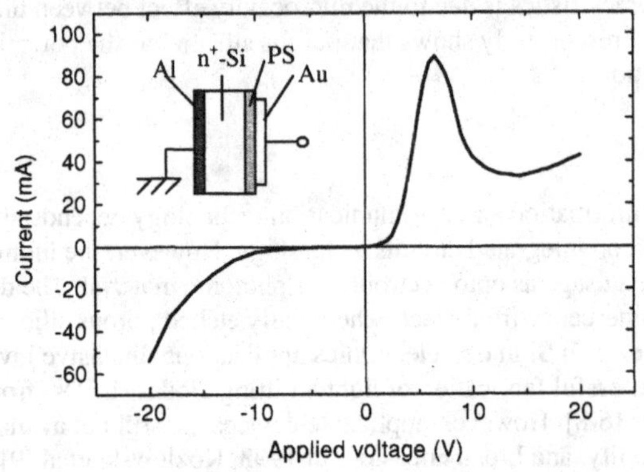

Figure 1: A typical I-V characteristics curve of the light emitting diode. The threshold voltage for negative resistance to occur is about 7 V.

Shown in Fig. 2 are the EL spectra under different applied voltage. Whitish orange colored EL could be readily observed by naked eyes under weak room illumination. The orange EL peaked at about 630 nm with broad featureless spectrum similar to photoluminescence spectrum of the device. The whitish EL came from sharp and much intensive near UV lines that have the strongest peak of the UV EL occurred at 337 nm. A higher resolution spectrum of the UV emissions is shown as inset in Figure 2. The sharp emission lines superimposed on the red EL were the second orders of the UV emissions, and this gives direct comparison of the intensities between the visible band and the second orders UV emission. These sharp UV emissions have the characteristic of atomic lines and are the same as that reported by Kozlowski et al. [9].

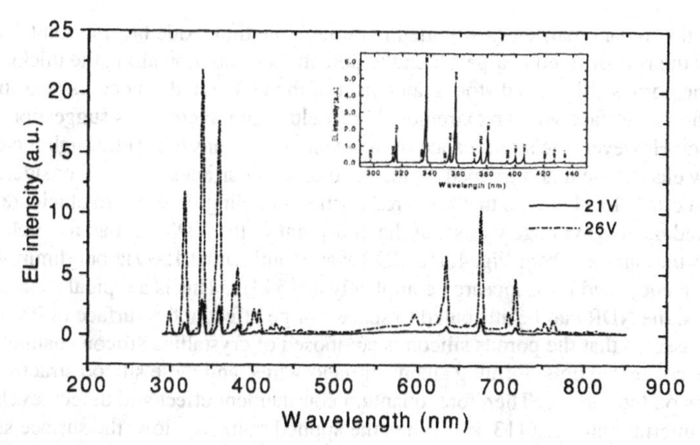

Figure 2: EL spectra of a typical PS LED under different bias voltage. The inset is a higher resolution spectrum of the emission.

In order to demonstrate the intensity of the PS LED UV EL, a single crystal of Tb-dipicolinic acid, was placed directly in front of the LED. When the crystal was pumped by the LED, the Tb 4*f* transition spectrum of the crystal was clearly observed (Figure 3). For comparison, a PL spectrum of the crystal obtained from using He-Cd laser as excitation source is shown as inset. Fig. 3 shows convincingly that the intensity of the UV EL is sufficient to act as a pump source.

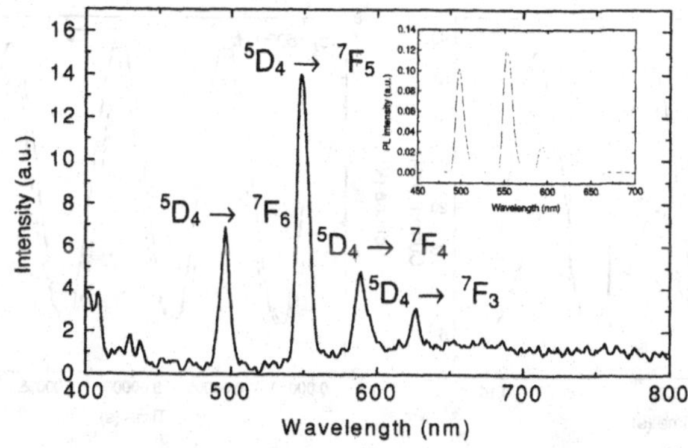

Figure 3: PL of Tb-dipicolinic acid single crystal using PS LED as the excitation source. The LED is driven at 25V, 0.1A. Inset is the PL spectrum of the crystal obtained from using 325 nm line of He-Cd laser as excitation source with 1 mW power.

To understand the I-V characteristics of the device, a brief review of previous works in this area is desirable. Ueno and Koshida [11] proposed that the NDR is due to the injection of electrons into the Si nano particles by tunneling through the thin oxide layer around. The injected electrons play the role of fixed charges, then the potential distribution along the thickness direction of the porous silicon is distorted and most of the potential drop occurs near the Au/porous silicon interface with an extremely high-field region there. This suggestion explains the NDR effect. However, the significance of the excited electrons tunnel through the electronic barriers at low electric field and trapped by the nano-silicon particles was not considered. This trapping could create local charge that hindered further tunneling. In order to elucidate the nature of NDR, a fixed d.c. bias voltage was set at the mid-point of the NDR, and an a.c. voltage was applied across the sample. From Fig. 4, the NDR was prominent at 150 Hz but diminishing with increasing frequency, and it disappeared completely at 15 kHz. This is a typical capacitive behavior. Thus, the NDR can be attributed to space charge effect at the surface of PS layer. It is reasonable to assume that the porous silicon is composed of crystalline silicon channels in the forms of inter-connected dots, forming chains of nano-wires, and these silicon structures has a layer of oxides on the surface. Therefore, quantum confinement effect and defect levels at the oxide/silicon interface co-exist [13-14]. When the applied voltage is low, the surface states provide effective ways for the electrons flowing from the bulk silicon substrate to the Au electrode. When the voltage increased above a threshold value, the electrical field in the porous silicon layer will become high enough for electrons to be injected into the oxide or defect states at the silicon-oxide interfaces and trapped there. The trapped electrons acted as static charges, as a result, space charging areas began to appear at the interface. This local charge effectively blocked the surface states from conducting electrons. Thus, the conduction path switched and the electrons conducted via the core instead. Hence, NDR reflects the conducting path switching from via surface states to via the core of Si wires.

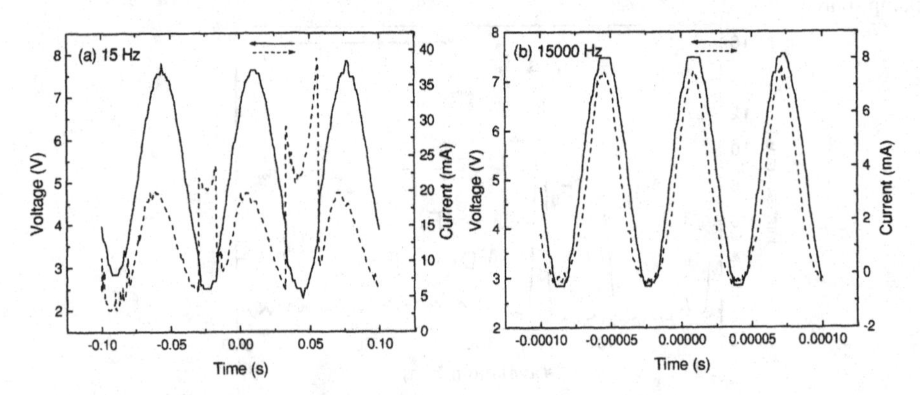

Figure 4: The NDR as a function of bias frequency. It was prominent at (a) 15 Hz but disappeared as the bias frequency increased to (b) 15 kHz.

In our experiment, the spacing between the top Au electrode and the bulk silicon is two microns. At an applied voltage of about 20 V, the field inside the porous silicon would exceed 10^5 V/cm, which is in the same order for breakdown of air at atmospheric pressure. For the discharge to take place, the microcavities should exist in the range of one micron or less, and it is essential that there should be fissures in the top layer of the samples. However, it is likely that the condition for plasma generation is not so critical and it should be investigated thoroughly. The topology and cross-section of our samples was examined under scanning electron microscope (Fig. 5), and there were no observable pores larger than µm in dimension. This raises doubt that the UV emission is really related to microsized N_2 plasma, as proposed by Kozlowski et al [9]. To further clarify the argument, several samples were annealed in Ar ambient at 150 °C for 20 minutes. This annealing procedure was sufficient to replace the N_2 in the pores with Ar.

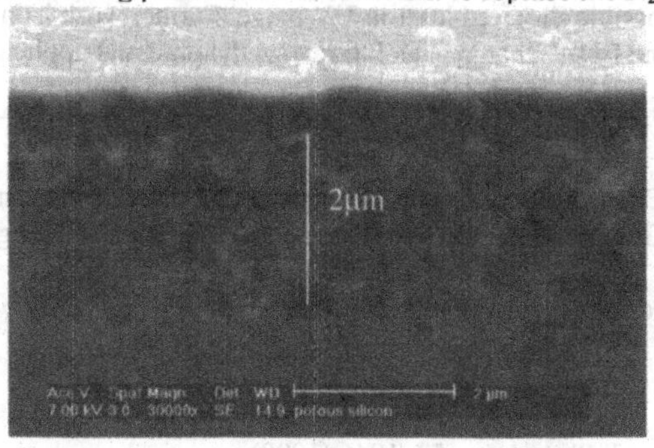

Figure 5: The Scanning Electron Micrograph of PS cross-section, showing the layer was about 2 µm thick and the pores were no larger than µm in dimension.

However there was no observable EL spectrum change under continuous Ar gas purging. We also performed similar process in different gases (N_2, O_2, and dry air), and similarly there was no obvious spectrum change. Thus, it appeared that the UV EL was not due to microsized plasma discharge. Furthermore, no oxygen plasma lines were observed although oxygen content in air is about 21%.

The result obtained showed that UV EL is very much linked to the PS surface states. It is possible to relate the emission energy to quantum confinement [14]. Narrow emission lines have been observed from Si tips [15]. This was attributed to quantum confined energy levels formed from oxygen passivated silicon narrow wires. Another possibility is the microcavity effect, which takes the PS as the active medium and the Au layer and Si substrate as losy reflectors. Calculation of such a system did produce a series of emission lines close to the observed spectra [16]. Furthermore, UV PL in the same energy range was observed by Chen et. al., [17] on Mn passivated PS. They argued that the PL was due to excited electrons confined in higher energy states in the conduction band and recombined there. The confinement was due to Mn passivated surface states that normally allow the electrons to tunnel across. Although their UV PL did not exhibit similar emission lines, the emission energy is in the same range. The sharp emission lines can be attributed to microcavity effect mentioned earlier. Thus, the result, here and elsewhere [14-17], indicates that the emission was probably due to PS core recombination when the surface states were passivated.

CONCLUSION

In conclusion, atomic line shaped UV and broad orange EL from porous silicon diodes has been achieved. The threshold voltage for the UV EL can be as low as 19 V. The UV EL could be related to the core recombination of PS when surface states were sufficiently passivated. However, the micro-process of the UV EL is yet to be elucidated. Nevertheless, the potential of the LED as an UV pump source was fully demonstrated. This more efficient process has the potential in applications such as small flat panel display units, gas sensors or micro-ozone generators for chemical processes in microsystems. In addition, if it is possible to combine recent demonstration of optical gain in silicon nano-crystals [18] and the result reported here, it would brings a step closer to possible Si-based laser in UV range. Further work is in progress to elucidate the details of the EL mechanism, and to realize the potential application.

ACKNOWLEDGEMENT

We thank S. C. Chen and P. N. Wang for many useful discussions. This project is supported by Research Grant Council of Hong Kong and FRG Grant of Hong Kong Baptist University.

* to whom all correspondence should be addressed to.

REFERENCES:

1. L. T. Canham, Appl. Phys. Lett., **57**, 1046 (1990).
2. S. Lazarouk, P. Jaguiro, S. Katsouba, G. Masini, S. LaMonica, G. Maiello and A. Ferrari, Appl. Phys. Lett., **68**, 2108 (1996).
3. L. Tsybeskov, S. P. Duttagupa, K. D. Hirschman, and P. M. Fauchet, Appl. Phys. Lett., **68**, 2058 (1996).
4. B. Gelloz, T. Nakagawa, and N. Koshida, Appl. Phys. Lett., **73**, 2021 (1998).
5. L. Pavesi, R. Chierchia, P. Bellutti, A. Lui, F. Fuso, M. Labardi, L. Pardi, F. Sbrana, M. Allegrini, S. Trusso, C. Vasi, P. J. Ventura, L. C. Costa, and O. Bisi, J. Appl. Phys., **86**, 6474 (1999).
6. G. F. Bai, Y. P. Qiao, Z. C. Ma, W. H. Zong, and G. G. Qin, Appl. Phys. Lett., **72**, 3408 (1998).
7. A. G. Nassiopoulou, V. I. Sougleridis, P. Photopoulos, A. Travlos, V. Tsakiri, and D. Papadimitriou, Phys. Stat. Sol., **165**, 79 (1998).
8. Y. Q. Wang, T. P. Zhao, J. Liu, and G. G. Qin, Appl. Phys. Lett., **74**, 3815 (1999).
9. F. Kozlowski, P. Steiner, M. Sauter, and W. Lang, J. luminescence, **57**, 185 (1993).
10. T. Oguro, H. Koyama, T. Ozaki, and N. Koshida, J. Appl. Phys., **81**, 1407 (1997).
11. K. Ueno and N. Koshida, Jpn. J. Appl. Phys. Part 1, **37**, 1096 (1998).
12. M. L. Gong, J. X. Shi, W. K. Wong, K. K. Shiu, W. H. Zheng and K. W. Cheah, Appl. Phys. A, **68**, 107 (1999).
13. K. W. Cheah, L.C. Ho, Jian-Bai Xia, J. Li, W. H. Zheng, W. R. Zhuang and Q.M. Wang, Appl. Phys. A, **60**, 601 (1995).
14. Jian-Bai Xia and K. W. Cheah, Phys. Rev. B **55**, 15688 (1997).
15. W. H. Zheng, Jian-bai Xia, S. D. Lam, K. W. Cheah, M. R. Rakhshandehroo and S. W. Pang, Appl. Phys. Lett. **74**, 386 (1999).
16. H. L. Lam, S. C. Chen and K. W. Cheah, unpublished.
17. Qianwang Chen, D. L. Zhu, and Y. H. Zhang, Appl. Phys. Lett. **77**, 854 (2000).
18. L. Pavesi, L. Dal Negro, C. Mazzoleni, G, Franzo and F. Priolo, Nature, **408**, 440 (2000).

AUTHOR INDEX

SUBJECT INDEX

Printed in the United States
By Bookmasters